Everyday Mathematics®

The University of Chicago School Mathematics Project

Everyday Mathematics®

The University of Chicago School Mathematics Project

Student Reference Book

McGraw Hill Education

Chicago, IL • Columbus, OH • New York, NY

UCSMP Elementary Materials Component

Max Bell, Director, UCSMP Elementary Materials Component; Director, *Everyday Mathematics* First Edition
James McBride, Director, *Everyday Mathematics* Second Edition
Andy Isaacs, Director, *Everyday Mathematics* Third Edition
Amy Dillard, Associate Director, *Everyday Mathematics* Third Edition
Rachel Malpass McCall, Associate Director, *Everyday Mathematics* Common Core State Standards Edition

Authors

Max Bell, Jean Bell, John Bretzlauf, Amy Dillard, James Flanders,
Robert Hartfield, Andy Isaacs, Deborah Arron Leslie, James McBride,
Kathleen Pitvorec, Peter Saecker

Assistants

Lance Campbell (Research), Adam Fischer (Editorial),
John Saller (Research)

Technical Art

Diana Barrie

 The *Student Reference Book* is based upon work supported by the National Science Foundation under Grant No. ESI-9252984. Any opinions, findings, conclusions, or recommendations expressed in this material are those of the authors and do not necessarily reflect the views of the National Science Foundation.

everyday**math**.com

 **Education**

STEM McGraw-Hill is committed to providing instructional
materials in Science, Technology, Engineering, and Mathematics
(STEM) that give all students a solid foundation, one that
prepares them for college and careers in the 21st century.

Send all inquiries to:
McGraw-Hill Education
STEM Learning Solutions Center
P.O. Box 812960
Chicago, IL 60681

ISBN: 978-0-07-657651-7
MHID: 0-07-657651-5

Printed in the United States of America.

4 5 6 7 8 9 QVR 17 16 15 14 13 12

Contents

Whole Numbers   1

Decimals and Percents   25

Contents

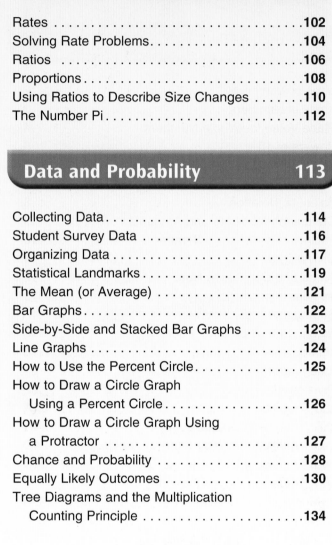

Mathematics ... Every Day
Space Travel　　　　　　　　　**95**

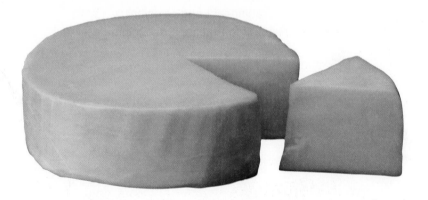

Geometry and Constructions 135

Mathematics ... Every Day
Mathematics and Architecture 175

Measurement 181

Contents

Games 293

American Tour 337

About the *Student Reference Book*

A reference book is organized to help people find information quickly and easily. Dictionaries, encyclopedias, atlases, cookbooks, and even telephone books are examples of reference books. Unlike novels and biographies, which are usually read in sequence from beginning to end, reference books are read in small segments to find specific information at the time it is needed.

You can use this *Student Reference Book* to look up and review information on topics in mathematics. It consists of the following sections:

♦ A **table of contents** that lists the topics covered and shows how the book is organized.

♦ Essays on **mathematical topics,** such as whole numbers, decimals, percents, fractions, data analysis, geometry, measurement, algebra, and problem solving.

♦ A collection of **photo essays** called **Mathematics... Every Day,** which show in words and pictures some of the ways that mathematics is used.

♦ Descriptions of how to use a **calculator** to perform various mathematical operations and functions.

♦ Directions on how to play **mathematical games** that will help you practice math skills.

♦ A set of **tables and charts** that summarize information, such as a place-value chart, prefixes for names of large and small numbers, tables of equivalent measures and of equivalent fractions, decimals, and percents, and a table of formulas.

♦ A **glossary** of mathematical terms consisting of brief definitions of important words.

♦ An **answer key** for every Check Your Understanding problem in the book.

♦ An **index** to help you locate information quickly.

This reference book also contains an **American Tour.** It is a collection of numerical information about the history, people, and environment of the United States.

How to Use the *Student Reference Book*

Suppose you are asked to solve a problem and you know that you have solved problems like it before. But at the moment, you are having difficulty remembering how to do it. This is a perfect time to use the *Student Reference Book*.

You can look in the **table of contents** or the **index** to find the page that gives a brief explanation of the topic. The explanation will often show a step-by-step sample solution.

There is a set of problems at the end of most essays, titled **Check Your Understanding.** It is a good idea to solve these problems and then turn to the **answer key** at the back of the book. Check your answers to make sure that you understand the information presented in the section you have been reading.

Always read mathematical text with paper and pencil in hand. Take notes, draw pictures and diagrams to help you understand what you are reading. Work the examples. If you get a wrong answer in the Check Your Understanding problems, try to find your mistake by working back from the correct answer given in the answer key.

It is not always easy to read text about mathematics, but the more you use the *Student Reference Book,* the better you will become at understanding this kind of material. You should find that your skills as an independent problem-solver are improving. We are confident that these skills will serve you well as you undertake more advanced mathematics courses.

Whole Numbers

Uses of Numbers

It is hard to live even one day without using or thinking about numbers. Numbers are used on clocks, calendars, license plates, rulers, scales, and so on. The major ways that numbers are used are listed here.

◆ Numbers are used for **counting.**

Examples Students sold 129 tickets to the school play.

The first U.S. Census counted 3,929,326 people.

The population of Hope is 10,290.

◆ Numbers are used for **measuring.**

Examples Ivan swam the length of the pool in 34.5 seconds.

The package is 31 inches long and weighs $4\frac{3}{8}$ pounds.

◆ Numbers are used to show where something is in a **reference system.**

Examples

Situation	Type of Reference System
Normal room temperature is 21°C.	Celsius temperature scale
David was born on June 22, 1997.	Calendar
The time is 10:08 A.M.	Clock time
Detroit is located at 42°N and 83°W.	Earth's latitude and longitude system

◆ Numbers are used to **compare** measures or counts.

Examples The cat weighs $\frac{1}{2}$ as much as the dog.

There were 4 times as many boys as girls at the game.

◆ Numbers can be used for **identification** and as **codes.**

Examples phone number: (709) 555-1212   ZIP code: 60637

driver's license number: M286-423-2061

bar code (to identify product and manufacturer):
6 96936 39883 2

car license plate: 02-5492

Kinds of Numbers

The **counting numbers** are the numbers used to count things. The set of counting numbers is 1, 2, 3, 4, and so on.

The **whole numbers** are any of the numbers 0, 1, 2, 3, 4, and so on. The whole numbers include all of the counting numbers and the number zero (0).

Counting numbers are useful for counting, but they often do not work for measures. Most measures fall between whole numbers. **Fractions** and **decimals** were invented to keep track of such in-between measures.

Fractions are often used in recipes for cooking and for measures in carpentry and other building trades. Decimals are used for almost all measures in science and industry. Money amounts are usually written as decimals.

Did You Know?

In 1498, Johann Widmann wrote the first book that used + and − signs. Traders and shopkeepers had used both of the signs long before this. They used + to show they had too much of something. They used − to show they had too little of something.

> **Examples** The calculator weighs 5.8 ounces.
>
> The recipe calls for $2\frac{1}{4}$ cups of flour.
>
> The window sill is 2 feet $5\frac{3}{4}$ inches above the floor.

Negative numbers are used to describe some locations when there is a zero point.

> **Examples** A temperature of 10 degrees below zero is written as −10°F or −10°C.
>
> A depth of 147 feet below sea level is written as −147 feet.

APPLE FRITTERS
$2\frac{1}{4}$ cups flour
12 eggs, beaten
$\frac{1}{2}$ tsp. saffron
$\frac{1}{4}$ tsp. black pepper
9 apples, peeled and diced

Negative numbers are also used to indicate changes in quantities.

> **Examples** A weight loss of $7\frac{1}{2}$ pounds is recorded as $-7\frac{1}{2}$ pounds.
>
> A decrease in income of $1,500 is recorded as −$1,500.

Place Value for Whole Numbers

Any number, no matter how large or small, can be written using one or more of the **digits** 0, 1, 2, 3, 4, 5, 6, 7, 8, and 9. A **place-value chart** is used to show how much each digit in a number is worth. The **place** for a digit is its position in the number. The **value** of a digit is how much it is worth according to its place in the number.

Study the place-value chart below. Look at the numbers that name the places. As you move from right to left along the chart, each number is **10 times as large as the number to its right.**

10,000s	1,000s	100s	10s	1s
ten thousands	thousands	hundreds	tens	ones
3	8	9	0	5

Example The number 38,905 is shown in the place-value chart above.

The value of the 3 is 30,000 (3 * 10,000).   The value of the 0 is 0 (0 * 10).
The value of the 8 is 8,000 (8 * 1,000).   The value of the 5 is 5 (5 * 1).
The value of the 9 is 900 (9 * 100).

38,905 is read as "thirty-eight thousand, nine hundred five."

In larger numbers, groups of 3 digits are separated by commas. Commas help identify the thousands, millions, billions, trillions, and so on.

Example The number 135,246,015,808,297 is shown in the place-value chart.

trillions				billions				millions				thousands				ones		
100	10	1	,	100	10	1	,	100	10	1	,	100	10	1	,	100	10	1
1	3	5	,	2	4	6	,	0	1	5	,	8	0	8	,	2	9	7

Read from left to right. Read "trillion" at the first comma. Read "billion" at the second comma. Read "million" at the third comma. And read "thousand" at the last comma.

This number is read as "135 **trillion,** 246 **billion,** 15 **million,** 808 **thousand,** 297."

Check Your Understanding

Read each number to yourself. What is the value of the 5 in each number?

1. 25,308   **2.** 74,546,002   **3.** 643,057   **4.** 2,450,609

Check your answers on page 433.

Powers of 10

Numbers like 10, 100, and 1,000 are called **powers of 10.** They are numbers that can be written as products of 10s.

100 can be written as $10 * 10$ or 10^2. 1,000 can be written as $10 * 10 * 10$ or 10^3.

The raised digit is called the **exponent.** The exponent tells how many 10s are multiplied.

A number written with an exponent, like 10^3, is in **exponential notation.** A number written in the usual place-value way, like 1,000, is in **standard notation.**

The chart below shows powers of 10 from ten through one billion.

Note

10^2 is read "10 to the second power" or "10 squared." 10^3 is read "10 to the third power" or "10 cubed." 10^4 is read "10 to the fourth power."

Powers of 10		
Standard Notation	Product of 10s	Exponential Notation
10	10	10^1
100	10*10	10^2
1,000 (1 thousand)	10*10*10	10^3
10,000	10*10*10*10	10^4
100,000	10*10*10*10*10	10^5
1,000,000 (1 million)	10*10*10*10*10*10	10^6
10,000,000	10*10*10*10*10*10*10	10^7
100,000,000	10*10*10*10*10*10*10*10	10^8
1,000,000,000 (1 billion)	10*10*10*10*10*10*10*10*10	10^9

Did You Know?

Raised numbers have been used to indicate powers since at least 1484.

Example $1,000 * 1,000 = ?$

Use the table above to write 1,000 as 10*10*10.
$$1,000 * 1,000 = (10 * 10 * 10) * (10 * 10 * 10)$$
$$= 10^6$$
$$= 1 \text{ million}$$

So, $1,000 * 1,000 = 1$ million.

Example 1,000 millions = ?

Write $1,000 * 1,000,000$ as $(10*10*10) * (10*10*10*10*10*10)$. This is a product of nine 10s, or 10^9.

1,000 millions = 1 billion

Exponential Notation

A **square array** is an arrangement of objects into rows and columns that form a square. All rows and columns must be filled; and the number of rows must equal the number of columns. A counting number that can be shown as a square array is called a **square number.** Any square number can be written as the product of a counting number with itself.

two square arrays

> **Example** 16 is a square number. It can be represented by an array consisting of 4 rows and 4 columns.
> $16 = 4 * 4$

Here is a shorthand way to write the square number 16: $16 = 4 * 4 = 4^2$. 4^2 is read as "4 times 4," "4 squared," or "4 to the second power." The raised 2 is called the **exponent.** It tells that 4 is used as a factor 2 times (two 4s are multiplied). The 4 is called the **base.** Numbers written with an exponent are said to be in **exponential notation.**

exponent

4^2

base

Exponents are also used to show that a factor is used more than twice.

> **Example** $2^3 = 2 * 2 * 2$
> The number 2 is used as a factor 3 times.
> 2^3 is read "2 cubed" or "2 to the third power."
>
> $9^5 = 9 * 9 * 9 * 9 * 9$
> The number 9 is used as a factor 5 times.
> 9^5 is read "9 to the fifth power."
>
> Any number raised to the first power is equal to itself. For example, $5^1 = 5$.

Some calculators have special keys for renaming numbers written in exponential notation as standard numerals.

> **Example** Use a calculator. Find the value of 2^4.
> To rename 2^4 on Calculator A, press 2 $\boxed{\wedge}$ 4 $\boxed{\text{Enter}}$. Answer: 16
> To rename 2^4 on Calculator B, press 2 $\boxed{x^y}$ 4 $\boxed{=}$. Answer: 16
> $2^4 = 16$ You can verify this by keying in 2 $\boxed{\times}$ 2 $\boxed{\times}$ 2 $\boxed{\times}$ 2 $\boxed{=}$.

Check Your Understanding

Write each number without exponents. Do not use a calculator to solve Problems 1–4.

1. 7^2 **2.** 3^3 **3.** 10^6 **4.** 6^1 **5.** 614^2 **6.** 15^4

Check your answers on page 433.

Positive and Negative Exponents

Positive exponents tell how many times to use the base as a factor.

2^6	64
2^5	32
2^4	16
2^3	8
2^2	4
2^1	2

Examples
$$5^3 = 5 * 5 * 5 \qquad 10^4 = 10 * 10 * 10 * 10$$
$$25^2 = 25 * 25 \qquad 2^6 = 2 * 2 * 2 * 2 * 2 * 2$$

Positive exponents are helpful for writing large numbers. As people used positive exponents, they noticed patterns like the one in the top table at the right. Note that as the exponents in the left-hand column become 1 less, the numbers in the right-hand column are divided in half. This pattern suggests what expressions like 2^0 or 2^{-3} might mean.

The bottom table at the right continues the pattern of the top table. It can be used to define powers of 2 for the exponent 0 and for negative exponents. Each time you move down by one row, the exponent becomes 1 less and the number in the right-hand column is $\frac{1}{2}$ of the number above it.

2^0	1
2^{-1}	$\frac{1}{2}$
2^{-2}	$\frac{1}{4}$
2^{-3}	$\frac{1}{8}$
2^{-4}	$\frac{1}{16}$
2^{-5}	$\frac{1}{32}$

Many people have trouble understanding how 2^0 equals 1. It's hard to see how you multiply 2s if there are no 2s to multiply. But mathematicians like patterns, and they have decided that 2^0 equals 1 because that fits the pattern in the tables.

Similar patterns hold for powers of other numbers. This is why we say that any number (except 0) raised to the 0 power equals 1.

Examples
$$4^0 = 1 \qquad 8^0 = 1 \qquad 12.893^0 = 1 \qquad 1^0 = 1$$

Note

For all numbers n (except 0), $n^0 = 1$.

Negative exponents are related to positive exponents. A number raised to a negative power is equal to the fraction 1 over the number raised to the positive power.

Examples
$$2^{-3} = \frac{1}{2^3} = \frac{1}{2 * 2 * 2} = \frac{1}{8} \qquad\qquad 10^{-2} = \frac{1}{10^2} = \frac{1}{10 * 10} = \frac{1}{100}$$
$$2^{-5} = \frac{1}{2^5} = \frac{1}{2 * 2 * 2 * 2 * 2} = \frac{1}{32} \qquad\qquad 5^{-3} = \frac{1}{5^3} = \frac{1}{5 * 5 * 5} = \frac{1}{125}$$

Check Your Understanding

Solve.

1. 5^{-2}
2. 10^{-3}
3. 6^0
4. 4^{-1}
5. 2^5
6. 10^0

Check your answers on page 433.

Scientific Notation

The population of the world is about 6 billion people. The number 6 billion can be written as 6,000,000,000 or as $6 * 10^9$.

The number 6,000,000,000 is written in **standard notation.**

The number $6 * 10^9$ is written in **scientific notation.** $6 * 10^9$ is read "six times ten to the ninth power."

Look at 10^9. 10^9 is the product of 10 used as a factor 9 times:

$$10^9 = 10 * 10 * 10 * 10 * 10 * 10 * 10 * 10 * 10$$
$$= 1,000,000,000$$
$$= 1 \text{ billion}$$

So $6 * 10^9 = 6 * 1,000,000,000$
$$= 6,000,000,000$$
$$= 6 \text{ billion}$$

The world's population is about $6 * 10^9$.

A number in scientific notation is written as the product of a number that is at least 1 and less than 10 and a power of 10. We often change numbers from standard notation to scientific notation so that they are easier to write and to work with.

Did You Know?

Scientific notation was not widely used until the 20th century.

Examples Write in scientific notation.

7,000,000 = ?
7,000,000 = 7 * 1,000,000
1,000,000 = 10 * 10 * 10 * 10 * 10 * 10 = 10^6
So, 7,000,000 = $7 * 10^6$.

240,000 = ?
240,000 = 2.4 * 100,000
100,000 = 10 * 10 * 10 * 10 * 10 = 10^5
So, 240,000 = $2.4 * 10^5$.

Examples Write in standard notation.

$4 * 10^3$ = ?
10^3 = 10 * 10 * 10 = 1,000
So, $4 * 10^3$ = 4 * 1,000 = 4,000.

$56 * 10^7$ = ?
10^7 = 10 * 10 * 10 * 10 * 10 * 10 * 10 = 10,000,000
So, $56 * 10^7$ = 56 * 10,000,000 = 560,000,000.

Check Your Understanding

Write each number in standard notation.

1. 4^2 2. 3^3 3. 7^1 4. $8 * 10^6$ 5. $76 * 10^4$

Write each number in scientific notation.

6. 500   7. 44,000   8. 600,000,000

Check your answers on page 433.

Comparing Numbers and Amounts

When two numbers or amounts are compared, there are two possible results: They are equal, or they are not equal because one is larger than the other. Different symbols are used to show that numbers and amounts are equal or not equal.

♦ Use an **equal sign** (=) to show that the numbers or amounts *are equal*.

♦ Use a **not-equal sign** (≠) to show that they are *not equal*.

♦ Use a **greater-than symbol** (>) or a **less-than symbol** (<) to show that they are *not equal* and to show which is larger.

Examples

Symbol	=	≠	>	<
Meaning	"equals" or "is the same as"	"is not equal"	"is greater than"	"is less than"
Examples	$\frac{1}{2} = 0.5$ 4 cm = 40 mm 2 * 5 = 9 + 1	2 ≠ 3 3^2 ≠ 6 1 m ≠ 100 mm	9 > 5 16 ft 9 in. > 15 ft 11 in. 10^3 > 100	3 < 5 98 minutes < 3 hours 3 * (3 + 4) < 5 * 6

When you compare amounts that include units, use the *same unit* for both amounts.

Example   Compare 30 yards and 60 feet.

The units are different—yards and feet. Change yards to feet, then compare.

1 yd = 3 ft, so 30 yd = 30 * 3 ft, or 90 ft. Now compare feet. 90 ft > 60 ft

Therefore, 30 yd > 60 ft.

Or, change feet to yards, and then compare. 1 ft = $\frac{1}{3}$ yd, so 60 ft = 60 * $\frac{1}{3}$ yd, or 20 yd.

Now compare yards. 30 yd > 20 yd

Therefore, 30 yd > 60 ft.

Check Your Understanding

True or false?

1. 6^2 < 13   **2.** 37 in. > 4 ft   **3.** 7 * 4 ≠ 90 / 2   **4.** 13 + 2 > 16 − 2

Check your answers on page 433.

Factors of a Counting Number

A **rectangular array** is an arrangement of objects into rows and columns that form a rectangle. All rows and columns must be filled. Each row has the same number of objects. Each column has the same number of objects. A multiplication **number model** can represent a rectangular array.

Example This rectangular array has 15 red dots.

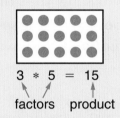

It has 3 rows with 5 dots in each row.

$3 * 5 = 15$ is a number model for this array.

3 and 5 are counting-number **factors** of 15.

15 is the **product** of 3 and 5.

3 and 5 are a **factor pair** for 15.

$$3 * 5 = 15$$
factors   product

Counting numbers can have more than one factor pair. 1 and 15 are another factor pair for 15 because $1 * 15 = 15$.

To test whether a counting number a is a **factor** of another counting number b, divide b by a. If the result is a counting number and the remainder is 0, then a is a factor of b.

Examples 4 is a factor of 12 because 12 / 4 gives 3 with a remainder of 0.

6 is *not* a factor of 14 because 14 / 6 gives 2 with a remainder of 2.

One way to find all the factors of a counting number is to find all the factor pairs for that number.

Example Find all the factors of the number 24.

	Number Models	Factor Pairs
The factors of 24 are 1, 2, 3, 4, 6, 8, 12, and 24.	$24 = 1 * 24$	1, 24
	$24 = 2 * 12$	2, 12
	$24 = 3 * 8$	3, 8
	$24 = 4 * 6$	4, 6

Check Your Understanding

List all the factors of each number.

1. 15   **2.** 8   **3.** 28   **4.** 36   **5.** 11   **6.** 100

Check your answers on page 433.

Divisibility

When one counting number is divided by another counting number and the quotient is a counting number with a remainder of 0, then the first number is **divisible by** the second number. If the quotient is a whole number with a non-zero remainder, then the first number is *not divisible by* the second number.

Note

Division by 0 is *never* permitted.

Examples 124 / 4 → 31 R0 The remainder is 0, so 124 *is* divisible by 4.

88 / 5 → 17 R3 The remainder is not 0, so 88 is *not* divisible by 5.

For some counting numbers, even large ones, it is possible to test for divisibility without dividing.

Here are **divisibility tests** that make it unnecessary to divide:
- All counting numbers are **divisible by 1.**
- Counting numbers with a 0, 2, 4, 6, or 8 in the ones place are **divisible by 2.** They are the **even numbers.**
- Counting numbers with 0 in the ones place are **divisible by 10.**
- Counting numbers with 0 or 5 in the ones place are **divisible by 5.**
- If the sum of the digits in a counting number is divisible by 3, then the number is **divisible by 3.**
- If the sum of the digits in a counting number is divisible by 9, then the number is **divisible by 9.**
- A counting number divisible by both 2 and 3 is **divisible by 6.**

Note

We say that the number 0 is divisible by any counting number a because $0 / a = 0$ R0.

Examples Tell some numbers that 216 is divisible by.

216 is divisible by:
2 because 6 in the ones place is an even number.
3 because the sum of its digits is 9, which is divisible by 3.
9 because the sum of its digits is divisible by 9.
6 because it is divisible both by 2 and by 3.
216 is not divisible by 10 or by 5 because it does not have a 0 or a 5 in the ones place.

Did You Know?

Six is the only counting number that is the product of three counting numbers and the sum of the same three numbers:
$6 = 1 * 2 * 3$ and
$6 = 1 + 2 + 3$.

Check Your Understanding

Which numbers are divisible by 2? By 3? By 5? By 6? By 9? By 10?

1. 705 **2.** 4,470 **3.** 616 **4.** 621 **5.** 14,580

Check your answers on page 433.

Prime and Composite Numbers

A **prime number** is a counting number greater than 1 that has exactly two *different* factors: 1 and the number itself. A prime number is divisible only by 1 and itself.

A **composite number** is a counting number that has more than two different factors.

> **Note**
>
> The only factor of 1 is 1 itself. So, the number 1 is neither prime nor composite.

Examples   11 is a prime number because its only factors are 1 and 11.

4 is a composite number because it has more than two different factors. Its factors are 1, 2, and 4.

Every composite number can be renamed as a product of prime numbers. This is called the **prime factorization** of that number.

> **Note**
>
> The prime factorization of a prime number is that number. For example, the prime factorization of 11 is 11.

Example   Find the prime factorization of 48. The number 48 can be renamed as the product $2 * 2 * 2 * 2 * 3$.
The prime factorization of 48 can be written as $2^4 * 3$.

One way to find the prime factorization of a composite number is to make a **factor tree.** First, write the number. Then, below the number, write any two factors whose product is that number. Repeat the process for these two factors. Continue until all the factors are prime numbers

Example   Find the prime factorization of 24.

No matter which two factors are used to start the tree, the tree will always end with the same prime factors.

$24 = 2 * 2 * 2 * 3$

The prime factorization of 24 is $2 * 2 * 2 * 3$, or $24 = 2^3 * 3$.

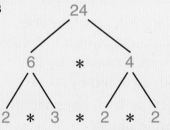

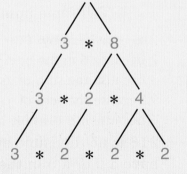

Check Your Understanding

Make a factor tree to find the prime factorization of each number.

1. 15   2. 20   3. 40   4. 36   5. 32   6. 100

Check your answers on page 433.

Addition Algorithms

Partial-Sums Method

The **partial-sums method** is used to find sums mentally or with paper and pencil.

To use the partial-sums method, add from left to right, one column at a time. Then add the partial sums.

Example $348 + 177 = ?$

	100s	10s	1s
	3	4	8
+	1	7	7

		100s	10s	1s
Add the 100s.	$300 + 100 \rightarrow$	4	0	0
Add the 10s.	$40 + 70 \rightarrow$	1	1	0
Add the 1s.	$8 + 7 \rightarrow$		1	5
Add the partial sums.	$400 + 110 + 15 \rightarrow$	5	2	5

$348 + 177 = 525$

Note

Larger numbers with 4 or more digits can be added in the same way.

Column-Addition Method

The **column-addition method** can be used to find sums with paper and pencil, but it is not a good method for finding sums mentally. To add numbers using the column-addition method:

♦ Draw lines to separate the 1s, 10s, 100s, and any other places.

♦ Add the numbers in each column. Write each sum in its column.

♦ If the sum of any column is a 2-digit number, adjust that column sum. Trade part of the sum into the column to the left.

Example $359 + 298 = ?$

	100s	10s	1s
	3	5	9
+	2	9	8

	100s	10s	1s
Add the numbers in each column. Adjust the 1s and 10s:	5	14	17
17 ones = 1 ten and 7 ones			
Trade the 1 ten into the tens column.	5	15	7
Adjust the 10s and 100s:			
15 tens = 1 hundred and 5 tens			
Trade the 1 hundred into the hundreds column.	6	5	7

$359 + 298 = 657$

Did You Know?

In 1642, at age 21, Blaise Pascal invented one of the first mechanical adding machines. The machine had 8 dials and was operated with a pointed, pen shaped tool. Pascal invented the machine as an aid to his father, who was a tax collector.

A Short Method

This addition method was taught to most adults in the United States. Add from right to left. Add one column at a time without displaying the partial sums.

Example   248 + 187 = ?

Step 1:	**Step 2:**	**Step 3:**
Add the ones.	Add the tens.	Add the hundreds.

Step 1:	**Step 2:**	**Step 3:**

1
   248
+ 187
─────
       5
	1 1
	   248
	+ 187
	─────
	    3 5
		1 1
		   248
		+ 187
		─────
		  4 3 5

8 ones + 7 ones = 15 ones = 1 ten + 5 ones	1 ten + 4 tens + 8 tens = 13 tens = 1 hundred + 3 tens	1 hundred + 2 hundreds + 1 hundred = 4 hundreds

248 + 187 = 435

The Opposite-Change Rule

Here is the **opposite-change rule:** If you subtract a number from one addend and add the same number to the other addend, the sum is the same.

Use this rule to make a problem easier by changing either of the addends to a number that has 0 in the ones place. Make the *opposite change* to the other addend.

Note

Addends are numbers that are added. In 8 + 4 = 12, the numbers 8 and 4 are addends.

Example   59 + 26 = ?

One way: Add and subtract 1.

59	(add 1)	60
+ 26	(subtract 1)	+ 25
		────
		85

59 + 26 = 85

Another way: Subtract and add 4.

59	(subtract 4)	55
+ 26	(add 4)	+ 30
		────
		85

Check Your Understanding

Add.

1.   355
    + 532

2. 46
   + 87

3. 277
   + 44

4. 678
   + 345

5. 329 + 534

6. 751 + 79

Check your answers on page 433.

Subtraction Algorithms

Trade-First Subtraction Method

The **trade-first method** is similar to the method for subtracting that most adults in the United States were taught.

♦ If each digit in the top number is greater than or equal to the digit below it, subtract separately in each column.

♦ If any digit in the top number is less than the digit below it, adjust the top number before doing any subtracting. Adjust the top number by "trading."

Example   Subtract 275 from 463 using the trade-first method.

100s	10s	1s
4	6	3
− 2	7	5

Look at the 1s place.
You cannot remove 5 ones from 3 ones.

100s	10s	1s
	5	13
4	6̸	3̸
− 2	7	5

So trade 1 ten for 10 ones.
Now look at the 10s place.
You cannot remove 7 tens from 5 tens.

100s	10s	1s
	15	
3	5̸	13
4̸	6̸	3̸
− 2	7	5
1	8	8

So trade 1 hundred for 10 tens.
Now subtract in each column.

$$463 - 275 = 188$$

Larger numbers with 4 or more digits are subtracted in the same way.

Check Your Understanding

Subtract.
1. $75 - 37$ 2. $853 - 471$ 3. $651 - 285$ 4. $704 - 442$ 5. $7{,}345 - 3{,}066$

Check your answers on page 433.

Counting-Up Method

You can subtract two numbers by counting up from the smaller number to the larger number. The first step is to count up to the nearest multiple of 10. Next, count up by 10s and 100s. Then count up to the larger number.

Example $425 - 48 = ?$

Write the smaller number, 48.

As you count from 48 up to 425, circle each number that you count up.

Add the numbers you circled:
$2 + 50 + 300 + 25 = 377$

You counted up by 377.

$$
\begin{array}{rl}
4\ 8 & \\
+\quad ②\ \ & \text{Count up to the nearest 10.} \\
\hline
5\ 0 & \\
+\ ⑤\ ⑩\ \ & \text{Count up to the nearest 100.} \\
\hline
1\ 0\ 0 & \\
+\ ③⓪⓪\ \ & \text{Count up to the largest possible hundred.} \\
\hline
4\ 0\ 0 & \\
+\quad ②⑤\ \ & \text{Count up to the larger number.} \\
\hline
4\ 2\ 5 & \\
\end{array}
$$

$425 - 48 = 377$

Left-to-Right Subtraction Method

Starting at the left, subtract column by column.

Examples

$932 - 356 = ?$

$$
\begin{array}{r}
9\ 3\ 2 \\
-\ 3\ 0\ 0 \\
\hline
6\ 3\ 2 \\
-\quad 5\ 0 \\
\hline
5\ 8\ 2 \\
-\qquad 6 \\
\hline
5\ 7\ 6 \\
\end{array}
$$

Subtract the 100s.

Subtract the 10s.

Subtract the 1s.

$932 - 356 = 576$

$782 - 294 = ?$

$$
\begin{array}{r}
7\ 8\ 2 \\
-\ 2\ 0\ 0 \\
\hline
5\ 8\ 2 \\
-\quad 9\ 0 \\
\hline
4\ 9\ 2 \\
-\qquad 4 \\
\hline
4\ 8\ 8 \\
\end{array}
$$

$782 - 294 = 488$

Check Your Understanding

Subtract.

1. $426 - 63$ 2. $936 - 777$ 3. $363 - 147$ 4. $505 - 262$

Check your answers on page 433.

Partial-Differences Method

1. Subtract left to right, one column at a time.

2. In some cases, the larger number is on the bottom and the smaller number is on the top. When this happens and you subtract, the difference will be a negative number.

Example $846 - 363 = ?$

$$\begin{array}{r} 8\ 4\ 6 \\ -\ 3\ 6\ 3 \\ \hline \end{array}$$

Subtract the 100s.	$800 - 300 \rightarrow$	5 0 0
Subtract the 10s.	$40 - 60 \rightarrow$	− 2 0
Subtract the 1s.	$6 - 3 \rightarrow$	3
Find the total.	$500 - 20 + 3 \rightarrow$	4 8 3

$846 - 363 = 483$

Same-Change Rules

Here are the **same-change rules** for subtraction problems:

♦ If you add the same number to both numbers in the problem, the answer is the same.

♦ If you subtract the same number from both numbers in the problem, the answer is the same.

Use this rule to change the second number in the problem to a number that has 0 in the ones place. Make the *same change* to the first number. Then subtract.

Example $83 - 27 = ?$

One way: Add 3.

$$\begin{array}{rl} 8\ \ 3 & \text{(add 3)} \\ -\ 2\ \ 7 & \text{(add 3)} \\ \hline \end{array} \qquad \begin{array}{r} 8\ \ 6 \\ -\ 3\ \ 0 \\ \hline 5\ \ 6 \end{array}$$

$83 - 27 = 56$

Another way: Subtract 7.

$$\begin{array}{rl} 8\ \ 3 & \text{(subtract 7)} \\ -\ 2\ \ 7 & \text{(subtract 7)} \\ \hline \end{array} \qquad \begin{array}{r} 7\ \ 6 \\ -\ 2\ \ 0 \\ \hline 5\ \ 6 \end{array}$$

Check Your Understanding

Subtract.

1. $518 - 62$ 2. $744 - 227$ 3. $435 - 152$ 4. $3{,}125 - 417$

Check your answers on page 433.

Extended Multiplication Facts

Numbers such as 10, 100, and 1,000 are called **powers of 10.**

It is easy to multiply a whole number, n, by a power of 10.
To the right of the number n, write as many zeros as there are
zeros in the power of 10.

Examples

10 * 63 = 630	10 * 50 = 500	100 * 160 = 16,000
100 * 63 = 6,300	100 * 50 = 5,000	10,000 * 23 = 230,000
1,000 * 63 = 63,000	1,000 * 50 = 50,000	1,000,000 * 8 = 8,000,000

If you have memorized the basic multiplication facts, you can
solve problems such as 8 * 70 and 5,000 * 3 mentally.

Examples

8 * 70 = ?	5,000 * 3 = ?
Think: 8 [7s] = 56	*Think:* 5 [3s] = 15
Then 8 [70s] is 10 times as much.	Then 5,000 [3s] is 1,000 times as much.
8 * 70 = 10 * 56 = 560	5,000 * 3 = 1,000 * 15 = 15,000

You can use a similar method to solve problems such as 40 * 50
and 300 * 90 mentally.

Examples

40 * 50 = ?	300 * 90 = ?
Think: 4 [50s] = 200	*Think:* 3 [90s] = 270
Then 40 [50s] is 10 times as much.	Then 300 [90s] is 100 times as much.
40 * 50 = 10 * 200 = 2,000	300 * 90 = 100 * 270 = 27,000

Check Your Understanding

Solve these problems in your head.

1. 9 * 100   **2.** 1,000 * 36   **3.** 5 * 500   **4.** 4,000 * 8   **5.** 80 * 50   **6.** 700 * 80

Check your answers on page 433.

Multiplication Algorithms

The symbols × and * are both used to indicate multiplication.

In this book, the symbol * is used more often.

Partial-Products Method
In the **partial-products method,** you must keep track of the place value of each digit. It may help to write 1s, 10s, and 100s above the columns. Each partial product is either a basic multiplication fact or an extended multiplication fact.

Example 4 * 236 = ?

	100s	10s	1s
	2	3	6
*			4

Think of 236 as 200 + 30 + 6.

Multiply each part of 236 by 4.

4 * 200 → 8 0 0 ⎫ extended multiplication facts
4 * 30 → 1 2 0 ⎬
4 * 6 → 2 4 basic multiplication fact

Add the three partial products. 9 4 4

4 * 236 = 944

Example 43 * 26 = ?

	100s	10s	1s
		2	6
*		4	3

Think of 26 as 20 + 6.

Think of 43 as 40 + 3.

Multiply each part of 26 by each part of 43.

40 * 20 → 8 0 0 ⎫
40 * 6 → 2 4 0 ⎬ extended multiplication facts
3 * 20 → 6 0 ⎭
3 * 6 → 1 8 basic multiplication fact

Add the four partial products. 1,1 1 8

43 * 26 = 1,118

Check Your Understanding

Multiply. Write each partial product. Then add the partial products.
1. 265 * 3   2. 42 * 67   3. 40 * 58   4. 83 * 54   5. 372 * 50

Check your answers on page 433.

Lattice Method

The **lattice method** for multiplying has been used for hundreds of years. It is very easy to use if you know the basic multiplication facts.

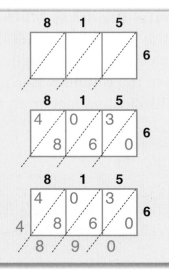

Example | 6 * 815 = ?

The box with cells and diagonals is called a **lattice**.
Write 815 above the lattice.
Write 6 on the right side of the lattice.

Multiply 6 * 5. Then multiply 6 * 1. Then multiply 6 * 8.
Write the answers as shown.

Add the numbers along each diagonal, starting at the right.

Read the answer. 6 * 815 = 4,890

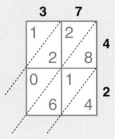

Example | 42 * 37 = ?

Write 37 above the lattice.
Write 42 on the right side of the lattice.

Multiply 4 * 7. Then multiply 4 * 3.
Multiply 2 * 7. Then multiply 2 * 3.
Write the answers as shown.

Add the numbers along each diagonal, starting at the right.

When the numbers along a diagonal add up to 10 or more:
• record the ones digit in the sum.
• add the tens digit to the sum along the next diagonal above.

Read the answer. 42 * 37 = 1,554

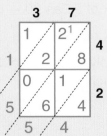

Check Your Understanding

Draw a lattice for each problem. Then multiply.

1. 6 * 78   **2.** 55 * 25   **3.** 77 * 89   **4.** 8 * 444   **5.** 357 * 6

Check your answers on pages 433 and 434.

Extended Division Facts

Numbers such as 10, 100, and 1,000 are called **powers of 10.**

In the examples below, use this method to divide a whole number that ends in zeros by a power of 10:

◆ Cross out zeros in the number, starting in the ones place.

◆ Cross out as many zeros as there are zeros in the power of 10.

Examples

40,000 / **10** = 40000	52,000 / **10** = 52000	790,000 / **10,000** = 790000
40,000 / **100** = 40000	52,000 / **100** = 52000	6,000,000 / **100,000** = 6000000
40,000 / **1,000** = 40000	52,000 / **1,000** = 52000	

If you know the basic division facts, you can solve problems such as 240 / 4 and 15,000 / 3 mentally.

Examples

240 / 4 = ?

Think: 24 / 4 = 6

Then 240 / 4 is 10 times as much.

240 / 4 = 10 * 6 = 60

15,000 / 3 = ?

Think: 15 / 3 = 5

Then 15,000 / 3 is 1,000 times as much.

15,000 / 3 = 1,000 * 5 = 5,000

You can use a similar method to solve problems such as 18,000 / 30 mentally.

Example

18,000 / 30 = ?

Think: 18 / 3 = 6

Try 6 as the answer: 6 * 30 = 180
You want 18,000, or 100 times 180.

Try 100 * 6 = 600 as the answer: 600 * 30 = 18,000

So, 18,000 / 30 = 600.

Check Your Understanding

Solve these problems mentally.

1. 45,000 / 1,000
2. 32,000 / 4
3. 48,000 / 8
4. 5,300 / 10
5. 4,900 / 70
6. 63,000 / 70

Check your answers on page 434.

Division Algorithms

Different symbols may be used to indicate division. For example, "94 divided by 6" may be written as $94 \div 6$, $6\overline{)94}$, $94 / 6$, or $\frac{94}{6}$.

- The number that is being divided is called the **dividend.**

- The number that divides the dividend is called the **divisor.**

- The answer to a division problem is called the **quotient.**

- Some numbers cannot be divided evenly. When this happens, the answer includes a quotient and a **remainder.**

Four ways to show "123 divided by 4"	
$123 \div 4$	$123 / 4$
$4\overline{)123}$	$\frac{123}{4}$
123 is the dividend. 4 is the divisor.	

Partial-Quotients Method

In the **partial-quotients method,** it takes several steps to find the quotient. At each step, you find a partial answer (called a **partial quotient**). These partial answers are then added to find the quotient.

Study the example below. To find the number of 6s in 1,010 first find partial quotients and then add them. Record the partial quotients in a column to the right of the original problem.

Example $1,010 / 6 = ?$

Write partial quotients in this column.

```
6)1,010    ↓     Think: How many [6s] are in 1,010? At least 100.
 − 600   100     The first partial quotient is 100. 100 * 6 = 600
   410           Subtract 600 from 1,010. At least 50 [6s] are left in 410.
 − 300    50     The second partial quotient is 50. 50 * 6 = 300
   110           Subtract. At least 10 [6s] are left in 110.
 −  60    10     The third partial quotient is 10. 10 * 6 = 60
    50           Subtract. At least 8 [6s] are left in 50.
 −  48     8     The fourth partial quotient is 8. 8 * 6 = 48
     2   168     Subtract. Add the partial quotients.
     ↑     ↑
Remainder  Quotient
```

The answer is 168 R2. Record the answer as $6\overline{)1,010}$ with 168 R2
or write $1,010 / 6 \rightarrow 168$ R2.

The partial-quotients method works the same whether you divide by a 2-digit or a 1-digit divisor. It often helps to write down some easy facts for the divisor.

Example   Divide 600 by 22.

Some facts for 22
(to help find partial quotients)

$$1 * 22 = 22$$
$$2 * 22 = 44$$
$$5 * 22 = 110$$
$$10 * 22 = 220$$

```
22)600
 - 440 | 20
   160 |        (20 [22s] in 600)
 - 110 | 5
    50 |        (5 [22s]  in 160)
 -  44 | 2
     6 | 27     (2 [22s] in 50)
```

Record the answer as $22)\overline{600}$ (27 R6)   or write $600 / 22 \rightarrow 27$ R6.

There are different ways to find partial quotients when you use the partial-quotients method. Study the example below. The answer is the same for each way.

Did You Know?

In 1820, Charles de Colmar invented a calculating machine called the *Arithmometer* that could both multiply and divide numbers. It was the first mass-produced calculator.

Example $381 / 4 = ?$

One way:
```
4)381  |
 - 200 | 50
   181 |
 - 120 | 30
    61 |
 -  40 | 10
    21 |
 -  20 | 5
     1 | 95
```

A second way:
```
4)381  |
 - 200 | 50
   181 |
 - 160 | 40
    21 |
 -  20 | 5
     1 | 95
```

A third way:
```
4)381  |
 - 360 | 90
    21 |
 -  20 | 5
     1 | 95
```

The answer, 95 R1, is the same for each way.

Check Your Understanding

Divide.

1. $4)\overline{71}$ **2.** $735 / 5$ **3.** $342 \div 4$ **4.** $3)\overline{674}$

Check your answers on page 434.

Column-Division Method

The best way to understand the **column-division method** is to think of a division problem as a money-sharing problem. In the example below, think of sharing $763 among 5 people.

Example $5\overline{)763} = ?$

1. Draw lines to separate the digits in the dividend (the number being divided).
   Work left to right. Begin in the left column.

2. Think of the 7 in the hundreds column as 7 $100 bills to be shared by 5 people.
   Each person gets 1 $100 bill. There are 2 $100 bills remaining.

3. Trade the 2 $100 bills for 20 $10 bills.
   Think of the 6 in the tens column as 6 $10 bills. That makes 20 + 6 = 26 $10 bills.

4. If 5 people share 26 $10 bills, each person gets 5 $10 bills. There is 1 $10 bill remaining.

5. Trade the 1 $10 bill for 10 $1 bills.
   Think of the 3 in the ones column as 3 $1 bills. That makes 10 + 3 = 13 $1 bills.

6. If 5 people share 13 $1 bills, each person gets 2 $1 bills. There are 3 $1 bills remaining.

Record the answer as 152 R3.
Each person receives $152 and $3 are left over.

U.S. Traditional Addition

You can add numbers with **U.S. traditional addition.**

Did You Know?

Writing numbers above the addends in U.S. traditional addition is sometimes called "carrying." Numbers that are carried, such as the 1 in the 10s column in Step 1 below, are sometimes called "carry marks" or "carries."

Example   7,945 + 8,438 = ?

Step 1: Start with the 1s: 5 + 8 = 13.
13 ones = 1 ten + 3 ones
Write 3 in the 1s place below the line.
Write 1 above the numbers in the 10s place.

```
    1
  7 9 4 5
+ 8 4 3 8
─────────
        3
```

Step 2: Add the 10s: 1 + 4 + 3 = 8.
There are 8 tens.
Write 8 in the 10s place below the line.

```
    1
  7 9 4 5
+ 8 4 3 8
─────────
      8 3
```

Step 3: Continue adding through the 1,000s place.

```
  1   1
  7 9 4 5
+ 8 4 3 8
─────────
1 6 3 8 3
```

7,945 + 8,438 = 16,383

Check Your Understanding

Add.

1. 4,569 + 1,032 = ?   **2.** 1,390 + 8,926 = ?   **3.** 9,429
                                                              + 683

4. 38,843
       + 6,309

Check your answers on page 442.

U.S. Traditional Subtraction

You can subtract numbers with **U.S. traditional subtraction.**

Example   572 − 385 = ?

Step 1:
Start with the 1s.
Since 5 > 2,
you need to regroup.
Trade 1 ten for 10 ones:
572 = 5 hundreds +
6 tens + 12 ones:
Subtract the 1s:
12 − 5 = 7.

```
      6  12
   5  7̸  2̸
 − 3  8  5
 ───────────
         7
```

Step 2:
Go to the 10s.
Since 8 > 6,
you need to regroup.
Trade 1 hundred for 10 tens:
572 = 4 hundreds +
16 tens + 12 ones.
Subtract the 10s:
16 − 8 = 8.

```
        16
   4  6̸  12
   5̸  7̸  2̸
 − 3  8  5
 ───────────
      8  7
```

Step 3:
Go to the 100s.
Subtract the 100s:
4 − 3 = 1.

```
        16
   4  6̸  12
   5̸  7̸  2̸
 − 3  8  5
 ───────────
   1  8  7
```

572 − 385 = 187

Example   904 − 385 = ?

Step 1: Start with the 1s. Since 5 > 4, you need to regroup.
There are no tens in 904, so trade 1 hundred for 10 tens and then
trade 1 ten for 10 ones: 904 = 8 hundreds + 9 tens + 14 ones.
Subtract the 1s: 14 − 5 = 9.

```
         9
   8  10̸ 14
   9̸  0̸  4̸
 − 3  8  5
 ───────────
         9
```

Step 2: Go to the 10s. Subtract the 10s: 9 − 8 = 1.

Step 3: Go to the 100s. Subtract the 100s: 8 − 3 = 5.

```
         9
   8  10̸ 14
   9̸  0̸  4̸
 − 3  8  5
 ───────────
   5  1  9
```

904 − 385 = 519

Check Your Understanding

Subtract.

1. 616 − 598 = ?   **2.** 540 − 278 = ?   **3.** 7,402 − 3,805 = ?   **4.** 5,071
− 2,986

Check your answers on page 442.

U.S. Traditional Multiplication

You can use **U.S. traditional multiplication** to multiply.

Example   5 * 629 = ?

Step 1: Multiply the ones.
5 * 9 ones = 45 ones = 4 tens + 5 ones
Write 5 in the 1s place below the line.
Write 4 above the 2 in the 10s place.

```
      4
  6 2 9
*     5
_____
      5
```

Step 2: Multiply the tens.
5 * 2 tens = 10 tens
Remember the 4 tens from Step 1.
10 tens + 4 tens = 14 tens in all
14 tens = 1 hundred + 4 tens
Write 4 in the 10s place below the line.
Write 1 above the 6 in the 100s place.

```
  1 4
  6 2 9
*     5
_____
    4 5
```

Step 3: Multiply the hundreds.
5 * 6 hundreds = 30 hundreds
Remember the 1 hundred from Step 2.
30 hundreds + 1 hundred = 31 hundreds in all
31 hundreds = 3 thousands + 1 hundred
Write 1 in the 100s place below the line.
Write 3 in the 1,000s place below the line.

```
  1 4
  6 2 9
*     5
_____
3 1 4 5
```

5 * 629 = 3,145

Did You Know?

Writing the numbers above the factors in the U.S. traditional multiplication method is sometimes called "carrying." The numbers are called "carry marks" or "carries."

You can use the U.S. traditional multiplication method to multiply by two-digit numbers.

Example

$73 * 826 = ?$

Step 1: Multiply 826 by the 3 in 73, as if the problem were 3 * 826.

```
        1
    8   2   6
*       7   3
─────────────
2   4   7   8   ← The partial product
                  3 * 826 = 2,478
```

Step 2: Multiply 826 by the 7 in 73, as if the problem were 7 * 826.

The 7 in 73 stands for 7 tens, so write this partial product one place to the left.

Write a 0 in the 1s place to show you are multiplying by tens.

Write the new carries above the old carries.

```
        1   4
            1
        8   2   6
*           7   3
─────────────
    2   4   7   8
5   7   8   2   0   ← 70 * 826 = 57,820
```

Step 3: Add the two partial products to get the final answer.

```
            1   4
                1
            8   2   6
*               7   3
─────────────
1   1
    2   4   7   8
+ 5 7   8   2   0
─────────────
6   0   2   9   8   ← 73 * 826 = 60,298
```

$73 * 826 = 60,298$

Check Your Understanding

Multiply.

1. $38 * 368 = ?$

2.
```
    829
*    52
```

3.
```
  2,943
*    78
```

4. $63 * 5,278 = ?$

Check your answers on page 442.

U.S. Traditional Long Division

U.S. traditional long division is a method you can use to divide.

Example Share $845 among 3 people.

Step 1: Share the $100 s.

$$\begin{array}{r} 2 \\ 3\overline{)845} \\ -6 \\ \hline 2 \end{array}$$

← Each person gets 2 $100 s.

← 2 $100 s each for 3 people

← 2 $100 s are left.

Step 2: Trade 2 $100 s for 20 $10 s.

That makes 24 $10 s in all.

$$\begin{array}{r} 2 \\ 3\overline{)845} \\ -6 \\ \hline 24 \end{array}$$

← 24 $10 s are to be shared.

Step 3: Share the $10 s.

$$\begin{array}{r} 28 \\ 3\overline{)845} \\ -6 \\ \hline 24 \\ -24 \\ \hline 0 \end{array}$$

← Each person gets 8 $10 s.

← 8 $10 s each for 3 people

← 0 $10 s are left.

Example

Step 4: Share the $1 s.

```
       281   ← Each person gets 1  $1 .
    3)845
     −6
      24
     −24
       05   ← 5  $1 s are to be shared.
       −3   ← 1  $1  each for 3 people
        2   ← 2  $1 s are left.
```

Each person gets $281; $2 is left over.

$845 / 3 → $281 R$2

Check Your Understanding

Divide.

1. $780 / 6

2. $973 / 4

3. $729 / 6

4. $828 / 8

Check your answers on page 442.

U.S. traditional long division is not limited to dividing money.

Example   9,427 / 7 → ?

Think about the problem as dividing 9,427 into 7 equal shares.

Step 1: Start with the thousands.

$$
\begin{array}{r}
1 \\
7{\overline{\smash{)}9427}} \\
-7 \\
\hline
2
\end{array}
$$

← Each share gets 1 thousand.

← 1 thousand * 7 shares

← 2 thousands are left.

Step 2: Trade 2 thousands for 20 hundreds. Share the hundreds.

$$
\begin{array}{r}
13 \\
7{\overline{\smash{)}9427}} \\
-7 \\
\hline
24 \\
-21 \\
\hline
3
\end{array}
$$

← Each share gets 3 hundreds.

← 20 hundreds + 4 hundreds

← 3 hundreds * 7 shares

← 3 hundreds are left.

Step 3: Trade 3 hundreds for 30 tens. Share the tens.

$$
\begin{array}{r}
134 \\
7{\overline{\smash{)}9427}} \\
-7 \\
\hline
24 \\
-21 \\
\hline
32 \\
-28 \\
\hline
4
\end{array}
$$

← Each share gets 4 tens.

← 30 tens + 2 tens

← 4 tens * 7 shares

← 4 tens are left.

Example *continued*

Step 4: Trade 4 tens for
40 ones.
Share the ones.

$$
\begin{array}{r}
1346 \\
7\overline{)9427} \\
-7 \\
\hline
24 \\
-21 \\
\hline
32 \\
-28 \\
\hline
47 \\
-42 \\
\hline
5
\end{array}
$$

← Each share gets 6 ones.

← 40 ones + 7 ones
← 6 ones * 7 shares
← 5 ones are left.

9,427 / 7 → 1,346 R5

Check Your Understanding

Divide.

1. 3,896 / 8

2. 3)6,789

3. 5)6,725

4. 4,304 / 4

Check your answers on page 442.

U.S. Traditional Long Division: Multidigit Divisors

You can use U.S. traditional long division to divide by larger numbers.

Example | Share $681 among 21 people.

Make a table of easy multiples of the divisor.
This can help you decide how many to share at each step.

1 * 21	21	
2 * 21	42	Double 21.
3 * 21	63	Add 2 * 21 and 1 * 21.
4 * 21	84	Double 2 * 21.
5 * 21	105	Halve 10 * 21.
6 * 21	126	Double 3 * 21.
8 * 21	168	Double 4 * 21.
10 * 21	210	Move decimal point one place to the right.

Step 1: There are not enough [$100]s to share 21 ways,

so trade 6 [$100]s for 60 [$10]s.

Share the 68 [$10]s.

```
        3     ← Each person gets 3 [$10]s.
  21)681      ← There are 68 [$10]s to share.
    −63       ← 3 [$10]s * 21 shares
      5       ← 5 [$10]s are left.
```

Step 2: Trade the 5 [$10]s for 50 [$1]s.

Share the 51 [$1]s.

```
       32     ← Each person gets 2 [$1]s.
  21)681
    −63
      51      ← 50 [$1]s + 1 [$1]
     −42      ← 2 [$1]s * 21 shares
       9      ← 9 [$1]s are left.
```

$681 / 21 → $32 R$9

Example   7720 / 25 → ?

Make a table of easy multiples of the divisor.

1 * 25	25
2 * 25	50
3 * 25	75
4 * 25	100
5 * 25	125
6 * 25	150
8 * 25	200
10 * 25	250

Double 25.
Add 2 * 25 and 1 * 25.
Double 2 * 25.
Halve 10 * 25.
Double 3 * 25.
Double 4 * 25.
Move decimal point one place to the right.

Step 1: There are not enough thousands to share 25 ways, so trade the thousands for hundreds. Share the hundreds.

$$
\begin{array}{r}
3 \\
25\overline{)7720} \\
-75 \\
\hline
2
\end{array}
$$

← Each share gets 3 hundreds.
← 77 hundreds
← 3 hundreds * 25 shares
← 2 hundreds are left.

Step 2: Trade the hundreds for tens. Share the tens.

$$
\begin{array}{r}
30 \\
25\overline{)7720} \\
-75 \\
\hline
22
\end{array}
$$

← There are not enough tens to share.

← 20 tens + 2 tens

Step 3: Trade the tens for ones. Share the ones.

$$
\begin{array}{r}
308 \\
25\overline{)7720} \\
-75 \\
\hline
220 \\
-200 \\
\hline
20
\end{array}
$$

← Each share gets 8 ones.

← 22 tens + 0 ones = 220 ones
← 8 ones * 25 shares
← 20 ones are left.

7720 / 25 → 308 R20

Check Your Understanding

Divide.

1. 359 / 57   **2.** 7,690 / 25   **3.** 15)6,730   **4.** 52)5,751

Check your answers on page 442.

Decimals and Percents

Decimals

Both decimals and fractions are used to write numbers that are between whole numbers. Decimals use the same base-ten place-value system as whole numbers. You can compute with decimals in the same way that you compute with whole numbers.

Examples   Some fractions between 1 and 2: $\frac{3}{2}$, $\frac{7}{4}$, $\frac{11}{8}$, $\frac{17}{16}$, $\frac{63}{32}$

Some decimals between 1 and 2: 1.5, 1.75, 1.375, 1.001, 1.999

Decimals are another way to write fractions. Many fractions have denominators of 10, 100, 1,000, and so on. It is easy to write the decimal names for fractions like these.

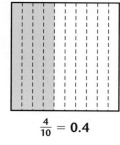

$\frac{4}{10} = 0.4$

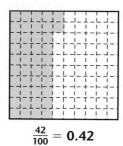

$\frac{42}{100} = 0.42$

This square is divided into 10 equal parts. Each part is $\frac{1}{10}$ of the square. The decimal name for $\frac{1}{10}$ is 0.1.

$\frac{4}{10}$ of the square is shaded. The decimal name for $\frac{4}{10}$ is 0.4.

This square is divided into 100 equal parts. Each part is $\frac{1}{100}$ of the square. The decimal name for $\frac{1}{100}$ is 0.01.

$\frac{42}{100}$ of the square is shaded. The decimal name for $\frac{42}{100}$ is 0.42.

Like mixed numbers, decimals can be used to name numbers greater than one.

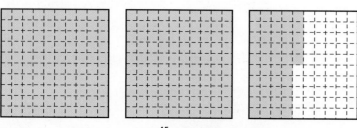

$2\frac{45}{100} = 2.45$

In a decimal, the dot is called the **decimal point.** It separates the whole-number part from the decimal part. A decimal with one digit after the decimal point names *tenths*. A decimal with two digits after the decimal point names *hundredths*. A decimal with three digits after the decimal point names *thousandths*.

decimal point

$$12{\cdot}105$$

whole-   decimal
number   part
part

Examples

tenths	hundredths	thousandths
$0.4 = \frac{4}{10}$	$0.34 = \frac{34}{100}$	$0.162 = \frac{162}{1,000}$
$0.8 = \frac{8}{10}$	$0.75 = \frac{75}{100}$	$0.003 = \frac{3}{1,000}$
$0.9 = \frac{9}{10}$	$0.03 = \frac{3}{100}$	$0.098 = \frac{98}{1,000}$

Did You Know?

Decimals were invented by the Dutch scientist Simon Stevin in 1585. But there is no single, worldwide form for writing decimals. For 3.25 (American notation), the British write 3·25, and the Germans and French write 3,25.

Reading Decimals

One way to read a decimal is to say it as you would a fraction or mixed number. For example, $0.001 = \frac{1}{1,000}$ and can be read as "one-thousandth." $7.9 = 7\frac{9}{10}$, so 7.9 can be read as "seven and nine-tenths."

You can also read decimals by first saying the whole number part, then saying "point," and finally saying the digits in the decimal part. For example, 6.8 can be read as "six point eight"; 0.15 can be read as "zero point one five." This way of reading decimals is often useful when there are many digits in the decimal.

Examples   0.18 is read as "18 hundredths" or "zero point one eight."

24.5 is read as "24 and 5 tenths" or "twenty-four point five."

0.008 is read as "8 thousandths" or "zero point zero zero eight."

Check Your Understanding

Write a decimal for each picture.

1.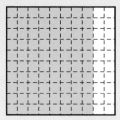

2.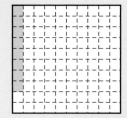

Read each decimal to yourself. Write each decimal as a fraction or mixed number.

3. 0.70   **4.** 4.506

5. 24.68   **6.** 0.014

Check your answers on page 434.

Extending Place Value to Decimals

The first systems for writing numbers were primitive. Ancient Egyptians used a stroke to record the number 1, a picture of an oxbow for 10, a coil of rope for 100, a lotus plant for 1,000, and a picture of a god supporting the sky for 1,000,000.

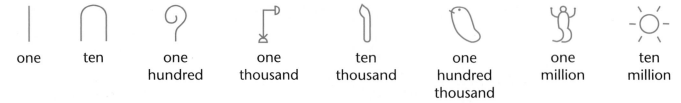

| one | ten | one hundred | one thousand | ten thousand | one hundred thousand | one million | ten million |

This is how an ancient Egyptian would write the number 54:

10 + 10 + 10 + 10 + 10 + 1 + 1 + 1 + 1

Our system for writing numbers was invented in India and later improved in Arabia. It is called a **base-ten** system. It uses only 10 symbols, which are called **digits**: 0, 1, 2, 3, 4, 5, 6, 7, 8, and 9. In this system, you can write any number using only these 10 digits.

For a number written in the base-ten system, each digit has a value that depends on its **place** in the number. That is why it is called a **place-value** system.

1,000s	100s	10s	1s
thousands	hundreds	tens	ones
7	0	8	6

In the number 7,086,

7 is in the **thousands** place; its value is 7 thousands, or 7,000.

0 is in the **hundreds** place; its value is 0.

8 is in the **tens** place; its value is 8 tens, or 80.

6 is in the **ones** place; its value is 6 ones, or 6.

The 0 in 7,086 serves a very important purpose: It "holds" the hundreds place so that the 7 can be in the thousands place. When used in this way, 0 is called a **placeholder.**

> **Note**
>
> It should come as no surprise that our number system uses exactly 10 symbols. People probably counted on their fingers when they first started using numbers.

The base-ten system works the same way for decimals as it does for whole numbers.

Examples

1,000s thousands	100s hundreds	10s tens	1s ones	.	0.1s tenths	0.01s hundredths	0.001s thousandths
		4	7	.	8	0	5
			4	.	3	6	0

In the number 47.805,

8 is in the **tenths** place; its value is 8 tenths, or $\frac{8}{10}$, or 0.8.

0 is in the **hundredths** place; its value is 0.

5 is in the **thousandths** place; its value is 5 thousandths, or $\frac{5}{1,000}$, or 0.005.

In the number 4.360,

3 is in the **tenths** place; its value is 3 tenths, or $\frac{3}{10}$, or 0.3.

6 is in the **hundredths** place; its value is 6 hundredths, or $\frac{6}{100}$, or 0.06.

0 is in the **thousandths** place; its value is 0.

Decimals can also be written in **expanded notation.** For the examples above,

$$47.805 = 40 * 10 + 7 * 1 + 8 * (\tfrac{1}{10}) + 0 * (\tfrac{1}{100}) + 5 * (\tfrac{1}{1,000})$$

$$4.360 \;\; = 4 * 1 + 3 * (\tfrac{1}{10}) + 6 * (\tfrac{1}{100}) + 0 * (\tfrac{1}{1,000})$$

Right to Left in the Place-Value Chart

Study the place-value chart below. Look at the numbers that name the places. As you move from *right to left* along the chart, each number is **10 times as large as the number to its right.**

Example

10 *	10 *	10 *	10 *	10 *	10 *

1,000s thousands	100s hundreds	10s tens	1s ones	.	0.1s tenths	0.01s hundredths	0.001s thousandths

one $\frac{1}{100}$ = ten $\frac{1}{1,000}$s   one 10 = ten 1s

one $\frac{1}{10}$ = ten $\frac{1}{100}$s   one 100 = ten 10s

one 1 = ten $\frac{1}{10}$s   one 1,000 = ten 100s

You use facts about the place-value chart each time you make trades using base-10 blocks.

Example | Suppose that a flat is worth 1. Then a long is worth $\frac{1}{10}$, or 0.1; and a cube is worth $\frac{1}{100}$, or 0.01

For this example:

A flat ▦ is worth 1.

A long ▯ is worth $\frac{1}{10}$ or 0.1.

A cube ▫ is worth $\frac{1}{100}$, or 0.01.

You can trade one long for ten cubes because one $\frac{1}{10}$ equals ten $\frac{1}{100}$s.

You can trade ten longs for one flat because ten $\frac{1}{10}$s equals one 1.

You can trade ten cubes for one long because ten $\frac{1}{100}$s equals one $\frac{1}{10}$.

Left to Right in the Place-Value Chart

Study the place-value chart below. Look at the numbers that name the places. As you move from *left to right* along the chart, each number is $\frac{1}{10}$ **as large as the number to its left.**

Example

$\frac{1}{10}$ * $\frac{1}{10}$ * $\frac{1}{10}$ * $\frac{1}{10}$ * $\frac{1}{10}$ * $\frac{1}{10}$ *

1,000s	100s	10s	1s	.	0.1s	0.01s	0.001s
thousands	hundreds	tens	ones		tenths	hundredths	thousandths

one 100 $= \frac{1}{10}$ of 1,000   one $\frac{1}{10}$ $= \frac{1}{10}$ of 1

one 10 $= \frac{1}{10}$ of 100   one $\frac{1}{100}$ $= \frac{1}{10}$ of $\frac{1}{10}$

one 1 $= \frac{1}{10}$ of 10   one $\frac{1}{1,000}$ $= \frac{1}{10}$ of $\frac{1}{100}$

Check Your Understanding

1. What is the value of the digit 2 in each of these numbers?

   a. 20,006.8

   b. 0.02

   c. 34.502

2. Using the digits 9, 3, and 5, what is

   a. the smallest decimal that you can write?

   b. the largest decimal less than 1 that you can write?

   c. the decimal closest to 0.5 that you can write?

Check your answers on page 434.

Powers of 10 for Decimals

Study the place-value chart. Look at the numbers across the top of the chart that name the places.

1,000s	100s	10s	1s		0.1s	0.01s	0.001s
thousands	hundreds	tens	ones	•	tenths	hundredths	thousandths
5	2	4	6	•	0	8	1
five thousand, two hundred forty-six			and		eighty-one thousandths		

A whole number that can be written using only 10s as factors is called a **power of 10.** A power of 10 can be written in exponential notation.

Powers of 10 (greater than 1)		
Standard Notation	**Product of 10s**	**Exponential Notation**
10	10	10^1
100	10 * 10	10^2
1,000	10 * 10 * 10	10^3
10,000	10 * 10 * 10 * 10	10^4
100,000	10 * 10 * 10 * 10 * 10	10^5

Note

A number written in the usual place-value way, like 100, is in **standard notation.** A number written with an exponent, like 10^2, is in **exponential notation.**

Decimals that can be written using only 0.1s as factors are also called powers of 10. They can be written in exponential notation with negative exponents.

Powers of 10 (less than 1)		
Standard Notation	**Product of 0.1s**	**Exponential Notation**
0.1	0.1	10^{-1}
0.01	0.1 * 0.1	10^{-2}
0.001	0.1 * 0.1 * 0.1	10^{-3}
0.0001	0.1 * 0.1 * 0.1 * 0.1	10^{-4}
0.00001	0.1 * 0.1 * 0.1 * 0.1 * 0.1	10^{-5}

Note

A number raised to a negative exponent is equal to the fraction 1 over the number raised to the positive exponent. For example, $10^{-2} = \frac{1}{10^2} = \frac{1}{10 * 10} = \frac{1}{100}$.

The number 1 is also called a power of 10 because $1 = 10^0$.

100,000s	10,000s	1,000s	100s	10s	1s		0.1s	0.01s	0.001s	0.0001s	0.00001s
10^5	10^4	10^3	10^2	10^1	10^0	•	10^{-1}	10^{-2}	10^{-3}	10^{-4}	10^{-5}

All of the numbers across the top of a place-value chart are powers of 10. Note the pattern: Each exponent is 1 less than the exponent in the place to its left. This is why mathematicians defined 10^0 to be equal to 1.

Comparing Decimals

One way to compare decimals is to model them with base-10 blocks. If you don't have the blocks, you can draw shorthand pictures of them.

Base-10 Block	Name	Shorthand Picture
▫	cube	▫
▯	long	|
▢	flat	▢
▨	big cube	◰

Example   Compare 2.3 and 2.16.

▢▢ ||| **2.3**   ▢▢ | ▫▫▫▫▫ **2.16**

2 flats and 3 longs are more than 2 flats, 1 long, and 6 cubes.

So, 2.3 is more than 2.16.
2.3 > 2.16

For this example:

A flat ▢ is worth 1.

A long \| is worth 0.1.

A cube ▫ is worth 0.01.

Sometimes the big cube is the whole, or ONE.

Example   Compare 2.23 and 2.174.

◰ ◰ ▢▢ ||| **2.23**   ◰ ◰ ▢ ||||||| ▫▫▫▫ **2.174**

Each picture shows 2 big cubes. 2 flats and 3 longs are more than 1 flat and 7 longs and 4 cubes.

So, 2.23 is more than 2.174.
2.23 > 2.174

For this example:

A big cube ◰ is worth 1.

A flat ▢ is worth 0.1.

A long \| is worth 0.01.

A cube ▫ is worth 0.001.

You can write a 0 at the end of a decimal without changing the value of the decimal: 0.7 = 0.70. Writing 0s is sometimes called "padding with 0s." Think of it as trading for smaller pieces.

For the examples on this page:

A flat ☐ is worth 1.

A long | is worth 0.1.

A cube ▫ is worth 0.01.

Example 0.3 = 0.30

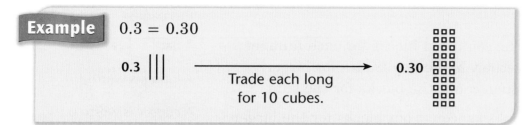

0.3 ||| Trade each long for 10 cubes. →   0.30

Padding with 0s makes comparing decimals easier.

Examples

Compare 0.3 and 0.06.

0.3 = 0.30 (Trade 3 longs for 30 cubes.)

30 cubes is more than 6 cubes.

30 hundredths is more than 6 hundredths.

0.30 > 0.06, so 0.3 > 0.06.

Compare 0.97 and 1.

1 = 1.00 (Trade 1 flat for 100 cubes.)

97 cubes is less than 100 cubes.

97 hundredths is less than 100 hundredths.

0.97 < 1.00, so 0.97 < 1.

A decimal place-value chart can be used to compare decimals.

Example Compare 4.825 and 4.862.

1s ones	.	0.1s tenths	0.01s hundredths	0.001s thousandths
4	.	8	2	5
4	.	8	6	2

The ones digits *are the same*. They are both worth 4.

The tenths digits *are the same*. They are both worth 8 tenths, or $\frac{8}{10}$, or 0.8.

The hundredths digits are *not* the same.

The 2 is worth 2 hundredths, or 0.02. The 6 is worth 6 hundredths, or 0.06. The 6 is worth more than the 2.

So, 4.862 is more than 4.825.

Did You Know❓

Imagine two points that are 100 meters apart. Sound would travel that distance in about 0.3 second. And a beam of light would travel that distance in about 0.0000003 second. Light travels about 1 million times as fast as sound.

Check Your Understanding

Compare the numbers in each pair.

1. 0.59, 0.059   **2.** 0.099, 0.1   **3.** $\frac{1}{4}$, 0.30   **4.** 0.99, 0.100

Check your answers on page 434.

Addition and Subtraction of Decimals

There are many ways to add and subtract decimals. One way is to use base-10 blocks. When working with decimals, we usually use a flat as the ONE.

To add with base-10 blocks, count out blocks for each number and put all the blocks together. Make any trades for larger blocks that you can, and then count the blocks for the sum.

To subtract with base-10 blocks, count out blocks for the larger number. Take away blocks for the smaller number, making trades as needed. Then count the remaining blocks.

Using base-10 blocks is a good idea, especially at first. However, drawing shorthand pictures is usually easier and faster.

For the examples on this page:

A flat ▢ is worth 1.

A long | is worth 0.1.

A cube ▫ is worth 0.01.

Example $1.63 + 3.6 = ?$

First, draw pictures for each number.

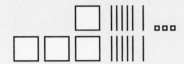

Next, draw a ring around 10 longs and trade them for a flat.

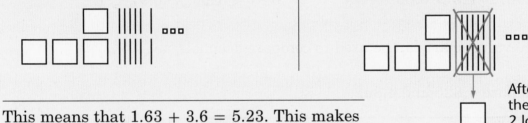

After the trade, there are 5 flats, 2 longs, and 3 cubes.

This means that $1.63 + 3.6 = 5.23$. This makes sense because 1.63 is near $1\frac{1}{2}$ and 3.6 is near $3\frac{1}{2}$. So, the answer should be near 5, which it is.

$1.63 + 3.6 = 5.23$

Example $3.07 - 2.6 = ?$

The picture for 3.07 does not show any longs.

You want to take away 2.6 (2 flats and 6 longs). To do this, trade 1 flat for 10 longs.

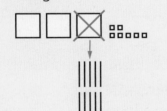

Now remove 2 flats and 6 longs (2.6).

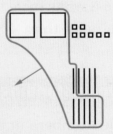

0 flats, 4 longs, and 7 cubes are left. These blocks show 0.47.

$3.07 - 2.6 = 0.47$

Most paper-and-pencil strategies for adding and subtracting whole numbers also work for decimals. The main difference is that you have to line up the places correctly, either by writing 0s at the end of the numbers or by lining up the ones places.

Examples $4.56 + 7.9 = ?$

Partial-Sums Method:

		1s	0.1s	0.01s
		4 .	5	6
	+	7 .	9	0

Add the ones.	$4 + 7 \rightarrow$	11 .	0	0
Add the tenths.	$0.5 + 0.9 \rightarrow$	1 .	4	0
Add the hundredths.	$0.06 + 0.00 \rightarrow$	0 .	0	6
Add the partial sums.	$11.00 + 1.40 + 0.06 \rightarrow$	**12** .	**4**	**6**

Column-Addition Method:

	1s	0.1s	0.01s
	4 .	5	6
+	7 .	9	0
	11 .	14	6

Add the numbers in each column.

Trade 14 tenths for one and 4 tenths.

Move the 1 one into the ones column.

	1s	0.1s	0.01s
	12 .	**4**	**6**

$4.56 + 7.9 = 12.46$, using either method.

Example $9.4 - 4.85 = ?$

Trade-First Method:
Write the problem in columns. Be sure to line up the places correctly. Since 4.85 has two decimal places, write 9.4 as 9.40.

1s	0.1s	0.01s
9 .	4	0
− 4 .	8	5

Look at the 0.01s place.

You cannot remove 5 hundredths from 0 hundredths.

1s	0.1s	0.01s
	3	10
9 .	4̸	0̸
− 4 .	8	5

So trade 1 tenth for 10 hundredths.

Now look at the 0.1s place. You cannot remove 8 tenths from 3 tenths.

1s	0.1s	0.01s
	13	
8	3̸	10
9̸ .	4̸	0̸
− 4 .	8	5
4 .	5	5

Trade 1 one for 10 tenths. Now subtract in each column.

$9.4 - 4.85 = 4.55$

Example $9.4 - 4.85 = ?$

Left-to-Right Subtraction Method:

Since 4.85 has two decimal places, write 9.4 as 9.40.

$$
\begin{array}{r}
\mathbf{9.40} \\
\text{Subtract the ones.} \quad -\ \mathbf{4.00} \\
\hline
5.40 \\
\text{Subtract the tenths.} \quad -\ 0.\mathbf{80} \\
\hline
4.60 \\
\text{Subtract the hundredths.} \quad -\ 0.0\mathbf{5} \\
\hline
\mathbf{4.55}
\end{array}
$$

$9.4 - 4.85 = 4.55$

Example $9.4 - 4.85 = ?$

Counting-Up Method:

Since 4.85 has two decimal places, write 9.4 as 9.40.
There are many ways to count up from 4.85 to 9.40. Here is one.

4.85 (+ 0.15) ———— 5.00 (+ 4.00) ———— 9.00 (+ 0.40) ———— 9.40	Add the numbers you circled and counted up by: 0.15 4.00 + 0.40 ———— 4.55 You counted up by 4.55.

$9.4 - 4.85 = 4.55$

Calculator:

If you use a calculator, it's important to check your answer by estimating because it's easy to press a wrong key accidentally.

Check Your Understanding

Add or subtract

1. $4.62 + 11.4$ **2.** $3.18 - 1.49$ **3.** $2.4 - 2.377$

Check your answers on page 434.

Multiplying by Powers of 10 Greater than 1

Multiplying decimals by a power of 10 greater than 1 is easy.
One way is to use **partial-products multiplication.**

Example | Solve 1,000 * 45.6 by partial-products
multiplication.

Step 1: Solve the problem as if there were no decimal point.

$$
\begin{array}{r}
1000 \\
*\quad 456 \\
\hline
\end{array}
$$

$$
\begin{array}{rr}
400 * 1000 \rightarrow & 400000 \\
50 * 1000 \rightarrow & 50000 \\
6 * 1000 \rightarrow & 6000 \\
\hline
& 456000
\end{array}
$$

Step 2: Estimate the answer to 1,000 * 45.6 and place the
decimal point where it belongs.

1,000 * 45 = 45,000, so 1,000 * 45.6 must be near 45,000.

So, the answer to 1,000 * 45.6 is 45,600.

Note

Some powers of 10
greater than 1:

$10^1 = 10$

$10^2 = 10 * 10 = 100$

$10^3 = 10 * 10 * 10$
$\quad = 1,000$

$10^4 = 10 * 10 * 10 * 10$
$\quad = 10,000$

Another way to multiply a number by a power of 10 greater than
one is to move the decimal point. Think of this as a *shortcut*.

Example | 1,000 * 45.6 = ?

Locate the decimal point in the power of 10. 1,000 = 1000.

Move the decimal point LEFT until you get to the number 1. 1.0 0 0.

Count the number of places you moved the decimal point. 3 places

Move the decimal point in the other factor
the same number of places, but to the RIGHT.
Insert 0s as needed. That's the answer. 4 5.6 0 0.

So, 1,000 * 45.6 = 45,600.

Check Your Understanding

Multiply.

1. 100 * 4.56   2. 0.28 * 10,000   3. 1,000 * $4.50   4. 1.04 * 10

Check your answers on page 434.

Multiplication of Decimals

You can use the same procedures for multiplying decimals that you use for whole numbers. The main difference is that with decimals you have to decide where to place the decimal point in the product.

One way to solve multiplication problems with decimals is to multiply as if both factors were whole numbers. Then adjust the product:

Step 1. Make a magnitude estimate of the product.

Step 2. Multiply as if the factors were whole numbers.

Step 3. Use the magnitude estimate to place the decimal point in the answer.

> **Note**
>
> A *magnitude estimate* is a very rough estimate that answers questions like: *Is the solution in the ones? Tens? Hundreds? Thousands?*

Example $15.2 * 3.6 = ?$

Step 1: Make a magnitude estimate.
- Round 15.2 to 20 and 3.6 to 4.
- Since $20 * 4 = 80$, the product will be in the tens. (*In the tens* means between 10 and 100.)

Step 2: Multiply as you would with whole numbers using the partial-products method. Work from left to right. Ignore the decimal points.

$$
\begin{array}{rcl}
& & 152 \\
& & *\ 36 \\
\hline
30 * 100 & \rightarrow & 3000 \\
30 * 50 & \rightarrow & 1500 \\
30 * 2 & \rightarrow & 60 \\
6 * 100 & \rightarrow & 600 \\
6 * 50 & \rightarrow & 300 \\
6 * 2 & \rightarrow & 12 \\
\hline
\text{Add the partial products.} & \rightarrow & 5472
\end{array}
$$

Step 3: Place the decimal point correctly in the answer. Since the magnitude estimate is in the tens, the product must be in the tens. Place the decimal point between the 4 and the 7 in 5472.

So, $15.2 * 3.6 = 54.72$.

> **Did You Know?**
>
> The price of a gallon of gasoline always includes an additional $\frac{9}{10}$ cent per gallon. Many people believe that this practice is deceptive.
>
> For example, suppose that gasoline costs $2.879 per gallon. A 10-gallon purchase would cost exactly $10 * \$2.879 = \$28.79$. A 9-gallon purchase should cost exactly $9 * \$2.879 = \$25.911$, but the buyer is charged $25.92.

Example $3.27 * 0.8 = ?$

Step 1: Make a magnitude estimate.
- Round 3.27 to 3 and 0.8 to 1.
- Since $3 * 1 = 3$, the product will be in the ones. (*In the ones* means between 1 and 10.)

Step 2: Multiply as you would with whole numbers. Ignore the decimal points.

$$
\begin{array}{r}
327 \\
*\quad 8 \\
\hline
\end{array}
$$

$8 * 300$	$\rightarrow$	2400
$8 * 20$	$\rightarrow$	160
$8 * 7$	$\rightarrow$	56
$2400 + 160 + 56$	$\rightarrow$	**2616**

Step 3: Place the decimal point correctly in the answer. Since the magnitude estimate is in the ones, the product must be in the ones. Place the decimal point between the 2 and the 6 in 2616.

So, $3.27 * 0.8 = 2.616$.

Note

Sometimes a magnitude estimate is on the "borderline" and you need to be more careful.

For example, a magnitude estimate for $18.5 * 5.2$ is $20 * 5 = 100$. The answer may be "in the 10s" or it may be "in the 100s." But the answer will be close to 100. Since $185 * 52 = 9620$, place the decimal point between the 6 and the 2: $18.5 * 5.2 = 96.20$.

There is another way to find where to place the decimal point in the product. This method is especially useful when the factors are less than 1 and have many decimal places.

Example $3.27 * 0.8 = ?$

Count the decimal places to the right of the decimal point in each factor.	2 decimal places in 3.27
	1 decimal place in 0.8
Add the number of decimal places. This is how many decimal places there will be in the product.	$2 + 1 = 3$
Multiply the factors as if they were whole numbers.	$327 * 8 = 2616$
Start at the right of the product. Move the decimal point LEFT the necessary number of decimal places.	$2.616.$

So, $3.27 * 0.8 = 2.616$.

Check Your Understanding

Multiply.

1. $1.7 * 5.7$ 2. $2.33 * 8.4$ 3. $0.61 * 4.04$ 4. $0.3 * 0.021$

Check your answers on page 434.

Lattice Multiplication with Decimals

Example Find 34.5 * 2.05 using lattice multiplication.

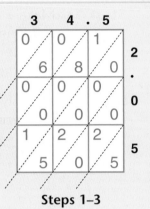

Step 1: Make a magnitude estimate. 34.5 * 2.05 ≈ 35 * 2 = 70
The product will be in the tens. (The symbol ≈ means *is about equal to*.)

Step 2: Draw the lattice and write the factors, including the decimal points, at the top and right side. In the factor above the grid, the decimal point should be above a column line. In the factor on the right side of the grid, the decimal point should be to the right of a row line.

Step 3: Find the products inside the lattice.

Step 4: Add along the diagonals, moving from right to left.

Steps 1–3

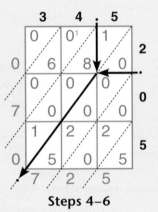

Step 5: Locate the decimal point in the answer as follows. Slide the decimal point in the factor above the grid down along the column line. Slide the decimal point in the factor on the right side of the grid across the row line. When the decimal points meet, slide the decimal point down along the diagonal line. Write a decimal point at the end of the diagonal line.

Step 6: Compare the result with the estimate.

The product, 70.725, is very close to the estimate of 70.

Steps 4–6

Example Find 73.4 * 10.5 using lattice multiplication.

A good magnitude estimate is
73.4 * 10.5 ≈ 73 * 10 = 730. (The symbol ≈ means *is about equal to*.)

The product, 770.70, is close to the estimate of 730.

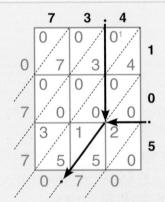

Did You Know?

The lattice method of multiplication was used by Persian scholars as long ago as the year 1010. It was often called the "grating" method.

Check Your Understanding

Draw a lattice for each problem and multiply.

1. 32.5 * 2.5   **2.** 4.02 * 17   **3.** 8.1 * 23.4

Check your answers on page 434.

Dividing by Powers of 10 Greater than 1

Here is one method for dividing by a power of 10 greater than 1.

Example 45.6 / 1,000 = ?

Step 1: Locate the decimal point in the power of 10.

1,000 = 1000.

Step 2: Move the decimal point LEFT until you get to the number 1.

1 . 0 0 0 .

Step 3: Count the number of places you moved the decimal point.

3 places

Step 4: Move the decimal point in the other number the same number of places to the LEFT. Insert 0s as needed.

0 . 0 4 5 . 6

45.6 / 1,000 = 0.0456

Note

Some powers of 10 greater than 1:

$10^1 = 10$

$10^2 = 100$

$10^3 = 1,000$

$10^4 = 10,000$

$10^5 = 100,000$

$10^6 = 1,000,000$

Examples

350 / 100 = ?	350 / 10,000 = ?	$290.50 / 1,000 = ?
100 = 100.	10,000 = 10000.	1,000 = 1000.
1 . 0 0 .	1 . 0 0 0 0 .	1 . 0 0 0 .
2 places	4 places	3 places
3 . 5 0 .	0 . 0 3 5 0 .	0 . 2 9 0 . 5 0
350 / 100 = 3.50	350 / 10,000 = 0.0350	$290.50 / 1,000 = $0.29 (rounded to the nearest cent)

Note: When the dividend (the number you are dividing) does not have a decimal point, you must locate the decimal point before moving it. For example, 350 = 350.

Check Your Understanding

Divide.

1. 56.7 / 10   **2.** 0.47 / 100   **3.** $290 / 1,000   **4.** 60 / 10,000

Check your answers on page 434.

forty-one   **41**

Division of Decimals

Here is one way to divide decimals:

Step 1: Make a magnitude estimate of the quotient.

Step 2: Divide as if the divisor and dividend were whole numbers.

Step 3: Use the magnitude estimate to place the decimal point in the answer.

Note

A *magnitude estimate* is a rough estimate of the size of an answer. A magnitude estimate tells whether an answer is in the ones, tens, hundreds, and so on.

Example   97.24 / 26 = ?

Step 1: Make a magnitude estimate.

- Since 26 is close to 25 and 97.24 is close to 100, the answer to 97.24 / 26 will be close to the answer to 100 / 25.

- Since 100 / 25 = 4, the answer to 97.24 / 26 should be in the ones. (*In the ones* means between 1 and 10.)

Step 2: Divide, ignoring the decimal point.

```
26)9724
  - 7800      300
    1924
  - 1040      40
     884
   - 780      30
     104
   - 104       4
       0     374
```

9724 / 26 = 374

Step 3: Decide where to place the decimal point. According to the magnitude estimate, the answer should be in the ones.

So, 97.24 / 26 = 3.74.

Note

Sometimes a magnitude estimate is on the "borderline" and you need to be more careful.

For example, a magnitude estimate for 2,890 / 3.4 is 3,000 / 3 = 1,000. This answer is "in the thousands." But the exact answer may be "in the hundreds." You should place the decimal point so that the answer is close to 1,000.

Since 2,890 / 34 = 85, you should attach one zero, followed by a decimal point: 2,890 / 3.4 = 850.

Check Your Understanding

Divide.

1. 148.8 / 6

2. 25.32 / 12

3. 4.55 / 3.5

Check your answers on page 434.

The answers to decimal divisions do not always come out even. When you divide as if the divisor and dividend were whole numbers, there may be a non-zero remainder. If the remainder is not zero:

1. Rewrite a remainder as a fraction:

   ♦ Make the remainder the *numerator* of the fraction.

   ♦ Make the divisor the *denominator* of the fraction.

2. Add this fraction to the quotient and round the sum to the nearest whole number.

3. Then use the magnitude estimate to place the decimal point in the answer.

The decimal division below does not come out even.

Example | 80.27 / 4 = ?

Make a magnitude estimate.
- Since 80.27 is close to 80, 80.27 / 4 ≈ 80 / 4.
- Since 80 / 4 = 20, the answer to 80.27 / 4 should be in the tens. (*In the tens* means between 10 and 100.)

Divide, ignoring the decimal point.

$$
\begin{array}{r|r}
4\overline{)8027} & \\
-\ 8000 & 2000 \\
\hline
27 & \\
-\ 24 & 6 \\
\hline
3 & 2006 \\
\end{array}
$$

8027 / 4 → 2006 R3. The quotient is 2006, and the remainder is 3.

Rewrite the remainder 3 as the fraction $\frac{3}{4}$.

Add this fraction to the quotient: 8027 / 4 = $2006\frac{3}{4}$.

Round this answer to the nearest whole number, 2007.

Decide where to put the decimal point. According to the magnitude estimate, the answer should be in the tens.

So, 80.27 / 4 ≈ 20.07.

Note

The symbol ≈ means *is about equal to.*

Check Your Understanding

Divide.

1. 8.8 / 3

2. 86.4 / 24

3. 45.2 / 3

Check your answers on page 434.

Column Division with Decimal Quotients

Column division can be used to find quotients that have a decimal part.
In the example below, think of sharing $15 equally among 4 people.

Example $4 \overline{)15} = ?$

1. Set the problem up. Draw a line to separate the digits in the dividend. Work left to right. Think of the 1 in the tens column as 1 $10 bill.

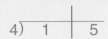

2. The 1 $10 bill cannot be shared by 4 people. So trade it for 10 $1 bills. Think of the 5 in the ones column as 5 $1 bills. That makes 10 + 5, or 15 $1 bills in all.

3. If 4 people share 15 $1 bills, each person gets 3 $1 bills. There are 3 $1 bills left over.

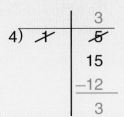

4. Draw a line and make decimal points to show amounts less than $1. Write 0 after the decimal point in the dividend to show there are 0 dimes. Then trade the 3 $1 bills for 30 dimes.

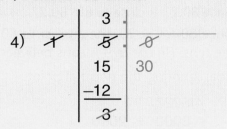

5. If 4 people share 30 dimes, each person gets 7 dimes. There are 2 dimes left over. Draw another line and another 0 in the dividend to show pennies.

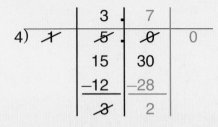

6. Trade the 2 dimes for 20 pennies.

7. If 4 people share 20 pennies, each person gets 5 pennies.

The column division shows that
15 / 4 = 3.75.

This means that $15 shared equally among 4 people is $3.75 each.

Rounding Decimals

Sometimes numbers have more digits than we need to use. This is especially true of decimals. A calculator display may show seven or more digits to the right of the decimal point, even when only one or two digits are needed.

Rounding is a way to get rid of unnecessary digits to the right of the decimal point. There are three basic ways to round numbers: A number may be rounded *down,* rounded *up,* or rounded to a *nearest place.* The examples on pages 45 and 46 involve rounding to hundredths, but rounding to any other place to the right of the decimal point is done in a similar way.

> **Note**
>
> See page 249 for examples that involve rounding to the *left* of the decimal point.

Rounding Down

To round a decimal down to a given place, just drop all the digits to the right of the desired place.

When a bank computes the interest on a savings account, the interest is calculated to the nearest tenth of a cent. But the bank cannot pay a fraction of one cent. So the interest is **rounded down,** and any fraction of one cent is ignored.

> **Did You Know?**
>
> *Truncate* means to shorten by cutting off a part. Rounding down a decimal by dropping digits to the right of the decimal point is often called *truncating.*

> **Example** The bank calculates the interest earned as $17.218. Round down to the next cent (the hundredths place).
>
> First, find the place you are rounding to: $17.2**1**8.
> Then, drop all the digits to the right of that place: $17.21.
>
> The bank pays $17.21 in interest.

Rounding Up

To round a decimal up to a given place, look at all the digits to the right of the desired place. If any digit to the right of the desired place is not 0, then add 1 to the digit in the place you are rounding to. (You will have to do some trading if there is a 9 in that place.) If all the digits to the right of the desired place are 0, then leave the digit unchanged. Finally, drop all the digits to the right of the desired place.

Running events at the Olympic Games are timed with automatic electric timers. The electric timer records a time to the nearest thousandth of a second and automatically **rounds up** to the next hundredth of a second. The rounded time becomes the official time.

1 minute 2.045 seconds
rounds up to 1 minute 2.05 seconds.

Examples   The winning time was 11.437 seconds.

Round up 11.437 seconds to the next hundredth of a second.

First, find the place you are rounding to: 11.4**3**7.
The digit to the right is not 0, so add 1 to the digit you are rounding to: 11.4**4**7.
Finally, drop all digits to the right of hundredths: 11.44.

The official winning time is 11.44 seconds.

11.431 seconds is rounded up to 11.44 seconds.

11.430 seconds is rounded up to 11.43 seconds because every digit to the right of the hundredths place is a 0. In this problem, rounding up does not change the number at all. 11.43 equals 11.430.

Rounding to the Nearest Place

Rounding to the nearest place is sometimes like rounding up and sometimes like rounding down. To round a decimal to the nearest place, follow these steps:

Step 1:   Find the digit to the right of the place you are rounding to.

Step 2:   If that digit is 5 or more, round up. If that digit is less than 5, round down.

Examples   Mr. Wilson is labeling the grocery shelves with unit prices so customers can compare the cost of items. To find a unit price, he divides the price by the quantity. Often, the quotient has more decimal places than are needed, so he **rounds to the nearest** cent (the nearest hundredth).

$1.2**3**422 is rounded down to $1.23.

$3.8**9**822 is rounded up to $3.90.

$1.8**6**5 is rounded up to $1.87.

Check Your Understanding

1. Round *down* to tenths.   **a.** 1.62   **b.** 36.592   **c.** 1.95

2. Round *up* to tenths.   **a.** 1.62   **b.** 36.592   **c.** 1.95

3. Round to the *nearest* tenth.   **a.** 1.62   **b.** 36.59   **c.** 10.95

Check your answers on page 434.

Percents

A percent is another way to name a fraction or decimal. Percent means *per hundred,* or *out of a hundred.* So, 1% has the same meaning as the fraction $\frac{1}{100}$ and the decimal 0.01. And 60% has the same meaning as $\frac{60}{100}$ and 0.60.

The statement "60% of students were absent" means that 60 out of 100 students were absent. This does *not* mean that there were exactly 100 students and that 60 of them were absent. It does mean that for *every* 100 students, 60 students were absent.

A percent usually represents a percent of something. The "something" is the whole (or ONE, or 100%). In the statement, "60% of the students were absent," the whole is the total number of students in the school.

> **Note**
>
> The word *percent* comes from the Latin *per centum: Per* means *for* and *centum* means *one hundred.*

> **Example** There are 250 students in Esmond School. One winter day, 60% of the students were absent. How many students were absent that day?
>
> *Think:* 250 = 100 + 100 + 50
>
> For every 100 students, 60 were absent.
>
> So, for every 50 students ($\frac{1}{2}$ of 100), 30 were absent ($\frac{1}{2}$ of 60).
>
> 60 + 60 + 30 = 150 students were absent that day.
>
> | 250 | = | 100 | + | 100 | + | 50 |
> |-----|---|-----|---|-----|---|----|
> | | | ↓ | | ↓ | | ↓ |
> | | | 60 absent | | 60 absent | | 30 absent |

Percents are used in many ways in everyday life:

♦ *Business:* "50% off" means that the price of an item will be reduced by 50 cents for every 100 cents the item usually costs.

♦ *Statistics:* "55% voter turnout" means that 55 out of every 100 registered voters will actually vote.

♦ *School:* An 80% score on a spelling test means that a student scored 80 out of 100 possible points for that test. One way to score 80% is to spell 80 words correctly out of 100. Another way to score 80% is to spell 8 words correctly out of 10.

♦ *Probability:* A "30% chance of showers" means that for every 100 days that have similar weather conditions, you would expect it to rain on 30 of the days.

> **SALE – 50% OFF
> Everything Must Go**

> Voter Turnout Pegged at 55% of Registered Voters

> For Wednesday, there is a 30% chance of showers.

Fractions, Decimals, and Percents

Percents may be used to rename both fractions and decimals.

Percents are another way of naming fractions with a denominator of 100.

> You can think of the fraction $\frac{25}{100}$ as 25 parts per hundred, or 25 out of 100, and write 25%.
>
> You can rename the fraction $\frac{1}{5}$ as $\frac{1*20}{5*20}$, or $\frac{20}{100}$, or 20%.
>
> 75% can be written as $\frac{75}{100}$, or $\frac{3}{4}$.

Percents are another way of naming decimals in terms of hundredths.

> Since 0.01 can be written as $\frac{1}{100}$, you can think of 0.01 as 1%. And you can think of 0.39 as $\frac{39}{100}$, or 39%.
>
> 58% means 58 hundredths, or 0.58.

Percents can also be used to name the whole.

> 100 out of 100 can be written as the fraction $\frac{100}{100}$, or 100 hundredths. This is the same as 1 whole, or 100%.

Examples

$$85\% = \frac{85}{100} = 0.85 \qquad 50\% = \frac{50}{100} = 0.5$$

$$200\% = \frac{200}{100} = 2 \qquad 37.5\% = \frac{37.5}{100} = 0.375$$

Example The amounts shown in the pictures below can be written as $\frac{1}{4}$, or 25%, or 0.25.

$$\frac{1}{4} = \frac{1*25}{4*25} = \frac{25}{100}$$

But $\frac{25}{100} = 0.25$. And $\frac{25}{100} = 25\%$.

So, $\frac{1}{4}$ and $\frac{25}{100}$ and 0.25 and 25% all name the same amount.

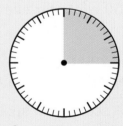
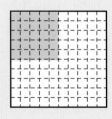

$\frac{1}{4} = 25\% = 0.25$

Finding a Percent of a Number

Finding a percent of a number is a basic problem that comes up over and over again. For example:

♦ A backpack that regularly sells for $60 is on sale for 20% off. What is the sale price?

♦ The sales tax on food is 5%. What is the tax on $80 worth of groceries?

♦ A borrower pays 10% interest on a car loan. If the loan is $6,000, how much is the interest?

There are many different ways of finding the percent of a number.

20% OFF
ALL CAMPING GEAR

Use a Fraction

Some percents are equivalent to "easy" fractions. For example, 25% is the same as $\frac{25}{100}$, or $\frac{1}{4}$. It is usually easier to find 25% of a number by thinking of 25% as $\frac{1}{4}$.

Some "easy" fractions and percents:		
$\frac{1}{2}$ =	$\frac{50}{100}$ =	50%
$\frac{1}{4}$ =	$\frac{25}{100}$ =	25%
$\frac{3}{4}$ =	$\frac{75}{100}$ =	75%
$\frac{1}{5}$ =	$\frac{20}{100}$ =	20%
$\frac{2}{5}$ =	$\frac{40}{100}$ =	40%
$\frac{3}{5}$ =	$\frac{60}{100}$ =	60%
$\frac{4}{5}$ =	$\frac{80}{100}$ =	80%
$\frac{1}{10}$ =	$\frac{10}{100}$ =	10%
$\frac{3}{10}$ =	$\frac{30}{100}$ =	30%
$\frac{7}{10}$ =	$\frac{70}{100}$ =	70%
$\frac{9}{10}$ =	$\frac{90}{100}$ =	90%

Example What is 25% of 48?

Think: 25% = $\frac{25}{100}$ = $\frac{1}{4}$, so 25% of 48 is the same as $\frac{1}{4}$ of 48.

Divide 48 into 4 equal groups. Each group is $\frac{1}{4}$ of 48, and each group has 12.

So, 25% of 48 is 12.

Example What is 20% of 60?

Think: 20% = $\frac{20}{100}$ = $\frac{1}{5}$, so 20% of 60 is the same as $\frac{1}{5}$ of 60.

Divide 60 into 5 equal groups. Each group is $\frac{1}{5}$ of 60, and each group has 12.

So, 20% of 60 is 12.

If a percent does not equal an "easy" fraction, you might find 1% first.

Example What is 7% of 300?

$1\% = \frac{1}{100}$, so 1% of 300 is the same as $\frac{1}{100}$ of 300.

If you divide 300 into 100 equal groups, there are 3 in each group.

1% of 300 is 3. Then 7% of 300 is 7 * 3.

So, 7% of 300 = 21.

Sometimes it is helpful to find 10% first.

Example What is 30% of 60?

$10\% = \frac{10}{100} = \frac{1}{10}$. 10% of 60 is $\frac{1}{10}$ of 60. If you divide 60 into 10 equal groups, each group has 6. 10% of 60 is 6. Then 30% of 60 is 3 * 6.

So, 30% of 60 = 18.

Use Decimal Multiplication

Finding a percent of a number is the same as multiplying the number by the percent. Usually, it's easiest to change the percent to a decimal and use a calculator.

Example What is 35% of 55?

$35\% = \frac{35}{100} = 0.35$

Change the percent to a decimal. Multiply using a calculator.

Key in: 0.35 ⊗ 55 ⊜   Answer: 19.25

If your calculator has a ⬭%⬭ key, you don't need to rename the percent as a decimal.

To find 35% of 55, key in 35 ⬭%⬭ ⊗ 55 ⏎ on Calculator A.
Or, key in 55 ⬭×⬭ 35 ⬭%⬭ ⬭=⬭ on Calculator B.

35% of 55 is 19.25.

1% of 55 means $\frac{1}{100}$ * 55
or 0.01 * 55.

7% of 55 means $\frac{7}{100}$ * 55
or 0.07 * 55.

35% of 55 means $\frac{35}{100}$ * 55
or 0.35 * 55.

The word *of* in problems like these means multiplication.

Check Your Understanding

Solve.

1. A $60 backpack is on sale for 20% off. What is the sale price?

2. The sales tax on food is 5%. What is the tax on $80 worth of groceries?

3. A borrower pays 10% interest on a $6,000 car loan. How much is the interest?

Check your answers on page 434.

Calculating a Discount

A **discount** is an amount taken off a regular price; it's the amount you save. Stores may display the regular price and percent discount and let customers figure out the sale price.

If the percent discount is equivalent to an "easy" fraction, then a good way to solve this kind of problem is by using the fraction.

> **Example** The list price for a desk lamp is $50, but it is on sale at a 20% discount (20% off the list price). What are the savings?
>
> Change 20% to a fraction: $20\% = \frac{20}{100} = \frac{1}{5}$.
>
> Since $20\% = \frac{1}{5}$, the discount is $\frac{1}{5}$ of $50.
>
> $\frac{1}{5}$ of $50 means $\frac{1}{5} * \$50$.
>
> $\frac{1}{5} * \$50 = \frac{1}{5} * \frac{\$50}{1} = \frac{1 * \$50}{5 * 1} = \frac{\$50}{5} = \$10$
>
> The discount is $10.

Note

For any fractions $\frac{a}{b}$ and $\frac{c}{d}$, $\frac{a}{b} * \frac{c}{d} = \frac{a * c}{b * d}$.

If the percent discount is not equivalent to an "easy" fraction, it's usually best to change the percent to a decimal first, and then to multiply with paper and pencil or a calculator.

> **Example** The list price for a radio is $45. The radio is sold at a 12% discount (12% off the list price). What are the savings? (*Reminder:* $12\% = \frac{12}{100} = 0.12$)
>
> **Paper and pencil:**
> 12% of 45 means $\frac{12}{100} * 45$, or $0.12 * 45$.
>
> 12% of $\$45 = \frac{12}{100} * \$45 = \frac{12}{100} * \frac{\$45}{1} = \frac{12 * \$45}{100 * 1} = \frac{\$540}{100} = \$5.40$
>
> **Calculator:**
> Key in: 0.12 ⊠ 45 ⊜. Interpret the answer 5.4 as $5.40.
>
> The discount is $5.40.

Did You Know?

Stores sometimes offer "33% off" sales. What they often mean is "$\frac{1}{3}$ off," but they do not want to write the discount as a fraction ($\frac{1}{3}$) or as a percent that includes a fraction ($33\frac{1}{3}\%$).

Check Your Understanding

Solve.

1. The list price of a pair of jeans is $30. The jeans are being sold at a 10% discount. What are the savings?

2. Movies cost $9.00, but shows before 4 P.M. are 25% off. How much cheaper are the early shows?

Check your answers on page 434.

Finding the Whole in Percent Problems

Sometimes you know a percent and how much it's worth, but you don't know what the ONE is.

Examples The sale price of a CD player is $120. It is on sale for 60% of its list price. What is the list price?

This problem can be solved in different ways.

Solution 1: Use fractions.

Find an "easy" fraction that is equivalent to 60%. $60\% = \frac{60}{100} = \frac{3}{5}$

This means that $\frac{3}{5}$ of the list price is $120.

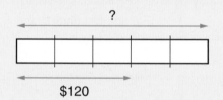

$120

Since $\frac{3}{5}$ of the list price is $120, $\frac{1}{5}$ of the list price is $40 ($\frac{\$120}{3} = \$40$).

Then $\frac{5}{5}$ of the list price is $5 * \$40 = \$200$.

Solution 2: Use percents.

60% is worth $120.

So, 1% is worth $2 ($\frac{\$120}{60} = \$2$), and 100% is worth $200 ($100 * \$2 = \$200$).

The list price is $200.

Examples A tea kettle is on sale for 80% of its list price. The sale price is $40. What is the list price?

Solution 1: Use fractions.

$80\% = \frac{80}{100} = \frac{4}{5}$ This means that $\frac{4}{5}$ of the list price is $40.

Since $\frac{4}{5}$ of the list price is $40, $\frac{1}{5}$ of the list price is $10 ($\frac{\$40}{4} = \$10$).

Then $\frac{5}{5}$ of the list price is $5 * \$10 = \$50$.

Solution 2: Use percents.

80% is worth $40.

So, 1% is worth $0.50 ($\frac{\$40}{80} = \$0.50$), and 100% is worth $50 ($100 * \$0.50 = \$50$).

The list price is $50.

Example In Alaska, there are about 103,000 Native Americans. These Native Americans make up about 16% of Alaska's population. What is the population of Alaska?

Use a 1% strategy. First find 1%. Then multiply by 100 to get 100%.

• Use your calculator to divide 103,000 by 16:

Key in: 103,000 ÷ 16 = Answer: 6437.5

• Multiply by 100:

Key in: 100 × 6437.5 = Answer: 643750

The total population of Alaska is 643,750, or about 650,000.

Check Your Understanding

Solve.

1. A bicycle is on sale for 50% of the list price. The sale price is $110. What is the list price?

2. A cellular telephone is on sale for 40% of the list price. The sale price is $60. What is the list price?

3. In Canada, there are about 6 million children aged 14 and younger. These children make up about 19% of Canada's population. What is the population of Canada?

Check your answers on page 435.

The Whole Circle

One full turn of a circle can be divided in various ways. One way is to divide it into four equal parts: a quarter turn, a half turn, three-fourths (or three-quarters) of a turn, and a full turn. A full turn can also be broken into 360 or 100 equal parts.

A Circle Protractor

There are 360 equally-spaced marks around the edge of the Circle Protractor. Each pair of side-by-side marks measures **one degree (1°)** of angle measure. That means there are 360 degrees (360°) in the whole circle.

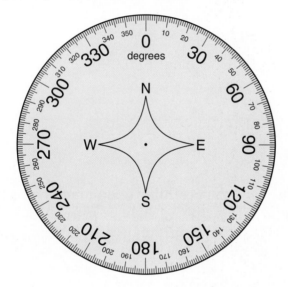

The Percent Circle

There are 100 equally-spaced marks around the edge of the Percent Circle. The marks define 100 thin, pie-shaped wedges. Each wedge contains **one percent (1%)** of the total area inside the circle. There are 100 wedges (100%) inside the whole circle.

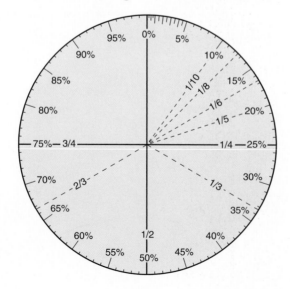

U.S. Traditional Addition: Decimals

You can use **U.S. traditional addition** to add decimals.

Did You Know?

When you add decimals, make sure you line up the places properly so that tenths are added to tenths, ones to ones, and so on. *For example:*

$$14.5$$
$$+\ \ 4.18$$

Example 5.29 + 4.96 = ?

Step 1: Start with the 0.01s: 9 + 6 = 15.
15 hundredths = 1 tenth + 5 hundredths

$$\begin{array}{r} {}^{1} \\ 5\ .\ 2\ 9 \\ +\ 4\ .\ 9\ 6 \\ \hline 5 \end{array}$$

Step 2: Add the 0.1s: 1 + 2 + 9 = 12.
12 tenths = 1 whole + 2 tenths

$$\begin{array}{r} {}^{1}\ \ \ {}^{1} \\ 5\ .\ 2\ 9 \\ +\ 4\ .\ 9\ 6 \\ \hline 2\ 5 \end{array}$$

Step 3: Add the 1s: 1 + 5 + 4 = 10.
10 ones = 1 ten + 0 ones
Remember to include the decimal point in the answer.

$$\begin{array}{r} {}^{1}\ \ \ {}^{1} \\ 5\ .\ 2\ 9 \\ +\ 4\ .\ 9\ 6 \\ \hline 1\ 0\ .\ 2\ 5 \end{array}$$

5.29 + 4.96 = 10.25

Check Your Understanding

Add.

1. 3.79 + 8.62 = ?

2. 74.50
+ 6.73

3. 8.189 + 348.2 = ?

4. 63.7
+ 31.84

Check your answers on page 442.

U.S. Traditional Subtraction: Decimals

You can use **U.S. traditional subtraction** to subtract decimals.

Example $7.83 - 2.89 = ?$

Step 1: Start with the 0.01s.
Since $9 > 3$, you need to regroup.
Trade 1 tenth for 10 hundredths:
$7.83 = 7$ ones $+ 7$ tenths $+ 13$ hundredths.
Subtract the 0.01s: $13 - 9 = 4$.

$$\begin{array}{r} 7\;\;13 \\ 7\,.\,\cancel{8}\;\cancel{3} \\ -\;2\,.\,8\;\;9 \\ \hline 4 \end{array}$$

Step 2: Go to the 0.1s.
Since $8 > 7$, you need to regroup.
Trade 1 one for 10 tenths:
$7.83 = 6$ ones $+ 17$ tenths $+ 13$ hundredths.
Subtract the 0.1s: $17 - 8 = 9$.

$$\begin{array}{r} 17 \\ 6\;\;\cancel{7}\;\;13 \\ \cancel{7}\,.\,\cancel{8}\;\cancel{3} \\ -\;2\,.\,8\;\;9 \\ \hline 9\;\;4 \end{array}$$

Step 3: Go to the 1s.
You don't need to regroup.
Subtract the 1s: $6 - 2 = 4$.
Remember to include the decimal point in the answer.

$$\begin{array}{r} 17 \\ 6\;\;\cancel{7}\;\;13 \\ \cancel{7}\,.\,\cancel{8}\;\cancel{3} \\ -\;2\,.\,8\;\;9 \\ \hline 4\,.\,9\;\;4 \end{array}$$

$7.83 - 2.89 = 4.94$

Check Your Understanding

Subtract.

1. $78.28 - 14.98 = ?$

2. $\begin{array}{r} 2.80 \\ -\;1.89 \\ \hline \end{array}$

3. $\begin{array}{r} 39.03 \\ -\;\;6.80 \\ \hline \end{array}$

4. $10.63 - 7.04 = ?$

Check your answers on page 442.

U.S. Traditional Multiplication: Decimals

You can use **U.S. traditional multiplication** with decimals. Here are two ways to use this method for multiplying decimals.

Method 1

One way to use U.S. traditional multiplication with decimals is to multiply them as though they were whole numbers and then use estimation to place the decimal point.

Step 1: Multiply as though both factors were whole numbers.

Step 2: Estimate the product.

Step 3: Use your estimate to place the decimal point in the answer.

Example $7 * 3.72 = ?$

Step 1: Multiply as though the factors were whole numbers.

Step 2: Estimate the product.
$7 * 3.72 \approx 7 * 4 = 28$

$$
\begin{array}{r}
{\scriptstyle 5 \quad 1} \\
3\ 7\ 2 \\
* \qquad 7 \\
\hline
2\ 6\ 0\ 4
\end{array}
$$

Step 3: Use the estimate to place the decimal point in the answer. The estimate is 28, so place the decimal point to make the answer close to 28: 26.04 is close to 28.

$7 * 3.72 = 26.04$

Method 2

The second method for multiplying decimals is useful when there are many decimal places in the factors and it becomes harder to estimate the answer.

Step 1: Multiply as though both factors were whole numbers.

Step 2: Count the total number of places to the RIGHT of the decimal point for both factors.

Step 3: Place the decimal point so that you have the same number of decimal places as the total in Step 2.

Did You Know?

When people talk about the number of decimal places in a number, they usually mean the number of places to the right of the decimal point. So, for example, we say the number 11.732 has three decimal places. The number 45.06 has two decimal places. The number 0.0078 has four decimal places.

Example   0.078 * 0.029 = ?

Step 1: Multiply as though the factors were whole numbers.

Step 2: Count the total number of decimal places in both factors. 0.078 has 3 decimal places; 0.029 has 3 decimal places. There are 6 decimal places in all.

Step 3: Place the decimal point 6 places from the right. Insert 0s as needed.

$$
\begin{array}{r}
6\,7 \\
2\ 9 \\
*\ \ \ 7\ 8 \\
\hline
2\ 3\ 2 \\
+\ 2\ 0\ 3\ 0 \\
\hline
2\ 2\ 6\ 2
\end{array}
$$

$$0.002262$$

0.078 * 0.029 = 0.002262

Check Your Understanding

Multiply.

1. 54.5 * 4 = ?   **2.** 7.05 * 0.25 = ?   **3.**   0.065   **4.** 12.05 * 6.71 = ?
                                                           *   1.04

Check your answers on page 442.

U.S. Traditional Long Division: Decimal Dividends

You can use U.S. traditional long division to divide money in dollars-and-cents notation.

Example Share $8.17 among 5 people.

Step 1: Share the 8 dollars.

$$
\begin{array}{r}
1 \\
5\overline{)\$8.17} \\
-5 \\
\hline
3
\end{array}
$$

← Each person gets 1 dollar.

← 1 dollar each for 5 people

← 3 dollars are left.

Step 2: Trade the dollars for dimes. Share the dimes.

$$
\begin{array}{r}
1.6 \\
5\overline{)\$8.17} \\
-5 \\
\hline
3\,1 \\
-3\,0 \\
\hline
1
\end{array}
$$

← Each person gets 6 dimes. Write a decimal point to show amounts less than a dollar.

← 30 dimes + 1 dime

← 6 dimes each for 5 people

← 1 dime is left.

Step 3: Trade the dime for pennies. Share the pennies.

$$
\begin{array}{r}
1.63 \\
5\overline{)\$8.17} \\
-5 \\
\hline
3\,1 \\
-3\,0 \\
\hline
17 \\
-15 \\
\hline
2
\end{array}
$$

← Each person gets 3 pennies.

← 10 pennies + 7 pennies

← 3 pennies each for 5 people

← 2 pennies are left.

Each person gets $1.63. There is 2¢ left.

$8.17 / 5 → $1.63 R2¢

Check Your Understanding

Divide.

1. $6.25 / 5   **2.** 5)$6.75   **3.** 8)$4.80   **4.** $3.85 / 7

Check your answers on page 442.

You can use U.S. traditional long division to divide decimals that
do not represent money.

Example   2.79 / 6 = ?

Step 1: Trade the ones for tenths and share the tenths.

```
    .4        ← Each share gets 4 tenths. Write a decimal point in the quotient.
6)2.79        ← 2 ones + 7 tenths = 27 tenths
 −2 4         ← 4 tenths * 6 shares = 24 tenths
    3         ← 3 tenths are left.
```

Step 2: Trade the remaining tenths for hundredths. Share the hundredths.

```
    .46       ← Each share gets 6 hundredths.
6)2.79
 −2 4
    39        ← 3 tenths + 9 hundredths = 39 hundredths
   −36        ← 6 hundredths * 6 shares = 36 hundredths
     3        ← 3 hundredths are left.
```

At this point, you can either round 0.46 to 0.5 and write 2.79 / 6 ≈ 0.5,
or you can continue dividing into the thousandths.

Step 3: Continue dividing into the thousandths. Add a 0 to the end of 2.79.
(Adding 0s or "padding" a decimal with 0s doesn't change its value.)

```
    .465      ← Each share gets 5 thousandths.
6)2.790
 −2 4
    39
   −36
    30        ← 3 hundredths + 0 thousandths = 30 thousandths
   −30        ← 5 thousandths * 6 shares = 30 thousandths
     0        ← No thousandths are left.
```

2.79 / 6 = 0.465

Check Your Understanding

Divide.

1. 6.29 / 4   **2.** 3)7.83   **3.** 4)8.37   **4.** 6.74 / 8

Check your answers on page 442.

U.S. Traditional Long Division: Decimal Divisors

To use U.S. traditional long division to divide by a decimal number, such as 0.7 or 4.5, you can find an equivalent problem that has no decimal in the divisor. The answer to the equivalent problem is the same as the answer to your original problem.

Step 1: Think of the division problem as a fraction.

Step 2: Use the multiplication rule to find an equivalent fraction that has no decimal in the denominator.

Step 3: Think of the equivalent fraction as a division problem.

Step 4: Solve the division problem. The answer to the equivalent problem is the same as the answer to the original problem.

Example 364 / 0.8 = ?

Step 1: Think of the division problem as a fraction.

$$364 / 0.8 = \frac{364}{0.8}$$

Step 2: Find an equivalent fraction with no decimal in the denominator.

$$\frac{364 * 10}{0.8 * 10} = \frac{3640}{8}$$

Step 3: Think of the equivalent fraction as a division problem.

$$\frac{3640}{8} = 3640 / 8$$

Step 4: Solve the equivalent division problem.

$$
\begin{array}{r}
455 \\
8\overline{)3640} \\
-32 \\
\hline
44 \\
-40 \\
\hline
40 \\
-40 \\
\hline
0
\end{array}
$$

Because $\frac{3640}{8}$ and $\frac{364}{0.8}$ are equivalent fractions, the division problems 3640 / 8 and 364 / 0.8 are equivalent. So the answer to 3640 / 8 is the same as the answer to 364 / 0.8.

364 / 0.8 = 455

U.S. Traditional Long Division: Decimal Divisors and Dividends

Sometimes *both* the divisor (the number you are dividing by) and the dividend (the number being divided) are decimal numbers. To use U.S. traditional long division in such cases, you can find an equivalent problem that has no decimal in the divisor. (Having a decimal part in the dividend is okay.) The answer to the equivalent problem is the same as the answer to your original problem.

Example   2.76 / 0.4 = ?

Step 1: Think of the division problem as a fraction.

$$2.76 / 0.4 = \frac{2.76}{0.4}$$

Step 2: Find an equivalent fraction with no decimal in the denominator.

$$\frac{2.76 * 10}{0.4 * 10} = \frac{27.6}{4}$$

Step 3: Think of the equivalent fraction as a division problem.

$$\frac{27.6}{4} = 27.6 / 4$$

Step 4: Solve the division problem.

```
      6.9
  4)27.6
   −24
     3 6
    −3 6
       0
```

Because $\frac{27.6}{4}$ and $\frac{2.76}{0.4}$ are equivalent fractions, the division problems 27.6 / 4 and 2.76 / 0.4 are equivalent. So the answer to 27.6 / 4 is the same as the answer to 2.76 / 0.4.

2.76 / 0.4 = 6.9

Check Your Understanding

Divide.

1. 687 / 0.3   **2.** 29.6 / 1.8   **3.** 3.95 / 0.05   **4.** 16.17 / 0.7

Check your answers on page 442.

SRB
54H

U.S. Traditional Long Division: Renaming Fractions as Decimals

U.S. traditional long division can be used to rename fractions as decimals.

Example Use U.S. traditional long division to rename $\frac{5}{8}$ as a decimal.

Step 1: Write $\frac{5}{8}$ as a division problem. Write 5 with several 0s after the decimal point: 5.000. (You can always add more 0s if you need them.)

$$8\overline{)5.000}$$

Step 2: Solve the division problem. Stop when the remainder is 0, or when you have enough precision for your purposes, or when you notice a repeating pattern.

$$
\begin{array}{r}
.625 \\
8\overline{)5.000} \\
-4\ 8 \\
\hline
20 \\
-16 \\
\hline
40 \\
-40 \\
\hline
0
\end{array}
$$

This division problem divided evenly in three decimal places.

$\frac{5}{8} = 0.625$

Example | Use U.S. traditional long division to rename $\frac{7}{11}$ as a decimal.

Step 1: Write $\frac{7}{11}$ as a division problem. Write 7 with several 0s after the decimal point: 7.000. (You can always add more 0s if you need them.)

$$11\overline{)7.000}$$

Step 2: Solve the division problem. Stop when the remainder is 0, or when you have enough precision for your purposes, or when you notice a repeating pattern.

```
        .636363
11)7.000000
   −6 6
      40
     −33
       70
      −66
        40
       −33
         70
        −66
          40
         −33
           7
```

The digits 6 and 3 in the quotient appear to repeat forever.

$$\frac{7}{11} = 0.636363... = 0.\overline{63}$$

Check Your Understanding

Use long division to rename these fractions as decimals.

1. $\frac{5}{6}$ 2. $\frac{2}{11}$ 3. $\frac{5}{9}$ 4. $\frac{7}{3}$

Check your answers on page 442.

Fractions

Fractions

Fractions were invented thousands of years ago to name numbers that are between whole numbers. People probably first used these in-between numbers for making more exact measurements.

Today most rulers and other measuring tools have marks to name numbers that are between whole measures. Learning how to read these in-between marks is an important part of learning to use these tools. Here are some examples of measurements that use fractions: $\frac{2}{3}$ cup, $\frac{3}{4}$ hour, $\frac{9}{10}$ kilometer, and $13\frac{1}{2}$ pounds.

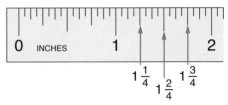

The $\frac{1}{4}$-in. marks between 1 and 2 are labeled.

Fractions are also used to name parts of wholes. The whole might be one single thing, like a stick of butter. Or, the whole might be a collection of things, like a box of cookies. The whole is sometimes called the ONE. In measurements, the whole is called the *unit*.

Whole
24 Cookies

The "whole" box names the ONE being considered.

To understand fractions you need to know what the ONE is. Half a box of crayons might be many crayons or just a few crayons depending on the size of the box. Half an inch is much less than half a mile.

Fractions are also used to show division, in rates and ratios, and in many other ways.

Naming Fractions

A fraction is written as $\frac{a}{b}$, with two whole numbers that are separated by a **fraction bar.** The top number is called the **numerator.** The bottom number is called the **denominator.** (The denominator cannot be 0.) In fractions that name parts of wholes, the denominator names the number of equal parts into which the whole is divided, and the numerator names the number of parts being considered.

$\frac{a}{b} \quad \begin{array}{l} \leftarrow \text{numerator} \\ \leftarrow \text{denominator} \end{array}$

$b \neq 0$

A fraction can be written in many ways. Fractions that name the same number are called **equivalent.** Multiplying or dividing a fraction's numerator and denominator by the same number (except 0) results in an equivalent fraction. Fractions also have decimal and percent names, which can be found by dividing the numerator by the denominator.

$\frac{1}{2} = \frac{2}{4} = \frac{3}{6} = 0.5 = 50\%$

$\frac{1}{3} = \frac{2}{6} = \frac{3}{9} = 0.\overline{3} = 33\frac{1}{3}\%$

Fractions have many equivalent names.

Two fractions can be compared, added, or subtracted by using equivalent fractions with the same denominator.

Fraction Uses

Fractions can be used in many ways.

Parts of Wholes Fractions are used to name a part of a whole object or a part of a collection of objects.

$\frac{5}{6}$ of the hexagon is shaded.

$\frac{6}{10}$ of the dimes are circled.

Points on Number Lines Fractions can name points on a number line that are between points named by whole numbers.

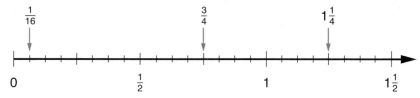

"In-Between" Measures Fractions can name measures that are between whole-number measures.

Division The fraction $\frac{a}{b}$ is another way of saying a divided by b.

The division problem 24 divided by 3 can be written in any of these ways: $24 \div 3$, $24 / 3$, or $\frac{24}{3}$.

Ratios Fractions are used to compare quantities with the *same unit*.

Curie won 7 games and lost 10 games during last year's basketball season. The fraction $\frac{\text{games won}}{\text{games lost}} = \frac{7}{10}$ compares quantities with the same unit (games and games).

We say that the ratio of wins to losses is 7 to 10.

PUBLIC-RED CENTRAL		
	Conf.	Overall
Dunbar	4−0	9−6
King	4−1	14−4
Robeson	3−2	8−9
Gage Park	2−3	8−10
Harper	2−3	8−7
Curie	1−3	7−10
Hubbard	1−4	8−9

Rates

Fractions are used to compare quantities with *different units*.

Bill's car can travel about 35 miles on 1 gallon of gasoline. The fraction $\frac{35 \text{ miles}}{1 \text{ gallon}}$ compares quantities with different units (miles and gallons). At this rate, it can travel about 245 miles on 7 gallons of gasoline.

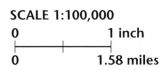

$$\frac{35 \text{ miles}}{1 \text{ gallon}} = \frac{245 \text{ miles}}{7 \text{ gallons}}$$

Scale Drawings, Scale Models, and Map Scales

Fractions are used to compare the size of a drawing or model to the size of the actual object.

A scale of 1:100,000 on a map means that every distance on the map is $\frac{1}{100,000}$ of the real-world distance. A 1-inch distance on the map stands for a real-world distance of 100,000 inches, or about $1\frac{1}{2}$ miles.

SCALE 1:100,000

```
0                    1 inch
├──────────┼──────────┤
0                    1.58 miles
```

1 inch represents 1.58 miles, or 100,000 inches.

Probabilities

Fractions are a way to describe the chance that an event will happen.

If one card is drawn from a well-shuffled deck of 52 playing cards, the chance of selecting the ace of spades is $\frac{1}{52}$, or about 2%. The chance of drawing any ace is $\frac{4}{52}$, or about 8%.

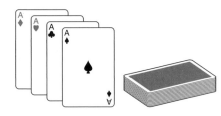

Miscellaneous

People use fractions in a variety of ways every day.

A film critic gave the new movie $3\frac{1}{2}$ stars out of 5.

Your half-birthday is 6 months after your birthday.

A half-baked idea is an idea that is not practical or has not been thought out properly.

A shoe size of $6\frac{1}{2}$ fits women whose feet are $9\frac{1}{2}$ inches long.

Did You Know?

Until recently, prices of shares of stock were reported in dollars and fractions of a dollar. In 2001, there was a shift to decimal pricing. Today, nearly all stock prices are reported in decimals.

Equivalent Fractions

Two or more fractions are called **equivalent fractions** if they name the same number. For example, the fractions $\frac{1}{2}$ and $\frac{2}{4}$ are equivalent because they both name the same part of a whole.

When you solve problems with fractions, it can often be easier to work with equivalent fractions instead of the given fractions.

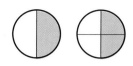

$\frac{1}{2}$ and $\frac{2}{4}$ are equivalent.

$$\frac{1}{2} = \frac{2}{4}$$

Using the Fraction-Stick Chart

To use the Fraction-Stick Chart to find equivalent fractions, first locate the fraction. Then use a vertical straightedge to find equivalent fractions. You can find a large Fraction-Stick Chart on page 399.

Example   Find equivalent fractions for $\frac{3}{4}$.

- Locate $\frac{3}{4}$ on the "fourths" stick.

- Place one edge of a straightedge at $\frac{3}{4}$.

- On the "eighths" stick, the straightedge touches the edge of the sixth piece, which is $\frac{6}{8}$.

So, $\frac{3}{4} = \frac{6}{8}$.

- On the "twelfths" stick, the straightedge touches the edge of the ninth piece, which is $\frac{9}{12}$.

So, $\frac{3}{4} = \frac{9}{12}$.

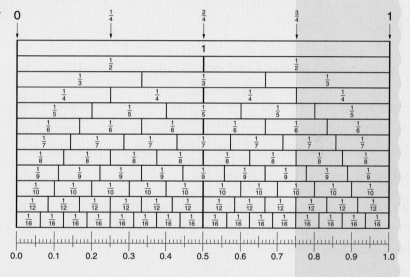

- On the "sixteenths" stick, the straightedge touches the edge of the twelfth piece, which is $\frac{12}{16}$.

So, $\frac{3}{4} = \frac{12}{16}$.

- The straightedge doesn't line up with an edge of any other sticks on the chart, so $\frac{3}{4}$ cannot be written as an equivalent fraction using those sticks (denominators).

Check Your Understanding

Use the Fraction-Stick Chart to find equivalent fractions.

1. $\frac{1}{2}$ 2. $\frac{2}{3}$ 3. $\frac{1}{6}$ 4. $\frac{4}{16}$ 5. $\frac{5}{7}$

Check your answers on page 435.

Methods for Finding Equivalent Fractions

Here are two methods for finding equivalent fractions.

Using Multiplication

If the numerator and the denominator of a fraction are both multiplied by the same number (not 0), the result is a fraction that is equivalent to the original fraction.

Example   Rename $\frac{3}{7}$ as a fraction with the denominator 21.

Multiply the numerator and the denominator of $\frac{3}{7}$ by 3.

In symbols, you can write $\frac{3}{7} = \frac{3 * 3}{7 * 3} = \frac{9}{21}$.

So, $\frac{3}{7}$ is equivalent to $\frac{9}{21}$.

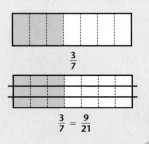

Using Division

If the numerator and the denominator of a fraction are both divided by the same number (not 0), the result is a fraction that is equivalent to the original fraction.

To understand why division works, use the example shown above. But start with $\frac{9}{21}$ this time and divide both numbers in the fraction by 3: $\frac{9 \div 3}{21 \div 3} = \frac{3}{7}$.

The division by 3 "undoes" the multiplication by 3 that we did before. Dividing both numbers in $\frac{9}{21}$ by 3 gives an equivalent fraction, $\frac{3}{7}$.

Examples $\frac{6}{15} = \frac{6 \div 3}{15 \div 3} = \frac{2}{5}$ $\frac{24}{60} = \frac{24 \div 4}{60 \div 4} = \frac{6}{15}$ $\frac{35}{100} = \frac{35 \div 5}{100 \div 5} = \frac{7}{20}$

Check Your Understanding

1. Use multiplication to find an equivalent fraction.

   a. $\frac{3}{4}$ **b.** $\frac{3}{8}$ **c.** $\frac{2}{5}$ **d.** $\frac{4}{7}$ **e.** $\frac{8}{3}$ **f.** $\frac{11}{12}$

2. Use division to find an equivalent fraction.

   a. $\frac{8}{12}$ **b.** $\frac{10}{25}$ **c.** $\frac{24}{36}$ **d.** $\frac{45}{60}$ **e.** $\frac{36}{24}$ **f.** $\frac{100}{8}$

Check your answers on page 435.

Table of Equivalent Fractions

The table below lists equivalent fractions. All the fractions in a row name the same number. For example, all the fractions in the last row are names for the number $\frac{11}{12}$.

Note

Every fraction in the first column is in simplest form. A fraction is in **simplest form** if there is no equivalent fraction with a smaller numerator and smaller denominator.

Every fraction is either in simplest form or is equivalent to a fraction in simplest form.

Lowest terms means the same as *simplest form*.

Simplest Name	Equivalent Fraction Name								
0 (zero)	$\frac{0}{1}$	$\frac{0}{2}$	$\frac{0}{3}$	$\frac{0}{4}$	$\frac{0}{5}$	$\frac{0}{6}$	$\frac{0}{7}$	$\frac{0}{8}$	$\frac{0}{9}$
1 (one)	$\frac{1}{1}$	$\frac{2}{2}$	$\frac{3}{3}$	$\frac{4}{4}$	$\frac{5}{5}$	$\frac{6}{6}$	$\frac{7}{7}$	$\frac{8}{8}$	$\frac{9}{9}$
$\frac{1}{2}$	$\frac{2}{4}$	$\frac{3}{6}$	$\frac{4}{8}$	$\frac{5}{10}$	$\frac{6}{12}$	$\frac{7}{14}$	$\frac{8}{16}$	$\frac{9}{18}$	$\frac{10}{20}$
$\frac{1}{3}$	$\frac{2}{6}$	$\frac{3}{9}$	$\frac{4}{12}$	$\frac{5}{15}$	$\frac{6}{18}$	$\frac{7}{21}$	$\frac{8}{24}$	$\frac{9}{27}$	$\frac{10}{30}$
$\frac{2}{3}$	$\frac{4}{6}$	$\frac{6}{9}$	$\frac{8}{12}$	$\frac{10}{15}$	$\frac{12}{18}$	$\frac{14}{21}$	$\frac{16}{24}$	$\frac{18}{27}$	$\frac{20}{30}$
$\frac{1}{4}$	$\frac{2}{8}$	$\frac{3}{12}$	$\frac{4}{16}$	$\frac{5}{20}$	$\frac{6}{24}$	$\frac{7}{28}$	$\frac{8}{32}$	$\frac{9}{36}$	$\frac{10}{40}$
$\frac{3}{4}$	$\frac{6}{8}$	$\frac{9}{12}$	$\frac{12}{16}$	$\frac{15}{20}$	$\frac{18}{24}$	$\frac{21}{28}$	$\frac{24}{32}$	$\frac{27}{36}$	$\frac{30}{40}$
$\frac{1}{5}$	$\frac{2}{10}$	$\frac{3}{15}$	$\frac{4}{20}$	$\frac{5}{25}$	$\frac{6}{30}$	$\frac{7}{35}$	$\frac{8}{40}$	$\frac{9}{45}$	$\frac{10}{50}$
$\frac{2}{5}$	$\frac{4}{10}$	$\frac{6}{15}$	$\frac{8}{20}$	$\frac{10}{25}$	$\frac{12}{30}$	$\frac{14}{35}$	$\frac{16}{40}$	$\frac{18}{45}$	$\frac{20}{50}$
$\frac{3}{5}$	$\frac{6}{10}$	$\frac{9}{15}$	$\frac{12}{20}$	$\frac{15}{25}$	$\frac{18}{30}$	$\frac{21}{35}$	$\frac{24}{40}$	$\frac{27}{45}$	$\frac{30}{50}$
$\frac{4}{5}$	$\frac{8}{10}$	$\frac{12}{15}$	$\frac{16}{20}$	$\frac{20}{25}$	$\frac{24}{30}$	$\frac{28}{35}$	$\frac{32}{40}$	$\frac{36}{45}$	$\frac{40}{50}$
$\frac{1}{6}$	$\frac{2}{12}$	$\frac{3}{18}$	$\frac{4}{24}$	$\frac{5}{30}$	$\frac{6}{36}$	$\frac{7}{42}$	$\frac{8}{48}$	$\frac{9}{54}$	$\frac{10}{60}$
$\frac{5}{6}$	$\frac{10}{12}$	$\frac{15}{18}$	$\frac{20}{24}$	$\frac{25}{30}$	$\frac{30}{36}$	$\frac{35}{42}$	$\frac{40}{48}$	$\frac{45}{54}$	$\frac{50}{60}$
$\frac{1}{8}$	$\frac{2}{16}$	$\frac{3}{24}$	$\frac{4}{32}$	$\frac{5}{40}$	$\frac{6}{48}$	$\frac{7}{56}$	$\frac{8}{64}$	$\frac{9}{72}$	$\frac{10}{80}$
$\frac{3}{8}$	$\frac{6}{16}$	$\frac{9}{24}$	$\frac{12}{32}$	$\frac{15}{40}$	$\frac{18}{48}$	$\frac{21}{56}$	$\frac{24}{64}$	$\frac{27}{72}$	$\frac{30}{80}$
$\frac{5}{8}$	$\frac{10}{16}$	$\frac{15}{24}$	$\frac{20}{32}$	$\frac{25}{40}$	$\frac{30}{48}$	$\frac{35}{56}$	$\frac{40}{64}$	$\frac{45}{72}$	$\frac{50}{80}$
$\frac{7}{8}$	$\frac{14}{16}$	$\frac{21}{24}$	$\frac{28}{32}$	$\frac{35}{40}$	$\frac{42}{48}$	$\frac{49}{56}$	$\frac{56}{64}$	$\frac{63}{72}$	$\frac{70}{80}$
$\frac{1}{12}$	$\frac{2}{24}$	$\frac{3}{36}$	$\frac{4}{48}$	$\frac{5}{60}$	$\frac{6}{72}$	$\frac{7}{84}$	$\frac{8}{96}$	$\frac{9}{108}$	$\frac{10}{120}$
$\frac{5}{12}$	$\frac{10}{24}$	$\frac{15}{36}$	$\frac{20}{48}$	$\frac{25}{60}$	$\frac{30}{72}$	$\frac{35}{84}$	$\frac{40}{96}$	$\frac{45}{108}$	$\frac{50}{120}$
$\frac{7}{12}$	$\frac{14}{24}$	$\frac{21}{36}$	$\frac{28}{48}$	$\frac{35}{60}$	$\frac{42}{72}$	$\frac{49}{84}$	$\frac{56}{96}$	$\frac{63}{108}$	$\frac{70}{120}$
$\frac{11}{12}$	$\frac{22}{24}$	$\frac{33}{36}$	$\frac{44}{48}$	$\frac{55}{60}$	$\frac{66}{72}$	$\frac{77}{84}$	$\frac{88}{96}$	$\frac{99}{108}$	$\frac{110}{120}$

Did You Know?

The ancient Romans used words instead of symbols to indicate parts of a whole. Each fraction had a special name, and the Romans usually used denominators of 12 or multiples of 12.

Check Your Understanding

1. True or false?
   a. $\frac{1}{3} = \frac{5}{15}$
   b. $\frac{4}{4} = \frac{8}{8}$
   c. $\frac{6}{10} = \frac{15}{25}$
   d. $\frac{5}{8} = \frac{40}{72}$

2. Use the table to list five fractions that are equivalent to $\frac{2}{3}$.

Check your answers on page 435.

Mixed Numbers and Improper Fractions

Numbers like $1\frac{1}{2}$, $2\frac{3}{5}$, and $4\frac{3}{8}$ are called **mixed numbers.**
A mixed number has a whole-number part and a fraction part.
In the mixed number $4\frac{3}{8}$, the whole-number part is 4 and the
fraction part is $\frac{3}{8}$. A mixed number is equal to the sum of the
whole-number part and the fraction part: $4\frac{3}{8} = 4 + \frac{3}{8}$. Mixed
numbers are used in many of the same ways as fractions.

An **improper fraction** is a fraction that is greater than or
equal to 1. Fractions like $\frac{4}{3}$, $\frac{5}{5}$, and $\frac{125}{10}$ are improper fractions.
In an improper fraction, the numerator is greater than or equal
to the denominator.

A **proper fraction** is a fraction that is less than 1. In a proper
fraction, the numerator is less than the denominator.

Note

Even though they are
called *improper,* there
is nothing wrong about
improper fractions.
Do not avoid them.

Renaming Mixed Numbers as Improper Fractions
Mixed numbers can be renamed as improper fractions. For
example, if a circle is the ONE, then $3\frac{1}{2}$ is three whole circles
and $\frac{1}{2}$ of a fourth circle.

$$3\frac{1}{2}$$

If you divide the three whole circles into halves, then you can
see that $3\frac{1}{2} = \frac{7}{2}$.

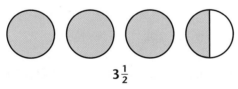

$$\frac{2}{2} \ + \ \frac{2}{2} \ + \ \frac{2}{2} \ + \ \frac{1}{2} \ = \ \frac{7}{2}$$

To change a mixed number to a fraction, rename the whole
number as a fraction with the same denominator as the fraction
part, and add the numerators. (See page 68.)

Examples

Rename $3\frac{1}{2}$ as a fraction.
$$3\frac{1}{2} = 3 + \frac{1}{2}$$
Rename 3 as $\frac{6}{2}$.
So, $3\frac{1}{2} = \frac{6}{2} + \frac{1}{2} = \frac{7}{2}$.

Rename $2\frac{3}{5}$ as a fraction.
$$2\frac{3}{5} = 2 + \frac{3}{5}$$
Rename 2 as $\frac{10}{5}$.
So, $2\frac{3}{5} = \frac{10}{5} + \frac{3}{5} = \frac{13}{5}$.

Renaming Improper Fractions as Mixed Numbers

An improper fraction can be renamed as a mixed number or a whole number.

Example Rename $\frac{19}{4}$ as a mixed number.

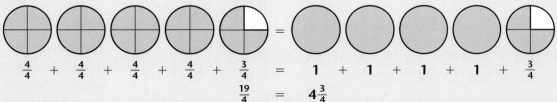

$$\frac{4}{4} + \frac{4}{4} + \frac{4}{4} + \frac{4}{4} + \frac{3}{4} = 1 + 1 + 1 + 1 + \frac{3}{4}$$

$$\frac{19}{4} = 4\frac{3}{4}$$

Shortcut:

Divide the numerator, 19, by the denominator, 4: 19 / 4 gives 4 R3.

$$\begin{array}{r} 4\overline{)19} \\ -16 \quad \lfloor 4 \end{array}$$

- The quotient, 4, is the whole-number part of the mixed number. It tells how many wholes there are in $\frac{19}{4}$.

- The remainder, 3, is the numerator of the fraction part of the mixed number. It tells how many fourths are left that cannot be made into a whole.

$$\frac{19}{4} = 4\frac{3}{4}$$

Some calculators have a special key for renaming fractions as mixed numbers or whole numbers.

Example Rename $\frac{19}{4}$ as a mixed number.

Key in on Calculator A: 19 ⬛n 4 ⬛d ⬛Enter Answer: $4\frac{3}{4}$

Key in on Calculator B: 19 ⬛b/c 4 ⬛d Answer: $4\frac{3}{4}$

See if you can do this on your calculator.

Note

The remainder in a division problem can be rewritten as a fraction. The remainder is the numerator and the divisor is the denominator. For example, $37 \div 5 \rightarrow 7$ R2, or $7\frac{2}{5}$.

Check Your Understanding

Write each mixed number as an improper fraction.

1. $4\frac{3}{4}$
2. $3\frac{2}{3}$
3. $5\frac{3}{5}$
4. $4\frac{5}{6}$
5. $1\frac{4}{3}$

Write each improper fraction as a mixed number.

6. $\frac{51}{4}$
7. $\frac{26}{3}$
8. $\frac{34}{5}$
9. $\frac{38}{8}$
10. $\frac{60}{16}$

Check your answers on page 435.

Least Common Multiples

Multiples

When you skip-count by a number, your counts are the **multiples** of that number. Since you can always count further, lists of multiples can go on forever.

Examples   Find multiples of 3 and 5.

Multiples of 3: 3, 6, 9, 12, 15, 18, 21, 24, 27, 30, 33, 36, …
Multiples of 5: 5, 10, 15, 20, 25, 30, 35, 40, 45, 50, 55, …

> **Note**
> The three dots, …, mean that a list can go on in the same way forever.

Common Multiples

Common multiples are on both lists of multiples.

Example   Find common multiples of 2 and 3.

Multiples of 2: 2, 4, **6**, 8, 10, **12**, 14, 16, **18**, 20, 22, **24**, …
Multiples of 3: 3, **6**, 9, **12**, 15, **18**, 21, **24**, 27, …
Common multiples of 2 and 3: 6, 12, 18, 24, …

Least Common Multiples

The **least common multiple** of two numbers is the smallest number that is a multiple of both numbers.

Example   Find the least common multiple of 6 and 8.

Multiples of 6: 6, 12, 18, **24**, 30, 36, 42, **48**, 54, …
Multiples of 8: 8, 16, **24**, 32, 40, **48**, 56, …

24 and 48 are common multiples. 24 is the smallest common multiple.

24 is the least common multiple of 6 and 8. It is the smallest number that can be divided evenly by both 6 and 8.

> **Did You Know?**
> The Maya were Indian people in Central America more than 1,000 years ago. They used two different calendars—one of 260 days and one of 365 days. Imagine that today is the first day on each calendar. The next time that the calendars are *both* on their first day will be 18,980 days from now. The least common multiple of 260 and 365 is 18,980.

Check Your Understanding

Find the least common mulitple of each pair of numbers.

1. 4 and 12   **2.** 6 and 10   **3.** 9 and 15   **4.** 6 and 14

Check your answers on page 435.

Common Denominators

Adding, subtracting, comparing, and dividing are easier with fractions that have the same denominator. If two fractions are renamed so that they have the same denominator, that denominator is called a **common denominator.**

There are different methods to rename fractions with a common denominator.

Examples Rename $\frac{3}{4}$ and $\frac{1}{6}$ with a common denominator.

Method 1: Equivalent Fractions Method

List equivalent fractions for $\frac{3}{4}$ and $\frac{1}{6}$.

$$\frac{3}{4} = \frac{6}{8} = \frac{9}{12} = \frac{12}{16} = \frac{15}{20} = \frac{18}{24} = \frac{21}{28} = \frac{24}{32} = \frac{27}{36} = \cdots$$

$$\frac{1}{6} = \frac{2}{12} = \frac{3}{18} = \frac{4}{24} = \frac{5}{30} = \frac{6}{36} = \frac{7}{42} = \frac{8}{48} = \frac{9}{54} = \cdots$$

Both $\frac{3}{4}$ and $\frac{1}{6}$ can be renamed as fractions with the common denominator 12.

$$\frac{3}{4} = \frac{9}{12} \text{ and } \frac{1}{6} = \frac{2}{12}$$

Method 2: The Multiplication Method

Multiply the numerator and the denominator of each fraction by the denominator of the *other* fraction.

$$\frac{3}{4} = \frac{3*6}{4*6} = \frac{18}{24} \qquad \frac{1}{6} = \frac{1*4}{6*4} = \frac{4}{24}$$

Method 3: Least Common Multiple Method

Find the least common multiple of the denominators.

Multiples of 4: 4, 8, **12**, 16, 20, …

Multiples of 6: 6, **12**, 18, 24, …

The least common multiple of 4 and 6 is 12.

Rename the fractions so that their denominator is the least common multiple.

$$\frac{3}{4} = \frac{3*3}{4*3} = \frac{9}{12} \qquad \frac{1}{6} = \frac{1*2}{6*2} = \frac{2}{12}$$

This method gives fractions with the **least common denominator.**

Note

The Multiplication Method gives what *Everyday Mathematics* calls the **quick common denominator.**

The quick common denominator can be used with variables, so it is common in algebra.

Note

The least common denominator is usually easier to use in complicated calculations, though finding it can often take more time.

Check Your Understanding

Rename each pair of fractions as fractions with a common denominator.

1. $\frac{1}{3}$ and $\frac{5}{6}$
2. $\frac{1}{2}$ and $\frac{4}{5}$
3. $\frac{3}{4}$ and $\frac{7}{10}$
4. $\frac{2}{3}$ and $\frac{5}{8}$

Check your answers on page 435.

Comparing Fractions

When you compare fractions, remember that the fractions are parts of the same whole, or ONE. Pay attention to both the numerators and denominators.

Like Denominators

Fractions with the same denominator are said to have **like denominators.** The fractions $\frac{1}{4}$ and $\frac{3}{4}$ have like denominators. To compare fractions that have like denominators, just compare the numerators. The fraction with the larger numerator is larger.

The denominator tells how many parts the whole has been divided into. If the denominators of two fractions are the same, then the parts are the same size.

Example   Compare $\frac{5}{8}$ and $\frac{3}{8}$.

For both fractions, the ONE has been divided into the same number of parts. Therefore, the parts for both fractions are the same size.

The fraction with the larger numerator has more parts shaded and so is larger.

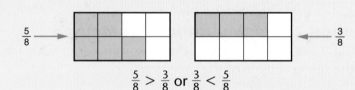

$$\frac{5}{8} > \frac{3}{8} \text{ or } \frac{3}{8} < \frac{5}{8}$$

So, $\frac{5}{8}$ is greater than $\frac{3}{8}$.

Like Numerators

Fractions with the same numerator are said to have **like numerators.** The fractions $\frac{2}{4}$ and $\frac{2}{5}$ have like numerators.

If the numerators of two fractions are the same, then the fraction with the smaller denominator is larger.

Example   Compare $\frac{2}{3}$ and $\frac{2}{5}$.

The ONE for $\frac{2}{3}$ has been divided into fewer parts than the ONE for $\frac{2}{5}$.

So, the parts in $\frac{2}{3}$ (the thirds) are larger than the parts in $\frac{2}{5}$ (the fifths).

Since thirds are larger than fifths, 2 thirds is larger than 2 fifths.

Therefore, $\frac{2}{3}$ is more than $\frac{2}{5}$.

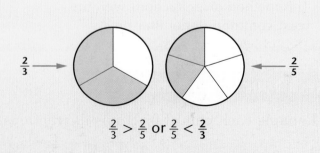

$$\frac{2}{3} > \frac{2}{5} \text{ or } \frac{2}{5} < \frac{2}{3}$$

Unlike Numerators and Unlike Denominators
Here are several strategies that can help when *both* the
numerators and the denominators are different.

Comparing to $\frac{1}{2}$ Compare $\frac{5}{7}$ and $\frac{3}{8}$.
Notice that $\frac{5}{7}$ is more than $\frac{1}{2}$ and
$\frac{3}{8}$ is less than $\frac{1}{2}$. So, $\frac{3}{8} < \frac{5}{7}$.

**Comparing to
0 or 1**   Compare $\frac{7}{8}$ and $\frac{3}{4}$.
$\frac{7}{8}$ is $\frac{1}{8}$ away from 1 and $\frac{3}{4}$ is $\frac{1}{4}$
away from 1. Since eighths
are smaller than fourths, $\frac{7}{8}$ is
closer to 1 than $\frac{3}{4}$ is. So, $\frac{7}{8} > \frac{3}{4}$.

**Using Common
Denominators**   Compare $\frac{5}{8}$ and $\frac{3}{5}$.
Rename both fractions with a common denominator.
The fraction with the larger numerator is larger.

A common denominator for $\frac{5}{8}$ and $\frac{3}{5}$ is 40.

$$\frac{5}{8} = \frac{5 * 5}{8 * 5} = \frac{25}{40} \qquad \frac{3}{5} = \frac{3 * 8}{5 * 8} = \frac{24}{40} \qquad \frac{25}{40} > \frac{24}{40}$$

So, $\frac{5}{8} > \frac{3}{5}$.

This way of comparing fractions will *always* work.

**Using Decimal
Equivalents**   Compare $\frac{2}{5}$ and $\frac{3}{8}$.
Use a calculator to rename both fractions
as decimals:

$\frac{2}{5}$: Key in: 2 ⌷÷⌷ 5 ⌷=⌷   Answer: 0.4

$\frac{3}{8}$: Key in: 3 ⌷÷⌷ 8 ⌷=⌷   Answer: 0.375

Since 0.4 > 0.375, you know that $\frac{2}{5} > \frac{3}{8}$.

This way of comparing fractions will *always* work.

> **Note**
> Remember that fractions
> can be used to show
> division problems.
> $$\frac{a}{b} = a / b$$

Check Your Understanding

Compare. Use <, >, or =.

1. $\frac{1}{7} \square \frac{1}{9}$ **2.** $\frac{5}{7} \square \frac{7}{9}$ **3.** $\frac{3}{4} \square \frac{9}{12}$ **4.** $\frac{5}{8} \square \frac{2}{3}$ **5.** $\frac{2}{5} \square \frac{5}{12}$

Check your answers on page 435.

Addition and Subtraction of Fractions

Like Denominators

To add or subtract fractions that have the same denominator,
just add or subtract the numerators. The denominator does not
change. After you find the sum or difference, you can use the
division rule to put it in simplest form.

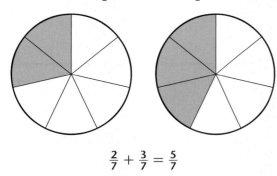

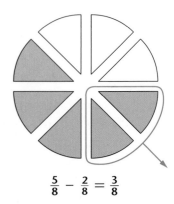

$$\frac{2}{7} + \frac{3}{7} = \frac{5}{7}$$

$$\frac{5}{8} - \frac{2}{8} = \frac{3}{8}$$

Examples

$$\frac{3}{9} + \frac{2}{9} = ?$$

$$\frac{3}{9} + \frac{2}{9} = \frac{3+2}{9} = \frac{5}{9}$$

$$\frac{7}{10} - \frac{3}{10} = ?$$

$$\frac{7}{10} - \frac{3}{10} = \frac{7-3}{10} = \frac{4}{10} = \frac{2}{5}$$

Unlike Denominators

To add or subtract fractions that have unlike denominators,
first rename the fractions with a common denominator.

Example $\frac{5}{12} + \frac{2}{3} = ?$

To rename $\frac{5}{12}$ and $\frac{2}{3}$ with a common denominator, you can multiply both the numerator
and the denominator of each fraction by the other fraction's denominator.

$$\frac{5}{12} = \frac{5 * 3}{12 * 3} = \frac{15}{36}$$

$$\frac{2}{3} = \frac{2 * 12}{3 * 12} = \frac{24}{36}$$

So, $\frac{5}{12} + \frac{2}{3} = \frac{15}{36} + \frac{24}{36} = \frac{39}{36}$.

Sometimes an answer like $\frac{39}{36}$ is fine as the solution to a problem. Other times you may
want to change it to a mixed number in simplest form.

$$\frac{39}{36} = 1\frac{3}{36} = 1\frac{1}{12}$$

Using a Slide Rule

In Fifth Grade *Everyday Mathematics,* you are given a paper slide rule to help you add and subtract some fractions.

Example   Find $\frac{3}{4} + \frac{1}{2}$.

Step 1: Place the fraction side of the slider inside the fraction side of the holder.

Step 2: Line up the 0-mark on the slider with the mark for $\frac{3}{4}$ on the holder.

Step 3: Find the mark for $\frac{1}{2}$ on the slider. It is lined up with the mark for $1\frac{1}{4}$ on the holder. That is the answer to the problem.

$$\frac{3}{4} + \frac{1}{2} = 1\frac{1}{4}$$

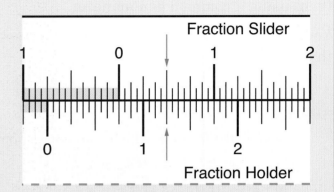

Example   Find $2\frac{1}{4} - \frac{1}{2}$.

Step 1: Line up the 0-mark on the slider with the mark for $2\frac{1}{4}$ on the holder.

Step 2: Find the mark for $\frac{1}{2}$ on the negative part of the slider. It is lined up with the mark for $1\frac{3}{4}$ on the holder. This is the answer to the problem.

$$2\frac{1}{4} - \frac{1}{2} = 1\frac{3}{4}$$

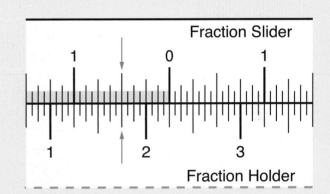

Using a Calculator

Some calculators can be used to add and subtract fractions.

Example $\frac{3}{8} + \frac{1}{4} = ?$

Key in on Calculator A: 3 ⊡n⊡ 8 ⊡d⊡ ⊕ 1 ⊡n⊡ 4 ⊡d⊡ ⊡Enter⊡   Answer: $\frac{5}{8}$

Key in on Calculator B: 3 ⊡b/c⊡ 8 ⊕ 1 ⊡b/c⊡ 4 ⊡=⊡   Answer: $\frac{5}{8}$

Check Your Understanding

Add or subtract.

1. $\frac{3}{8} - \frac{1}{3}$ **2.** $\frac{7}{8} - \frac{1}{4}$ **3.** $\frac{7}{12} - \frac{1}{4}$ **4.** $\frac{7}{12} + \frac{1}{4}$ **5.** $\frac{3}{8} + \frac{1}{3}$

Check your answers on page 435.

Addition of Mixed Numbers

One way to add mixed numbers is to add the whole numbers and the fractions separately. If the fractions in the mixed numbers have unlike denominators, you should first rename them using a common denominator.

You will have to rename the sum of the fractions if it is more than 1.

Example   Find $3\frac{3}{4} + 5\frac{2}{3}$.

Rename the fractions with a common denominator.

$$3\ \frac{3}{4} \qquad (3\tfrac{3}{4} = 3\tfrac{9}{12}) \qquad 3\ \frac{9}{12}$$
$$+\ 5\ \frac{2}{3} \qquad (5\tfrac{2}{3} = 5\tfrac{8}{12}) \qquad +\ 5\ \frac{8}{12}$$

Add the fractions.

$$3\ \frac{9}{12}$$
$$+\ 5\ \frac{8}{12}$$
$$\overline{\quad \frac{17}{12}}$$

Add the whole numbers.

$$3\ \frac{9}{12}$$
$$+\ 5\ \frac{8}{12}$$
$$\overline{8\ \frac{17}{12}}$$

Rename the sum.

$$8\tfrac{17}{12} = 8 + \tfrac{12}{12} + \tfrac{5}{12}$$
$$= 8 + 1 + \tfrac{5}{12}$$
$$= 9 + \tfrac{5}{12}$$
$$= 9\tfrac{5}{12}$$
$$3\tfrac{3}{4} + 5\tfrac{2}{3} = 9\tfrac{5}{12}$$

Some calculators have special keys for entering and renaming mixed numbers. See pages 259 and 260.

Check Your Understanding

Add.

1. $2\frac{1}{2} + 1\frac{1}{4}$ 2. $4\frac{1}{3} + 6\frac{3}{4}$ 3. $5\frac{5}{6} + 3\frac{2}{3}$ 4. $2\frac{1}{2} + 1\frac{3}{4} + 1\frac{5}{8}$

Check your answers on page 435.

Subtraction of Mixed Numbers

Here is one way to subtract a mixed number from a mixed number:

First, subtract the fraction parts of the mixed numbers.

♦ If the fractions have unlike denominators, rename them using a common denominator.

♦ If necessary, rename the larger mixed number so the fraction part is large enough to subtract from.

Second, subtract the whole-number parts.

Example   Find $5\frac{1}{4} - 3\frac{2}{3}$.

Rename the fractions using a common denominator.

$$5\frac{1}{4} \qquad \left(5\frac{1}{4} = 5\frac{3}{12}\right) \qquad 5\frac{3}{12}$$
$$-\,3\frac{2}{3} \qquad \left(3\frac{2}{3} = 3\frac{8}{12}\right) \qquad -\,3\frac{8}{12}$$

Rename the larger mixed number $\left(5\frac{3}{12}\right)$ so the fraction part is at least as large as $\frac{8}{12}$. (Remember that 1 can be renamed as a fraction with the same numerator and denominator.)

$$\mathbf{5\frac{3}{12}} = 4 + 1 + \frac{3}{12} = 4 + \frac{12}{12} + \frac{3}{12} = \mathbf{4\frac{15}{12}}$$
$$-\,3\frac{8}{12} \qquad\qquad\qquad\qquad\qquad\qquad -\,3\frac{8}{12}$$

Subtract the fractions.

$$4\frac{15}{12}$$
$$-\,3\frac{8}{12}$$
$$\overline{\frac{7}{12}}$$

Subtract the whole numbers.

$$\mathbf{4\frac{15}{12}}$$
$$-\,\mathbf{3\frac{8}{12}}$$
$$\overline{\mathbf{1\frac{7}{12}}}$$

$$5\frac{1}{4} - 3\frac{2}{3} = 1\frac{7}{12}$$

Example Find $5 - 2\frac{2}{3}$.

Rename the whole number as a mixed number.

$$5 = \quad 4 + 1 = 4 + \frac{3}{3} = \quad 4\frac{3}{3}$$
$$-2\frac{2}{3} \qquad\qquad\qquad -2\frac{2}{3}$$

Subtract the fractions. Subtract the whole numbers.

$$\begin{array}{r} 4\frac{3}{3} \\ -2\frac{2}{3} \\ \hline \frac{1}{3} \end{array} \qquad \begin{array}{r} 4\frac{3}{3} \\ -2\frac{2}{3} \\ \hline 2\frac{1}{3} \end{array}$$

$$5 - 2\frac{2}{3} = 2\frac{1}{3}$$

Another way to subtract (or add) mixed numbers is to rename the mixed numbers as improper fractions.

Example Find $4\frac{1}{6} - 2\frac{2}{3}$.

Rename the mixed numbers as fractions. $4\frac{1}{6} = \frac{25}{6}$ $2\frac{2}{3} = \frac{8}{3}$

Rename the fractions with a common denominator. Subtract.

$$\begin{array}{r} \frac{25}{6} \\ -\frac{8}{3} \end{array} \quad (\frac{8}{3} = \frac{16}{6}) \qquad \begin{array}{r} \frac{25}{6} \\ -\frac{16}{6} \\ \hline \frac{9}{6} \end{array}$$

Rename the result as a mixed number. $\frac{9}{6} = 1\frac{3}{6} = 1\frac{1}{2}$

So, $4\frac{1}{6} - 2\frac{2}{3} = 1\frac{1}{2}$.

Check Your Understanding

Subtract.

1. $4\frac{1}{5} - 2\frac{3}{5}$
2. $1\frac{1}{2} - \frac{2}{3}$
3. $4\frac{1}{2} - 1\frac{1}{8}$
4. $5\frac{1}{6} - 3\frac{3}{4}$

Check your answers on page 435.

Finding a Fraction of a Number

Many problems with fractions involve finding a fraction of a number.

Example   Find $\frac{2}{3}$ of 24.

Model the problem by using 24 pennies. Divide the pennies into 3 equal groups.

Each group has $\frac{1}{3}$ of the pennies. So $\frac{1}{3}$ of 24 pennies is 8 pennies. Since $\frac{1}{3}$ of 24 is 8, $\frac{2}{3}$ of 24 must be twice as much: $2 * 8 = 16$.

$\frac{1}{3}$ of 24 = 8, so $\frac{2}{3}$ of 24 = 16.

$\frac{2}{3}$ of 24 = $\frac{2}{3} * 24 = 16$.

Note

"$\frac{2}{3}$ of 24" has the same meaning as "$\frac{2}{3} * 24$." When you find the fraction of a number, you can replace the word *of* with a multiplication symbol.

Other examples:

$\frac{1}{6}$ of 18 means $\frac{1}{6} * 18$.

2 of 18 means $2 * 18$.

$\frac{3}{4}$ of 40 means $\frac{3}{4} * 40$.

34 of 40 means $34 * 40$.

Example   A jacket that sells for \$45 is on sale for $\frac{2}{3}$ of the regular price. What is the sale price?

To find the sale price, you have to find $\frac{2}{3}$ of \$45.

Step 1: Find $\frac{1}{3}$ of 45.

Step 2: Use the answer from Step 1 to find $\frac{2}{3}$ of 45.

Since $\frac{1}{3}$ of 45 is 15, $\frac{2}{3}$ of 45 is $2 * 15 = 30$.

The sale price is \$30.

$\frac{2}{3}$ of \$45 = $\frac{2}{3} * \$45 = \$30$.

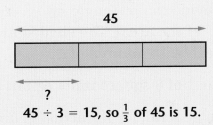

45

?

$45 \div 3 = 15$, so $\frac{1}{3}$ of 45 is 15.

Check Your Understanding

Solve each problem.

1. $\frac{1}{4}$ of 32   **2.** $\frac{3}{4}$ of 32   **3.** $\frac{3}{5} * 20$ **4.** $\frac{5}{6}$ of 48

5. Rita and Hunter earned \$20 raking lawns. Since Rita did most of the work, they decided that Rita should get $\frac{4}{5}$ of the money. How much does each person get?

Check your answers on page 435.

Multiplying Fractions and Whole Numbers

There are several ways to show the multiplication of a whole number and a fraction.

Number Line Model

One way to multiply a fraction and a whole number is to think about "hops" on a number line. The whole number tells how many hops to make, and the fraction tells how long each hop should be. For example, to find $4 * \frac{2}{3}$, imagine taking 4 hops, each $\frac{2}{3}$ unit long, on a number line.

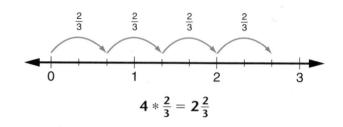

$$4 * \frac{2}{3} = 2\frac{2}{3}$$

Addition Model

You can use addition to multiply a fraction and a whole number. For example, to find $4 * \frac{2}{3}$, draw 4 models of $\frac{2}{3}$. Then add all of the fractions.

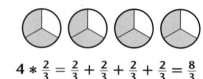

$$4 * \frac{2}{3} = \frac{2}{3} + \frac{2}{3} + \frac{2}{3} + \frac{2}{3} = \frac{8}{3}$$

Area Model

Think of the problem $\frac{2}{3} * 4$ as "What is $\frac{2}{3}$ of an area that has 4 square units?" Remember: $\frac{2}{3}$ of 4 means $\frac{2}{3} * 4$.

Draw 4 squares, each with an area of 1 square unit. The rectangle has an area of 4 square units.

4 squares

Divide the rectangle into 3 equal strips and shade 2 strips $\left(\frac{2}{3} \text{ of the area} \right)$. The shaded area equals $\frac{8}{3}$ (8 small rectangles, each with an area of $\frac{1}{3}$).

$$\frac{2}{3} \text{ of } 4 = \frac{2}{3} * 4 = \frac{8}{3}$$

So, $\frac{2}{3}$ of 4 square units equals $\frac{8}{3}$ square units. $\frac{2}{3} * 4 = \frac{8}{3}$

In general, for a fraction $\frac{a}{b}$, $b \neq 0$, and a whole number n,

$$\frac{a}{b} * n = \frac{a * n}{b}.$$

Example   Marie wants $\frac{2}{5}$ of the 20 tiles on her new floor to be blue. How many blue tiles does she need?

Marie needs $\frac{2}{5} * 20$ tiles $= \frac{2 * 20}{5}$ tiles $= 8$ tiles.

$$\frac{2}{5} * 20 = 8$$

Using a Unit Fraction to Find the Whole

A fraction with 1 in the numerator is called a **unit fraction.**
The fractions $\frac{1}{2}$, $\frac{1}{3}$, $\frac{1}{4}$, and $\frac{1}{5}$ are unit fractions. Unit fractions can often be helpful in solving problems with fractions.

Did You Know?

The ancient Egyptians used fractions, but generally only used unit fractions. For example, an Egyptian would have written $\frac{2}{5}$ as $\frac{1}{3} + \frac{1}{15}$.

Example Alex collects sports cards. Seventy of his cards feature basketball players. These 70 cards are $\frac{2}{3}$ of Alex's collection. How many sports cards does Alex have?

$\frac{2}{3}$ of the collection is 70 cards.

So, $\frac{1}{3}$ of the collection is 35 cards.

The whole collection ($\frac{3}{3}$) is
$3 * 35 = 105$.

Alex has 105 cards in his whole collection.

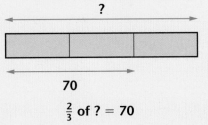

?

70

$\frac{2}{3}$ of ? = 70

Example Alicia baked cookies. She gave away 24 cookies, which was $\frac{3}{5}$ of the total she baked. How many cookies did Alicia bake?

$\frac{3}{5}$ of Alicia's cookies is 24 cookies.

So $\frac{1}{5}$ of Alicia's cookies is 8 cookies.

The whole batch of cookies ($\frac{5}{5}$) is
$5 * 8 = 40$ cookies.

Alicia baked 40 cookies.

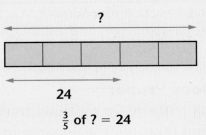

?

24

$\frac{3}{5}$ of ? = 24

Check Your Understanding

Solve each problem.

1. $\frac{1}{2}$ of a package is 22 cookies. How many cookies are in the whole package?

2. $\frac{2}{3}$ of a package is 6 cookies. How many cookies are in the whole package?

3. $\frac{3}{4}$ of a package is 15 cookies. How many cookies are in the whole package?

Check your answers on page 435.

Multiplying Fractions

When you multiply two fractions, thinking about area can help.

Example What is $\frac{3}{4}$ of $\frac{2}{3}$?

Step 1: *Think:* How much is $\frac{3}{4}$ of $\frac{2}{3}$ of this rectangular region?

Step 2: Shade $\frac{2}{3}$ of the region this way:

Step 3: Shade $\frac{3}{4}$ of the region this way:

Now $\frac{3}{4}$ of $\frac{2}{3}$ of the region is shaded both ways:

That's $\frac{6}{12}$ or $\frac{1}{2}$ of the whole region.

$\frac{3}{4}$ of $\frac{2}{3} = \frac{6}{12} = \frac{1}{2}$

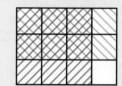

When you find a fraction of a number, you can replace the word *of* by a multiplication symbol. $\frac{3}{4}$ of $\frac{2}{3} = \frac{3}{4} * \frac{2}{3} = \frac{6}{12} = \frac{1}{2}$

Multiplication of Fractions Property

The example above shows a pattern: To multiply fractions, multiply the numerators and multiply the denominators.

This pattern can be expressed as follows:

$\frac{a}{b} * \frac{c}{d} = \frac{a * c}{b * d}$ (b and d may not be 0.)

Examples

$\frac{3}{4} * \frac{2}{3} = ?$

$\frac{3}{4} * \frac{2}{3} = \frac{3 * 2}{4 * 3} = \frac{6}{12} = \frac{1}{2}$

$\frac{2}{5} * \frac{6}{7} = ?$

$\frac{2}{5} * \frac{6}{7} = \frac{2 * 6}{5 * 7} = \frac{12}{35}$

$\frac{4}{1} * \frac{7}{9} = ?$

$\frac{4}{1} * \frac{7}{9} = \frac{4 * 7}{1 * 9} = \frac{28}{9} = 3\frac{1}{9}$

Check Your Understanding

Multiply.

1. $\frac{1}{2} * \frac{2}{3}$

2. $\frac{2}{5} * \frac{3}{4}$

3. $\frac{5}{6} * \frac{9}{10}$

4. $\frac{1}{4} * \frac{1}{3}$

Check your answers on page 435.

Multiplying Fractions, Whole Numbers, and Mixed Numbers

Multiplying Whole Numbers and Fractions

The **multiplication of fractions property** can be used to multiply a whole number and a fraction. First, rename the whole number as a fraction with 1 in the denominator.

Examples

Find $5 * \frac{2}{3}$.

$$5 * \frac{2}{3} = \frac{5}{1} * \frac{2}{3} = \frac{5 * 2}{1 * 3} = \frac{10}{3} = 3\frac{1}{3}$$

$$5 * \frac{2}{3} = 3\frac{1}{3}$$

Find $12 * \frac{3}{7}$.

$$12 * \frac{3}{7} = \frac{12}{1} * \frac{3}{7} = \frac{12 * 3}{1 * 7} = \frac{36}{7} = 5\frac{1}{7}$$

$$12 * \frac{3}{7} = 5\frac{1}{7}$$

Multiplying Mixed Numbers by Renaming Them as Improper Fractions

One way to multiply two mixed numbers is to rename each mixed number as an improper fraction and multiply the fractions. Then rename the product as a mixed number.

Example Find $3\frac{1}{4} * 1\frac{5}{6}$.

Rename the mixed numbers as fractions.
Multiply the fractions.

$$3\frac{1}{4} * 1\frac{5}{6} = \frac{13}{4} * \frac{11}{6}$$

$$= \frac{13 * 11}{4 * 6}$$

$$= \frac{143}{24}$$

Rename $\frac{143}{24}$ as a mixed number.

A fraction $\frac{a}{b}$ is another way of saying a divided by b.
So, $\frac{143}{24}$ can be written as $143 \div 24$. $\frac{143}{24} = 143 \div 24$

$$\begin{array}{r} 24\overline{)143} \\ -120 \quad \boxed{5} \\ \hline 23 \quad 5 \end{array}$$

So, $\frac{143}{24} = 143 \div 24 = 5 \text{ R}23 = 5\frac{23}{24}$.

$$3\frac{1}{4} * 1\frac{5}{6} = 5\frac{23}{24}$$

Note

The remainder in a division problem can be rewritten as a fraction. The remainder is the numerator and the divisor is the denominator.
For example,
$24\overline{)143} = 5 \text{ R}23 = 5\frac{23}{24}$.

Multiplying Mixed Numbers by Using Partial Products

Another way to multiply mixed numbers is to find partial products and add them.

Example   Find $7\frac{1}{2} * 2\frac{3}{5}$.

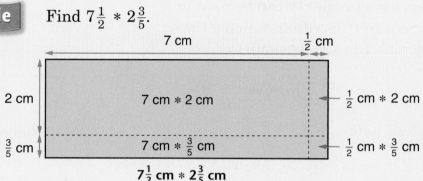

$7\frac{1}{2}$ cm $* 2\frac{3}{5}$ cm

Find the partial products. $7 * 2 = 14$

$$7 * \frac{3}{5} = \frac{7}{1} * \frac{3}{5} = \frac{21}{5} = 4\frac{1}{5}$$

$$\frac{1}{2} * 2 = \frac{1}{2} * \frac{2}{1} = \frac{2}{2} = 1$$

$$\frac{1}{2} * \frac{3}{5} = \frac{3}{10}$$

Add the partial products. $14 + 4\frac{1}{5} + 1 + \frac{3}{10} = 19 + \frac{1}{5} + \frac{3}{10}$

$$= 19 + \frac{2}{10} + \frac{3}{10}$$

$$= 19 + \frac{5}{10}$$

$$= 19\frac{5}{10}$$

$$= 19\frac{1}{2}$$

$$7\frac{1}{2} * 2\frac{3}{5} = 19\frac{1}{2}$$

Did You Know?

Until 1971, Great Britain had an interesting system of money, in which certain amounts were written as mixed numbers.

Pennies, half-pennies, and quarter-pennies (called farthings) were in use. This led to expressions of the form $2\frac{1}{2}$d (for two and one-half pennies) and $5\frac{3}{4}$d (for five and three-quarters pennies).

Check Your Understanding

Multiply.

1. $1\frac{1}{2} * \frac{1}{4}$ **2.** $2\frac{2}{3} * 6$ **3.** $3\frac{2}{5} * 2\frac{1}{2}$ **4.** $4\frac{5}{6} * \frac{1}{2}$

Check your answers on page 435.

Resizing (Scaling) by Multiplying Fractions

Makers of scale models, scale drawings, and maps often use fractions to describe how their work is **resized**, or **scaled**, compared to the actual-size objects. For example, the drawing below of the woodpecker is a $\frac{1}{4}$ scale *reduction* of the lengths of the actual bird, while the drawing of the aphid garden pest is a 10 times *enlargement* of the lengths of the actual bug. The numbers $\frac{1}{4}$ and 10 are **size-change factors**, or **scale factors**.

2 in.

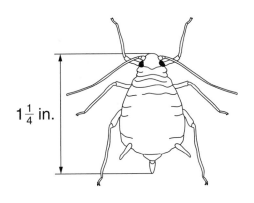

$1\frac{1}{4}$ in.

The $\frac{1}{4}$ scale factor of the woodpecker drawing means that every length in the drawing is $\frac{1}{4}$ of the length in the real bird. In general, if a positive scale factor is less than 1, the drawing is a reduction of the object. You can show this with multiplication.

Example Show that the 2-inch scale length of the woodpecker in the scale drawing is a $\frac{1}{4}$ reduction of the actual 8-inch woodpecker.

Each length in the drawing is $\frac{1}{4}$ of the length in the real bird.
So, 8 in. $* \frac{1}{4} = \frac{8}{4}$ in. $= 2$ in.

The drawing is a $\frac{1}{4}$ reduction of the actual bird length.

Example A desktop model car is a reduction of an actual car. Which scale factor might be used?

a. $\frac{2}{1}$ **b.** $\frac{4}{4}$ **c.** $\frac{1}{16}$ **d.** $\frac{10}{3}$

A reduction means multiplying by a scale factor less than 1, so the possible scale factor is $\frac{1}{16}$.

So, in general, when you multiply a given number by a positive number less than 1, the product is less than the given number.

If a scale factor is greater than 1, then the drawing is an enlargement of the object. You can also show this with multiplication.

Example Show that the $1\frac{1}{4}$-inch scale length of the aphid in the drawing on page 78A is 10 times the length of the actual $\frac{1}{8}$-inch long aphid.

Each length in the drawing is 10 times the length in the real aphid.

So $\frac{1}{8}$ in. $* 10 = \frac{10}{8}$ in.

$= \frac{5}{4}$ in.

$= 1\frac{1}{4}$ in.

The drawing is a 10-times enlargement of the actual bug length.

$\frac{1}{8}$ in. $* 10 = 1\frac{1}{4}$ in.

Example A biologist draws an enlargement of a red blood cell. Which scale factor might be used?

a. $\frac{1}{12,500}$ **b.** $\frac{12,500}{1}$ **c.** $\frac{12,500}{12,500}$

An enlargement means multiplying by a scale factor greater than 1, so the possible scale factor is $\frac{12,500}{1}$.

So, in general, when you multiply a given number by a number greater than 1, the product is greater than the given number.

If a drawing of an object has a scale factor of 1, then every length in the drawing is 1 times the length in the actual object. This means the drawing is exactly the same size as the object.

> **Note**
>
> A scale-factor-1 illustration of an object is commonly called a *life-size* or *full-size* illustration.

Multiplying a fraction by a scale factor of 1 is an easy way to find an equivalent fraction. For example, multiplying $\frac{5}{6}$ by the scale factor 1, renamed as $\frac{2}{2}$, gives the equivalent fraction:

$$\frac{5}{6} * \frac{2}{2} = \frac{5*2}{6*2} = \frac{10}{12}.$$

This scaling of $\frac{5}{6}$ by $\frac{2}{2}$ is illustrated in the drawing on the right.

The fraction $\frac{5}{6}$ is shown as 5 yellow sections out of 6 on the left. The fraction $\frac{5*2}{6*2} = \frac{10}{12}$ is shown as 10 yellow sections out of 12 on the right. The fractions are equivalent because the yellow shaded regions cover the same area.

In general, you can find an equivalent fraction to $\frac{a}{b}$ if you multiply by $\frac{n}{n}$, where $b \neq 0$ and $n \neq 0$.

$$\frac{a}{b} = \frac{n*a}{n*b}.$$

Using Fractions for Division

The fraction $\frac{a}{b}$ is another way of saying a divided by b. So another name for the fraction $\frac{2}{3}$ is $2 \div 3$. Another name for $24 \div 6$ is $\frac{24}{6} = \frac{4}{1} = 4$. The next two examples show how fractions can be used to solve equal-sharing division problems.

Example   Two bagels are equally shared among Rick, Sasha, and Kim. What part of a whole bagel does each person get?

Solve this equal-sharing problem by finding $2 \div 3$. Think of equally dividing each bagel into 3 pieces, 1 piece per person.

Each person gets $\frac{1}{3}$ of each bagel, or $2 * \frac{1}{3}$ bagel $= \frac{2}{3}$ of a bagel in all.

$\frac{2}{3} = 2 \div 3$

Example   If 4 people share a 25-pound bag of rice equally by weight, how many pounds of rice will each person get?

Solve this equal-sharing problem by dividing 25 by 4.

Write the division as a fraction. $25 \div 4 = \frac{25}{4}$

Rename as a mixed number. $= 6\frac{1}{4}$

Each person gets 6 and $\frac{1}{4}$ pounds of rice.

$25 \div 4 = 6\frac{1}{4}$

Fractions can also help you solve division problems involving rates.

Example   Katie completed 3 math problems in 8 minutes. On average, how many problems did she complete each minute?

A rate of 3 problems in 8 minutes is 3 problems $\div$ 8 minutes.
Write the division as a fraction.

$$3 \text{ problems} \div 8 \text{ minutes} = \frac{3 \text{ problems}}{8 \text{ minutes}}$$

$$= \frac{3}{8} \text{ problem per minute.}$$

$3 \div 8 = \frac{3}{8}$

Fractions can also help you solve area problems.

Example   Ben's dad wants to add a rectangular deck along the 24 foot length of their house. He wants an area of about 180 square feet. How wide will the deck need to be?

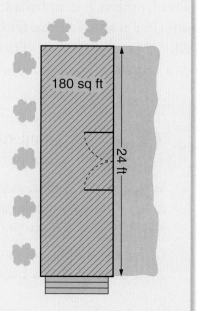

180 sq ft

24 ft

The area A of a rectangle is equal to its length l times its width w, or $A = l * w$.

For the deck, $180 = 24 * ?$. So $? = 180 \div 24$.

Write the division as a fraction. $180 \div 24 = \frac{180}{24}$

Rename. $= \frac{180 \div 12}{24 \div 12}$

$= \frac{15}{2}$

$= 7\frac{1}{2}$

The deck will be about $7\frac{1}{2}$ feet wide.

$180 \div 24 = 7\frac{1}{2}$

The next example shows how you can use fractions to divide decimals if you don't have a calculator.

Example   Calculate $5.6 \div 20$.

Write the division as a fraction. $5.6 \div 20 = \frac{5.6}{20}$

Multiply the numerator and denominator by 5. $= \frac{(5.6 * 5)}{(20 * 5)}$

$= \frac{28}{100}$

Rename. $= 0.28$

$5.6 \div 20 = 0.28$

Check Your Understanding

1. Write as a division problem. $\frac{9}{24}$ **2.** Write as a fraction. $111 \div 222$

Rewrite each division as a mixed number.

3. $8 \div 5$ **4.** $135 \div 40$

5. Three pizzas are shared equally among 4 friends. What part of a whole pizza will each friend get?

Check your answers on page 442.

Division of Fractions

Dividing a number by a fraction often gives a quotient that is larger than the dividend. For example, $4 \div \frac{1}{2} = 8$. To understand this, it's helpful to think about what division means.

There are 8 halves in 4 wholes.

Equal Groups

A division problem like $a \div b = ?$ is asking "How many bs are there in a?" For example, the problem $6 \div 3 = ?$ asks, "How many 3s are there in 6?" The figure at the right shows that there are two 3s in 6, so $6 \div 3 = 2$.

$6 \div 3 = 2$

A division problem like $6 \div \frac{1}{3} = ?$ is asking, "How many $\frac{1}{3}$s are there in 6?" The figure at the right shows that there are 18 thirds in 6, so $6 \div \frac{1}{3} = 18$.

$6 \div \frac{1}{3} = 18$

Example   Frank has 5 pounds of rice. A cup of rice is about $\frac{1}{2}$ pound.

How many cups of rice does Frank have?

This problem is solved by finding how many $\frac{1}{2}$s are in 5, which is the same as $5 \div \frac{1}{2}$.

$\frac{1}{2}$ lb $+ \frac{1}{2}$ lb $+ \frac{1}{2}$ lb $+ \frac{1}{2}$ lb $+ \frac{1}{2}$ lb $+ \frac{1}{2}$ lb $+ \frac{1}{2}$ lb $+ \frac{1}{2}$ lb $+ \frac{1}{2}$ lb $+ \frac{1}{2}$ lb $= 5$ lb

So, Frank has about 10 cups of rice.

Common Denominators

One way to solve a fraction division problem is to rename both dividend and divisor as fractions with a common denominator. Then divide the numerators and the denominators.

Example   Find $6 \div \frac{2}{3}$.

Rename 6 as $\frac{18}{3}$.

Divide the numerators and the denominators.

$6 \div \frac{2}{3} = 9$

$$6 \div \frac{2}{3} = \frac{18}{3} \div \frac{2}{3}$$
$$= \frac{18 \div 2}{3 \div 3}$$
$$= \frac{9}{1} \text{ or } 9$$

To see why this method works, imagine putting the 18 thirds in groups of $\frac{2}{3}$s each. There would be 9 groups. The figure shows nine $\frac{2}{3}$s in $\frac{18}{3}$.

Example Jake has 6 pounds of sugar. He wants to put it in packages that hold $\frac{3}{4}$ of a pound each. How many packages can he make?

Solve $6 \div \frac{3}{4}$.

Rename 6 as $\frac{24}{4}$. $6 \div \frac{3}{4} = \frac{24}{4} \div \frac{3}{4}$

Then divide. $= \frac{24 \div 3}{4 \div 4} = \frac{8}{1}$ or 8

The 24 fourths can be put in 8 groups of $\frac{3}{4}$s each.

So Jake can make 8 packages.

Missing Factors

A division problem is equivalent to a multiplication problem with a missing factor. A problem like $6 \div \frac{1}{2} = \square$ is equivalent to $\frac{1}{2} * \square = 6$. And $\frac{1}{2} * \square = 6$ is the same as asking "$\frac{1}{2}$ of what number equals 6?" Since $\frac{1}{2}$ of 12 is 6, you know that $\frac{1}{2} * 12 = 6$ and $6 \div \frac{1}{2} = 12$.

Example Find $8 \div \frac{2}{3}$.

$8 \div \frac{2}{3} = \square$ is equivalent to $\frac{2}{3} * \square = 8$, which means "$\frac{2}{3}$ of what number is 8?"

The diagram above shows that $\frac{2}{3}$ of the missing number is 8.

So $\frac{1}{3}$ of the missing number must be 4. The missing number is $3 * 4 = 12$.

So, $\frac{2}{3}$ of 12 = 8, which means that $\frac{2}{3} * 12 = 8$. (Remember: "$\frac{2}{3}$ of 12" has the same meaning as "$\frac{2}{3} * 12$.")

This means that $8 \div \frac{2}{3} = 12$.

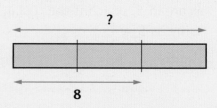

$\frac{2}{3}$ of ? = 8

Check Your Understanding

Solve. Write a division number model for each problem.

1. Richard has 7 pizzas. If each person can eat $\frac{1}{2}$ of a pizza, how many people can Richard serve?

2. Maya has 6 yards of plastic strips for making bracelets. She needs $\frac{1}{2}$ yard for each bracelet. How many bracelets can she make?

3. 8 is $\frac{1}{2}$ of a number. What is the number?

Check your answers on page 435.

Division with Unit Fractions

Dividing a fraction by a whole number gives a quotient that is smaller than the dividend. For example, $\frac{1}{3} \div 4 = \frac{1}{12}$. You can use a diagram to help you understand this.

Example A rope is $\frac{1}{2}$ meter long. If it is cut into 3 equal pieces, how long is each piece?

A number model for this problem is $\frac{1}{2} \div 3 = ?$

Draw a figure showing the $\frac{1}{2}$ meter of rope as part of a whole meter on a number line.

$\frac{1}{2}$ meter

0   1 meter

Then mark the rope showing it cut into 3 equal pieces.

$\frac{1}{6}$ meter

0   1 meter

Because 3 pieces make up $\frac{1}{2}$ of a meter, 6 pieces would make up 1 whole meter. So each piece of rope is $\frac{1}{6}$ of a meter long.

$\frac{1}{2} \div 3 = \frac{1}{6}$

Another way to divide a fraction by a whole number is to think of multiplying the fraction by a unit fraction. In the example above, $\frac{1}{2} \div 3$ means find one-third, or $\frac{1}{3}$, of $\frac{1}{2}$. Remember: "$\frac{1}{3}$ of $\frac{1}{2}$" means the same as "$\frac{1}{3} * \frac{1}{2}$." So $\frac{1}{2} \div 3 = \frac{1}{2} * \frac{1}{3} = \frac{1}{6}$.

Example One half of a loaf of cornbread is divided into 8 equal pieces. What fraction of the whole loaf is each piece?

To solve $\frac{1}{2} \div 8$, multiply $\frac{1}{2} * \frac{1}{8}$. Because $\frac{1}{2} * \frac{1}{8} = \frac{1}{16}$, each piece is $\frac{1}{16}$ of the whole loaf.

$\frac{1}{2} \div 8 = \frac{1}{16}$

The previous example shows that if the dividend in a division problem is a unit fraction $\frac{1}{a}$ and the divisor is n, where $a \neq 0$ and $n \neq 0$, then:

$$\frac{1}{a} \div n = \frac{1}{a} * \frac{1}{n} = \frac{1}{an}.$$

Here is another example of this type of problem.

Example Six teammates each run an equal part of a half-mile obstacle course. What part of a mile does each teammate run?

Solve $\frac{1}{2} \div 6 = ?$

$\frac{1}{2} \div 6 = \frac{1}{2} * \frac{1}{6} = \frac{1}{12}.$

So each teammate runs $\frac{1}{12}$ of a mile.

$\frac{1}{2} \div 6 = \frac{1}{12}$

In general, dividing *any* number by a whole number n, where $n \neq 0$, is the same as multiplying the dividend by $\frac{1}{n}$.

You can also use the missing factor meaning of division on page 80 to think about division of a unit fraction by a whole number.

Example Find $\frac{1}{4} \div 5$.

$\frac{1}{4} \div 5 = \square$ is equivalent to $5 * \square = \frac{1}{4}$, which means, "5 times what number is $\frac{1}{4}$?" Because $5 * \frac{1}{20} = \frac{1}{4}$, $\frac{1}{4} \div 5 = \frac{1}{20}$.

Check Your Understanding

Solve.

1. $\frac{1}{3} \div 5$

2. One eighth divided by four

3. $\frac{1}{100} \div 10$

4. Three friends equally share a half of a watermelon. How much of the whole watermelon does each friend get?

Check your answers on page 442.

Different Types of Numbers

Counting is almost as old as the human race and has been used in some form by every human society. Long ago, people found that the **counting numbers** (1, 2, 3, and so on) did not meet all their needs.

♦ Counting numbers cannot be used to express measures that are between two consecutive whole numbers, such as $2\frac{1}{2}$ inches and 1.6 kilometers.

♦ With the counting numbers, division problems such as 8/5 and 3/7 do not have an answer.

Fractions were invented to meet these needs. Fractions can also be renamed as decimals and percents. And most of the numbers you have seen—such as $\frac{1}{2}$, $5\frac{1}{6}$, 1.23, and 25%—are either fractions, or can be renamed as fractions. With the invention of fractions, it became possible to express rates and ratios, to name many more points on the number line, and to solve any division problem involving whole numbers (except division by 0).

> **Note**
>
> Every whole number can be renamed as a fraction. For example, 0 can be written as $\frac{0}{1}$ and 8 can be written as $\frac{8}{1}$.

However, even fractions did not meet every need. For example, problems such as $5 - 7$ and $2\frac{3}{4} - 5\frac{1}{4}$ have answers that are less than 0 and cannot be named as fractions. (Fractions, by the way they are defined, can never be less than 0.) This led to the invention of **negative numbers.** Negative numbers are numbers that are less than 0. The numbers $-\frac{1}{4}$, -3.25, and -100 are negative numbers. The number -3 is read as "negative 3."

> **Note**
>
> Since every whole number can be renamed as a fraction, every negative whole number can be renamed as a negative fraction. For example, $-7 = -\frac{7}{1}$.

Negative numbers serve several purposes:

♦ To express locations such as temperatures below zero on a thermometer and depths below sea level;

♦ To show changes such as yards lost in a football game and

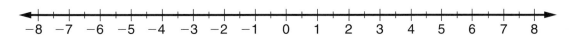

decreases in weight;

♦ To extend the number line to the left of 0;

♦ To calculate answers to many subtraction problems.

The **opposite** of every positive number is a negative number. And the opposite of every negative number is a positive number. The diagram shows this relationship. The number 0 is neither positive nor negative. 0 is its own opposite.

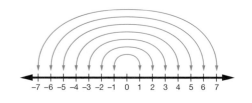

A Summary of Different Types of Numbers

The **counting numbers** are 1, 2, 3, and so on.

The **whole numbers** are 0, 1, 2, 3, and so on. The whole numbers are all the counting numbers, together with 0.

The **integers** are 0, 1, −1, 2, −2, 3, −3, and so on. So the integers are all the whole numbers, together with the opposites of all the whole numbers.

The **fractions** are all numbers that can be written as $\frac{a}{b}$, where a and b can be any whole numbers ($b \neq 0$).

♦ Every whole number can be renamed as a fraction. For example, 5 can be renamed as $\frac{5}{1}$ and 0 can be renamed as $\frac{0}{1}$.

♦ Every fraction can be renamed as a decimal and as a percent. For example, $\frac{3}{4}$ can be renamed as 0.75 and as 75%, and $\frac{9}{8}$ can be renamed as 1.125 and as 112.5%.

The **rational numbers** are all numbers that can be written or renamed as fractions or their opposites.

♦ Counting numbers, whole numbers, integers, fractions, and the opposites of these numbers are all rational numbers.

♦ Every number shown on the Probability Meter at the right is a rational number.

There are other numbers that are called **irrational numbers.** Some of these, like the number pi (π), you have used before. An irrational number cannot be renamed as a fraction or its opposite. You will learn more about irrational numbers when you study algebra.

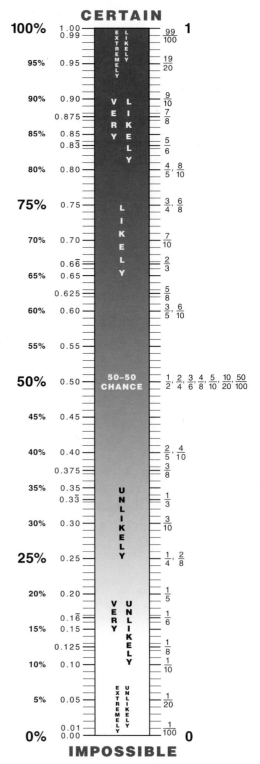

Renaming Fractions as Decimals

Fractions, decimals, and percents are different ways to write numbers. Sometimes it's easier to work with a decimal instead of a fraction, or with a percent instead of a decimal. So it's good to know how to rename fractions, decimals, and percents in whatever way you want.

Any fraction can always be renamed as a decimal. Sometimes the decimal will end after a certain number of places. However, sometimes the decimal will have one or more digits that repeat in a pattern forever. Decimals that end are called **terminating decimals;** those that repeat in a pattern forever are called **repeating decimals.** The fraction $\frac{1}{2}$ is equal to the terminating decimal 0.5. The fraction $\frac{1}{3}$ is equal to the repeating decimal 0.3333….

There are several ways to rename a fraction as a decimal. One way is just to memorize the decimal name: $\frac{1}{2} = 0.5$, $\frac{1}{4} = 0.25$, $\frac{1}{8} = 0.125$, and so on. But this only works for a few simple fractions, unless you have a very good memory.

Another way to find the decimal name for a fraction is to use logical thinking. For example, if $\frac{1}{8} = 0.125$, then $\frac{3}{8} = 0.125 + 0.125 + 0.125 = 0.375$. Logical thinking and a few memorized decimal names will help you rename many common fractions as decimals.

Sometimes you can rename a fraction as a decimal by finding an equivalent fraction with a denominator such as 10, 100, or 1,000. For example, $\frac{3}{5} = \frac{6}{10} = 0.6$ and $\frac{3}{20} = \frac{15}{100} = 0.15$.

For other fractions, you can find decimal names by using the Fraction-Stick Chart on page 399.

Remember that fractions can be used to show division problems: $\frac{a}{b} = a \div b$. So, you can *always* rename any fraction as a decimal by dividing the numerator by the denominator. You can do the division either with paper and pencil or with a calculator.

Equivalent Fractions Method

One way to rename a fraction as a decimal is to find an equivalent fraction with a denominator that is a power of 10, such as 10, 100, or 1,000. This method works only for some fractions.

Example Rename $\frac{3}{5}$ as a decimal.

The solid lines divide the square into 5 equal parts. Each part is $\frac{1}{5}$ of the square. $\frac{3}{5}$ of the square is shaded.

The dashed lines divide each fifth into 2 equal parts. Each of these smaller parts is $\frac{1}{10}$, or 0.1, of the square.

$\frac{6}{10}$, or 0.6, of the square is shaded.

$\frac{3}{5} = \frac{6}{10} = 0.6$

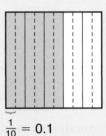

$\frac{1}{10} = 0.1$

Example Rename $\frac{3}{4}$ as a decimal.

The solid lines divide the square into 4 equal parts. Each part is $\frac{1}{4}$ of the square. $\frac{3}{4}$ of the square is shaded.

The dashed lines divide each fourth into 25 equal parts. Each of these smaller parts is $\frac{1}{100}$, or 0.01, of the square.

$\frac{75}{100}$, or 0.75, of the square is shaded.

$\frac{3}{4} = \frac{75}{100} = 0.75$

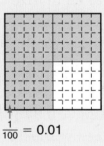

$\frac{1}{100} = 0.01$

Example Rename $\frac{6}{40}$ as a decimal.

$\frac{6}{40} = \frac{6 \div 2}{40 \div 2} = \frac{3}{20}$ Dividing numerator and denominator by the same number gives an equivalent fraction.

$\frac{3}{20} = \frac{3 * 5}{20 * 5} = \frac{15}{100}$ Multiplying numerator and denominator by the same number gives an equivalent fraction.

So, $\frac{6}{40} = \frac{15}{100} = 0.15$.

Check Your Understanding

Rename each fraction as a decimal.

1. $\frac{3}{4}$ **2.** $\frac{2}{5}$ **3.** $\frac{7}{2}$ **4.** $\frac{11}{20}$ **5.** $\frac{12}{15}$ **6.** $\frac{8}{25}$

Check your answers on page 435.

Using the Fraction-Stick Chart

The Fraction-Stick Chart below (and on page 399) can also be used to rename fractions as decimals. Note that the result is often not exact.

Example Rename $\frac{2}{3}$ as a decimal.

1. Locate $\frac{2}{3}$ on the "thirds" stick.
2. Place one edge of a straightedge at $\frac{2}{3}$.
3. Find where the straightedge crosses the number line.

The straightedge crosses the number line between 0.66 and 0.67.

So $\frac{2}{3}$ is equal to about 0.66 or 0.67.

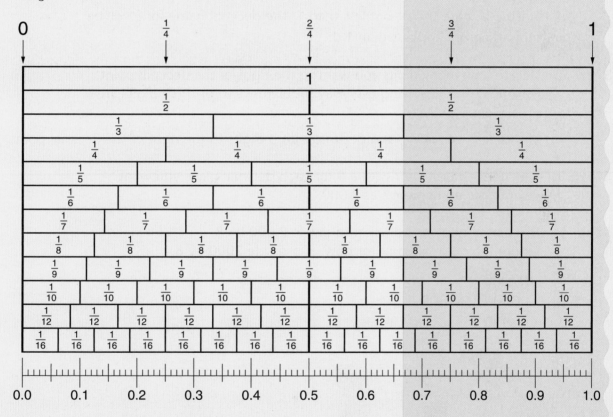

Check Your Understanding

Use the chart above to find an approximate decimal name for each fraction or mixed number.

1. $\frac{7}{10}$ 2. $\frac{3}{8}$ 3. $2\frac{1}{3}$ 4. $\frac{12}{16}$ 5. $\frac{6}{7}$

Check your answers on page 435.

Using Division

Fractions can be used to show division problems. For example, $\frac{7}{8}$ is another way to write $7 \div 8$. So, one way to rename $\frac{7}{8}$ as a decimal is to divide 7 by 8.

This method will *always* work. Any fraction may be renamed as a decimal by dividing its numerator by its denominator. The division can usually be done easily on a calculator. But it may also be done by computing with pencil and paper.

Example   Change $\frac{7}{8}$ to a decimal by division.

Step 1:   First, estimate the answer.

Since $\frac{7}{8}$ is more than $\frac{1}{2}$ but less than 1, the decimal name for $\frac{7}{8}$ will be more than 0.5 and less than 1.0.

Step 2:   Decide how many digits you want to the right of the decimal point. For measuring or solving everyday problems, two or three digits are usually enough.

In this case, rename $\frac{7}{8}$ as a decimal with 3 digits to the right of the decimal point.

Step 3:   Rewrite the numerator with a 0 for each decimal place you want. Rewrite the numerator 7 as 7.000.

Step 4:   Use partial-quotients division to divide 7.000 by 8. Ignore the decimal point for now, and divide 7000 by 8.

```
   8)7000
   - 6400   |  800
      600   |
   -  560   |   70
       40   |
   -   40   |    5
        0   |  875
```

Step 5:   Use the estimate from Step 1 to place the decimal point in the quotient.

Since $\frac{7}{8}$ is between 0.5 and 1.0, the decimal point should be placed before the 8, giving 0.875.

So, $\frac{7}{8} = 0.875$.

In the example above, there was no remainder.

When there is a remainder, you should round the quotient before placing the decimal point.

Example Rename $\frac{2}{3}$ as a decimal using division.

Step 1: Estimate the answer.

Since $\frac{2}{3}$ is more than $\frac{1}{2}$ and less than 1, the decimal name for $\frac{2}{3}$ should be between 0.5 and 1.0.

Step 2: Decide how many digits you want to the right of the decimal point.

In this example, rename $\frac{2}{3}$ as a decimal with two digits to the right of the decimal point.

Step 3: Rewrite the numerator with a 0 for each decimal place you want. Rewrite the numerator 2 as 2.00.

Step 4: Use partial-quotients division to divide 2.00 by 3. Ignore the decimal point for now, and divide 200 by 3.

$$
\begin{array}{r|r}
3)\overline{200} & \\
-180 & 60 \\
\hline
20 & \\
-18 & 6 \\
\hline
2 & 66 \\
\end{array}
$$

Division shows that $200 / 3 = 66\frac{2}{3}$, which you can round to 67.

Step 5: Use the estimate from Step 1 to place the decimal point in the quotient.

Since $\frac{2}{3}$ is between 0.5 and 1.0, the decimal point should be placed before the first 6, giving 0.67.

So, $\frac{2}{3} = 0.67$, rounded to the nearest hundredth.

Note

The remainder in a division problem can be rewritten as a fraction. The remainder is the numerator and the divisor is the denominator. For example, $200 \div 3 \rightarrow$ 66 R2, or $66\frac{2}{3}$.

Check Your Understanding

Use division to rename these fractions as decimals.

1. $\frac{3}{8}$ (to nearest thousandth)

2. $\frac{1}{6}$ (to nearest thousandth)

3. $\frac{5}{9}$ (to nearest hundredth)

Check your answers on page 435.

Using a Calculator

You can also rename a fraction as a decimal by dividing the numerator by the denominator using a calculator.

Examples Rename $\frac{2}{3}$ as a decimal.

Key in: 2 ÷ 3 =

$\frac{2}{3}$ = 0.6666666667 on Calculator A.

$\frac{2}{3}$ = 0.6666666 on Calculator B.

Rename $\frac{5}{6}$ as a decimal.

Key in: 5 ÷ 6 =

$\frac{5}{6}$ = 0.8333333333 on Calculator A.

$\frac{5}{6}$ = 0.8333333 on Calculator B.

In some cases, the decimal takes up the entire calculator display. If one or more digits repeat in a pattern, the decimal can be written by writing the repeating digit or digits just once, and putting a bar over whatever repeats.

Examples

Fraction	Key in:	Calculator Answer	Decimal
$\frac{1}{3}$	1 ÷ 3 =	0.3333333	$0.\overline{3}$
$\frac{4}{9}$	4 ÷ 9 =	0.4444444	$0.\overline{4}$
$\frac{6}{11}$	6 ÷ 11 =	0.5454545	$0.\overline{54}$
$\frac{7}{12}$	7 ÷ 12 =	0.5833333	$0.58\overline{3}$

Decimals in which one or more digits repeat according to a pattern are called **repeating decimals.** In a repeating decimal, the pattern of repeating digits goes on forever. For example, suppose you converted $\frac{4}{9}$ to a decimal using a calculator with a display that could show 1,000 digits. The display would show a 0 and 999 fours: 0.4444444....

A decimal that ends is called a **terminating decimal.** For example, 0.4, 1.25, and 0.625 are terminating decimals.

Note

On some calculators, the final digit for some repeating decimals may not follow the pattern. For example, a calculator may show $\frac{2}{3}$ = 2 ÷ 3 = 0.6666666667. The digit 6 really does repeat forever, but this calculator rounded the last digit that the calculator can display.

Check Your Understanding

Use a calculator to rename each fraction as a decimal.

1. $\frac{5}{8}$ 2. $\frac{2}{13}$ 3. $\frac{7}{8}$ 4. $\frac{4}{7}$ 5. $\frac{1}{12}$ 6. $\frac{8}{15}$

Check your answers on page 435.

Renaming Fractions, Decimals, and Percents

The previous pages describe several ways to rename fractions as decimals. Here we discuss how to change fractions to percents, decimals to fractions, and so on.

Renaming a Decimal as a Fraction

Any terminating decimal can be renamed as a fraction whose denominator is a power of 10. To change a terminating decimal to a fraction, use the place of the rightmost digit to help you write the denominator.

Example Write as fractions.

0.5   The rightmost digit is 5, which is in the tenths place.
So $0.5 = \frac{5}{10}$ or, in simplest form, $\frac{1}{2}$.

0.307 The rightmost digit is 7, which is in the thousandths place.
So $0.307 = \frac{307}{1,000}$.

4.75   The rightmost digit is 5, which is in the hundredths place.
So $4.75 = \frac{475}{100}$. You can simplify $\frac{475}{100}$ as $4\frac{75}{100}$ or $4\frac{3}{4}$.

Did You Know?

The examples at the left rename *terminating* decimals as fractions. But any *repeating* decimal can also be renamed as a fraction. Here are some simple examples.

$$0.111\ldots = \frac{1}{9}$$
$$0.2323\ldots = \frac{23}{99}$$
$$0.456456\ldots = \frac{456}{999}$$
$$0.78907890\ldots = \frac{7890}{9999}$$

Some calculators have a special key for renaming decimals as fractions.

Example Rename 0.32 as a fraction.

To rename 0.32 as a fraction using Calculator A, key in:
.32 (Enter) (F↔D) Answer: $\frac{32}{100}$

Or, using Calculator B, key in: .32 (F↔D) Answer: $\frac{8}{25}$

Both answers are correct since $\frac{8}{25} = \frac{32}{100}$. Calculator B has renamed the decimal with a fraction in simplest form.

Renaming Fractions as Percents

The best way to change a fraction to a percent is by memory. If you have a good memory, you may be able to remember almost all of the decimal and percent names for "easy" fractions given in the table on page 398.

If you can't remember the percent name for a fraction, the best thing to do is change the fraction to a decimal (by dividing). Then multiply the decimal by 100 and write the % symbol.

Example Use a calculator to rename $\frac{3}{8}$ as a percent.

Key in: 3 $\div$ 8 $\times$ 100 $=$ Answer: 37.5

Therefore, $\frac{3}{8} = 37.5\%$.

Renaming Percents as Fractions

A percent can always be written as a fraction with a denominator of 100. Simply remove the % symbol and write a fraction bar and a denominator of 100 below the number. The fraction can be renamed in simplest form.

Examples $40\% = \frac{40}{100} = \frac{2}{5}$ $85\% = \frac{85}{100} = \frac{17}{20}$ $150\% = \frac{150}{100} = \frac{3}{2} = 1\frac{1}{2}$

Renaming a Percent as a Decimal

Remove the % symbol and then divide by 100.

Examples $45\% = 45 / 100 = 0.45$ $120\% = 120 / 100 = 1.20$ $1\% = 1 / 100 = 0.01$

Renaming a Decimal as a Percent

Multiply by 100 and write the % symbol.

Examples $0.45 = (0.45 * 100)\% = 45\%$ $1.2 = (1.2 * 100)\% = 120\%$

Check Your Understanding

Copy and complete this table.

Fraction	Decimal	Percent
$\frac{1}{4}$	0.25	25%
		80%
$\frac{1}{2}$		
	0.35	
		10%
$\frac{5}{8}$		

Check your answers on page 436.

Uses of Negative Numbers

You have probably used positive and negative numbers before. For example, when using the Celsius scale, 20 degrees above zero can be written as +20°C and 5 degrees below zero as −5°C.

Many other real-world situations have zero as a starting point. Numbers go in opposite directions from zero. The numbers greater than zero are called **positive numbers;** the numbers less than zero are called **negative numbers.**

Examples

Situation	Negative (−)	Zero (0)	Positive (+)
Temperature	below zero	zero	above zero
Business	loss	break even	profit
Bank account	withdrawal	unchanged	deposit
Time	past; before	present; now	future; after
Calendar (Gregorian)	B.C. or B.C.E.	zero	A.D.
Game	behind	tied; even	ahead
Elevation	below sea level	at sea level	above sea level
Gauges, dials, dipsticks	below correct level	at correct level	above correct level

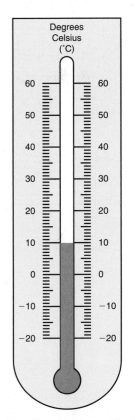

This thermometer is a vertical number line.

A number line may be used to show both positive and negative numbers. Some number lines may be horizontal, others vertical. A **timeline** is usually shown as a horizontal number line. A thermometer and the oil dipstick in an automobile can be thought of as vertical number lines.

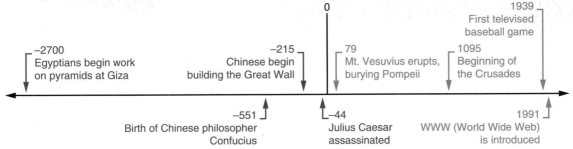

A timeline based on the Gregorian calendar

Addition and Subtraction of Positive and Negative Numbers

Number-Line Walking

One way to add and subtract positive and negative numbers is to imagine walking on a number line.

♦ The first number tells you where to start.

♦ The operation sign (+ or −) tells you which way to face:
   + means face toward the positive end of the number line.
   − means face toward the negative end of the number line.

♦ If the second number is negative (has a − sign), then you will walk backward. Otherwise, walk forward.

♦ The second number tells you how many steps to take.

♦ The number where you end up is the answer.

Example $-3 + 5 = ?$

Start at -3.

The operation sign is $+$, so face the positive end of the number line.

The second number is positive, so walk forward 5 steps.

You end up at 2.

So, $-3 + 5 = 2$.

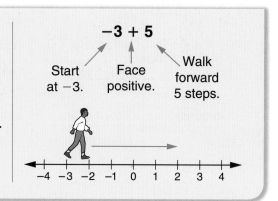

Example $-2 - (-4) = ?$

Start at -2.

The operation sign is $-$, so face the negative end of the number line.

The second number is negative, so walk backward 4 steps.

You end up at 2.

So, $-2 - (-4) = 2$.

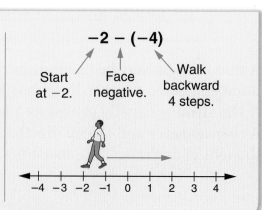

Check Your Understanding

Add or subtract.

1. $4 + (-6)$ **2.** $-5 - 3$ **3.** $-6 - (-7)$ **4.** $-2 + (-10)$

Check your answers on page 436.

Using a Slide Rule

In Fifth Grade *Everyday Mathematics* there is a special slide rule that can be used for adding and subtracting positive and negative numbers. (The other side of the slide rule can be used for adding and subtracting fractions; see page 69.)

Addition and subtraction with a slide rule is similar to number-line walking.

Addition with a Slide Rule

Examples

$-4 + 3 = ?$

Line up the 0-mark on the slider with -4 on the holder.

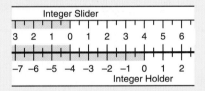

- Imagine facing in the positive direction on the slider.

- Go forward 3 on the slider.

- The 3 on the slider lines up with -1 on the holder.

- This is the answer to the problem.

$$-4 + 3 = -1$$

$1 + (-5) = ?$

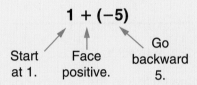

Line up the 0-mark on the slider with 1 on the holder.

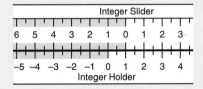

- Imagine facing in the positive direction on the slider.

- Go backward 5 on the slider. (So you are actually going in the negative direction on the slider.)

- The 5 on the slider lines up with -4 on the holder.

- This is the answer to the problem.

$$1 + (-5) = -4$$

Subtraction with a Slide Rule

Examples

$$2 - 3 = ?$$

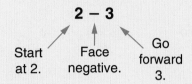

Start at 2. Face negative. Go forward 3.

Line up the 0-mark on the slider with 2 on the holder.

Integer Slider
5 4 3 2 1 0 1 2 3 4
-3 -2 -1 0 1 2 3 4 5 6
Integer Holder

- Imagine facing in the negative direction on the slider.

- Go forward 3 on the slider. (So you are actually going in the negative direction on the slider.)

- The 3 on the slider is lined up with −1 on the holder.

- This is the answer to the problem.

$$2 - 3 = -1$$

$$-4 - (-6) = ?$$

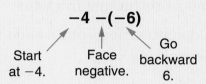

Start at −4. Face negative. Go backward 6.

Line up the 0-mark on the slider with −4 on the holder.

Integer Slider
2 1 0 1 2 3 4 5 6 7
-6 -5 -4 -3 -2 -1 0 1 2 3
Integer Holder

- Imagine facing in the negative direction on the slider.

- Go backward 6 on the slider. (So you are actually going in the positive direction on the slider.)

- The 6 on the slider is lined up with 2 on the holder.

- This is the answer to the problem.

$$-4 - (-6) = 2$$

Check Your Understanding

Add or subtract.

1. $-4 + (-6)$

2. $3 - (-9)$

3. $5 - 11$

4. $6 + (-16)$

Check your answers on page 436.

Space Travel

When we travel on Earth, we typically measure distances in miles or kilometers. In space, where the distances are much greater, astronomers use units of measure based on the fastest thing in the universe: light.

Traveling at the Speed of Light

A common unit of measure is the light-year. One **light-year** is the distance light travels in a year. Light travels 186,000 miles every second. In one year, light can travel nearly 6 trillion (6.0×10^{12}) miles.

To get a sense of the speed of light and the vastness of the universe, imagine you have hopped on a beam of light and are about to begin a journey in space. ▶

▲
Start with a quick warm-up on Earth. At light speed, you can do about 7 laps around the earth every second!

◀ Next, take a trip to the moon. At about 240,000 miles away, your light beam will take you there in about 1.3 seconds.

Our Solar System

The earth and other planets revolve around the sun in our solar system. Your next few trips will take you to places in our solar system.

A trip to the sun would take only 8 minutes, if you could survive the 10,000°F surface temperature. The distance from the sun to the earth is 93 million miles. Astronomers call that distance 1 Astronomical Unit (AU). This unit helps them to calculate and compare distances to other celestial bodies. ➤

◄ Now travel to Saturn, the 6th planet from the sun. Saturn is nearly 10 AU from the sun, so your light-speed trip will take about 10 times as long as the Sun-Earth trip, or about 80 minutes.

This illustration shows your final destination in our solar system, Pluto and one of its moons, Charon. Pluto is around 3.6 billion (3.6×10^9) miles from the sun. You can be there in a little more than 5 hours. ➤

Stars

Once you leave our solar system, the closest stars are not too
far away when you are traveling at light-speed.

◄ Sirius, the brightest star in the Northern
Hemisphere, is 8 years away on your light
beam. This may seem like a long journey,
but the same trip on a space shuttle traveling
at 18,000 miles per hour would take nearly
320,000 years, or 40,000 times as long.

Sirius

Polaris

A well-known star to visit is
Polaris, or the North Star.
Polaris is the last star in the
handle of the Little Dipper.
It is 300 light-years from the
earth, about 70 times farther
away than the closest star,
Proxima Centauri. ➤

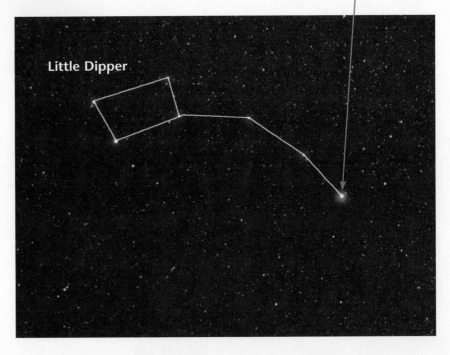

Little Dipper

Galaxies

Our sun, Sirius, Polaris, and all the other stars we can see are members of the Milky Way galaxy. The Milky Way contains at least 100 billion (1.0×10^{11}) stars. Scientists believe it is 1 of about 200 billion (2.0×10^{11}) galaxies in the entire universe.

◄ The Milky Way is a spiral galaxy like the Whirlpool galaxy in this photograph.

From Earth, the Milky Way looks like this. Our home is about $\frac{2}{3}$ of the way from the center of the galaxy. It will take you about 30,000 years to travel on your light beam to its center. ➤

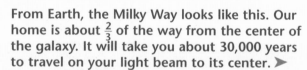

◄ The Small Magellanic Cloud is one of the galaxies closest to the Milky Way. It is named after the explorer, Ferdinand Magellan, who observed and noted this galaxy during his voyage around the world in the 1500s. On your beam of light, you can travel there in about 210,000 years.

◄ The Andromeda Galaxy is the closest spiral galaxy to ours. At the speed of light, it will take you about 2.5 million (2.5×10^6) years to get there. The Andromeda Galaxy, the Milky Way, and more than 30 other galaxies are part of a cluster of galaxies known as the **Local Group.**

▲ The Local Group is one of many clusters that make up the Local Supercluster. There are thousands of galaxies in the Local Supercluster. Traveling on your light beam, you could reach the center of the supercluster in around 60 million (6.0×10^7) years.

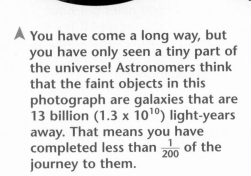

▲ You have come a long way, but you have only seen a tiny part of the universe! Astronomers think that the faint objects in this photograph are galaxies that are 13 billion (1.3×10^{10}) light-years away. That means you have completed less than $\frac{1}{200}$ of the journey to them.

Back to Reality

Unfortunately, we cannot ride a beam of light to the planets, stars, and galaxies. But technological achievements have enabled us to see and learn a great deal about space.

◄ The Hubble Space Telescope, first launched in 1990, orbits the earth once every 97 minutes. The Hubble can track and photograph planets, stars, galaxies, and other celestial bodies. It then transmits the photographs to the earth by radio.

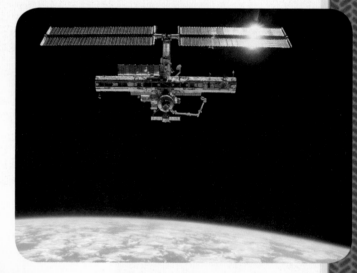

The U.S. space shuttle and the Russian Soyuz rocket bring supplies and crew to the International Space Station. In this photograph, the space shuttle Discovery is preparing to "dock" with the space station. ►

▲ The International Space Station is operated by the United States, Russia, and several other countries. The space station is a zero-gravity science laboratory that orbits the earth at over 17,000 miles per hour. Normally, a crew of three astronauts lives and works in the space station for several months.

Based on your "experiences" traveling through space, how much of the universe do you think humans will be able to visit in your lifetime?

Rates, Ratios, and Proportions

Rates

The easiest fractions to understand are fractions that name parts of wholes and fractions used in measurement. In working with such fractions, it's important to know what the ONE, or whole, is. For example, if a cake is the ONE, then 3/4 of a cake describes 3 parts of the cake after it has been divided into 4 equal parts.

Not all fractions name parts of wholes. Some fractions compare two different amounts, where one amount is not part of the other. For example, a store might sell apples at 3 apples for 75 cents, or a car's gas mileage might be 160 miles per 8 gallons. These comparisons can be written as fractions: $\frac{3 \text{ apples}}{75\cancel{c}}$; $\frac{160 \text{ miles}}{8 \text{ gallons}}$. These fractions do not name parts of wholes: The apples are *not* part of the money; the miles are *not* part of the gallons of gasoline.

Fractions like $\frac{160 \text{ miles}}{8 \text{ gallons}}$ show rates. A **rate** tells how many of one thing there are for a certain number of another thing. Rates often contain the word *per,* meaning "for each," "for every," or something similar.

Examples

Alex rode her bicycle 10 miles in 1 hour. Her rate was 10 miles per hour. This rate describes the distance Alex traveled and the time it took her. The rate "10 miles per hour" is often written as 10 mph. The fraction for this rate is $\frac{10 \text{ miles}}{1 \text{ hour}}$.

Here are some other rates:

typing speed	50 words per minute	$\frac{50 \text{ words}}{1 \text{ minute}}$
price	14 cents per ounce	$\frac{14\cancel{c}}{1 \text{ ounce}}$
scoring average	17 points per game	$\frac{17 \text{ points}}{1 \text{ game}}$
exchange rate	0.7792 Euros for each U.S. dollar	$\frac{0.7792 \text{ Euros}}{1 \text{ U.S. dollar}}$
allowance	$5.00 per week	$\frac{\$5.00}{1 \text{ week}}$
baby-sitting	$4.00 per hour	$\frac{\$4.00}{1 \text{ hour}}$

Per-Unit Rates

A **per-unit rate** is a rate with 1 in the denominator. Per-unit rates tell how many of one thing there are for a single one of another thing. We say that "2 dollars per gallon" is a **per-gallon rate,** "12 miles per hour" is a **per-hour rate,** and "4 words per minute" is a **per-minute rate.** The fractions for these per-unit rates each have a 1 in the denominator: $\frac{\$2}{1 \text{ gallon}}$, $\frac{12 \text{ miles}}{1 \text{ hour}}$, and $\frac{4 \text{ words}}{1 \text{ minute}}$.

Any rate can be renamed as a per-unit rate by dividing the numerator and the denominator by the denominator.

A gas pump displays the per-gallon rate, the number of gallons pumped, and the total cost of the gas.

Example Change each rate to a per-unit rate.

$$\frac{36 \text{ in.}}{3 \text{ ft}} = \frac{36 \text{ in.} \div 3}{3 \text{ ft} \div 3} \qquad \frac{72¢}{12 \text{ eggs}} = \frac{72¢ \div 12}{12 \text{ eggs} \div 12}$$

$$= \frac{12 \text{ in.}}{1 \text{ ft}} \qquad\qquad\qquad = \frac{6¢}{1 \text{ egg}}$$

A rate can also have 1 in the numerator. A food stand might sell apples at a rate of 1 apple for 25¢ or $\frac{1 \text{ apple}}{25¢}$. Conversions between inches and centimeters are at a rate of 1 inch to 2.54 centimeters or $\frac{1 \text{ in.}}{2.54 \text{ cm}}$. Rates with 1 in the numerator or in the denominator are often easier to work with than other rates.

Rate Tables

Rate information can be used to make a **rate table.**

Example Write a fraction and make a rate table for the statement, "A computer printer prints 4 pages per minute."

4 pages per minute = $\frac{4 \text{ pages}}{1 \text{ minute}}$

This is the per-unit rate.

pages	4	8	12	16	20	24	28
minutes	1	2	3	4	5	6	7

The table shows that if a printer prints 4 pages per minute, it will print 8 pages in 2 minutes, 12 pages in 3 minutes, and so on.

Equivalent rates are rates that make the same comparison. Each rate in a rate table is **equivalent** to each of the other rates in the table.

Check Your Understanding

Write each rate as a fraction and make a rate table showing 4 equivalent rates.

1. Joan baby-sits for 6 dollars per hour.   **2.** Water weighs about 8 pounds per gallon.

Check your answers on page 436.

Solving Rate Problems

In many problems that involve rates, a per-unit rate is given and you need to find an equivalent rate. These problems can be solved in more than one way.

Examples   Bill's car can travel 35 miles on 1 gallon of gasoline. At this rate, how far can the car travel on 7 gallons?

Solution 1: Using a rate table

First, set up a rate table and enter what you know. Write a question mark in place of what you are trying to find.

miles	35					?
gallons	1					7

Next, work from what you know to what you need to find. In this case, by doubling, you can find how far Bill could travel on 2 gallons, 4 gallons, and 8 gallons of gasoline.

miles	35	70	140	280			?
gallons	1	2	4	8			7

There are two different ways to use the rate table to answer the question. You will find that Bill can travel 245 miles:

- by adding the distances for 1 gallon, 2 gallons, and 4 gallons: 35 miles + 70 miles + 140 miles = 245 miles.

- by subtracting the distance for 1 gallon from the distance for 8 gallons: 280 miles − 35 miles = 245 miles.

Solution 2: Using multiplication

If the car can travel 35 miles on 1 gallon, then it can travel 7 times as far on 7 gallons.

7 * 35 = 245, so the car can travel 245 miles on 7 gallons of gas.

Check Your Understanding

Solve.

1. There are 3 feet in 1 yard. How many feet are in 5 yards? In 14 yards?

2. Angie's heart rate is 70 beats per minute. How many times does her heart beat in 9 minutes? In 20 minutes?

Check your answers on page 436.

Sometimes, a rate is given and you need to find the equivalent per-unit rate. You can use a rate table or division.

Examples Keisha receives an allowance of $20 for 4 weeks. At this rate, how much does she get per week?

Solution 1: Using a rate table

First, set up a table and enter what you know.

Next, work from what you know to what you need to find. By halving $20, you can find how much Keisha gets for 2 weeks. By halving again, you can find what she gets for 1 week.

allowance	$20				?
weeks	4				1

So, Keisha gets $5 for 1 week.

allowance	$20	$10	$5		?
weeks	4	2	1		1

Solution 2: Using division

If Keisha receives $20 for 4 weeks, she receives $\frac{1}{4}$ of $20 for 1 week. If you divide $20 into 4 equal parts, each part has $5. $(20 \div 4 = 5)$

So, Keisha receives $5 per week.

Sometimes a rate that is not a per-unit rate is given and you need to find an equivalent rate that is not a per-unit rate.

♦ First find the equivalent per-unit rate.

♦ Then use the per-unit rate to find the rate asked for in the problem. A rate table can help you organize your work.

Example A gray whale's heart beats 24 times in 3 minutes. At this rate, how many times does it beat in 2 minutes?

If the whale's heart beats 24 times in 3 minutes, it beats $\frac{1}{3}$ of 24 times in 1 minute (24 / 3 = 8). Double this to find how many times it beats in 2 minutes (2 * 8 = 16).

heartbeats	24	8	16		?
minutes	3	1	2		2

The whale's heart beats 16 times in 2 minutes.

Check Your Understanding

Solve.

1. Ashley baby-sat for 5 hours. She was paid $30. How much did she earn per hour?
2. Bob saved $420 last year. How much did he save per month?
3. A carton of 12 eggs costs $1.80 (180 cents). At this rate, how much do 8 eggs cost?

Check your answers on page 436.

Ratios

A **ratio** is a comparison of two counts or measures that have the *same unit*. Ratios can be expressed as fractions, decimals, percents, words, or with a colon.

Some ratios compare part of a collection of things to the total number of things in the collection.

> **Note**
>
> A *rate* is also a comparison of two counts or measures, but the counts or measures have *different units*.

Example The statement "1 out of 6 students in the class is absent" compares the number of absent students to the total number of students in the class. This ratio can be expressed in many ways.

In words: For every 6 students enrolled in the class, 1 student is absent. *One in 6 students is absent.* The ratio of absent students to all students is *1 to 6*.

As a fraction: $\frac{\text{number absent}}{\text{total number}}$ is the fraction of students in the class that is absent. $\frac{1}{6}$ of the students are absent.

As a percent: $\frac{1}{6} = 0.166\ldots = 16.6\ldots\%$. So, about 16.7% of the students are absent.

With a colon: The ratio of absent students to all students is 1:6 (read as: "one to six"). A colon is written between the numbers that are being compared.

Some ratios compare part of a quantity to the total quantity.

Example The statement "Seth has biked the first 16 miles in a 25-mile race" compares the number of miles Seth has biked with the total number of miles for this race.

In words: *16 of 25* miles have been biked. The ratio of miles biked to total miles is *16 to 25*.

As a fraction: $\frac{\text{miles biked}}{\text{total miles}}$ is the fraction of total miles that have been biked. $\frac{16}{25}$ of the total distance has been biked.

As a percent: $\frac{16}{25} = \frac{64}{100} = 0.64 = 64\%$. So, 64% of the total distance has been biked.

With a colon: The ratio of miles biked to total distance is 16:25.

Ratios are similar to rates in some ways:

♦ Ratios and rates are each used to compare two different amounts.
♦ Ratios and rates can each be written as fractions.

Ratios and rates are different in one important way.

♦ Rates compare amounts that have *different units,* while ratios compare amounts that have the *same unit.*

So, you must always mention both units when you name a rate. But a ratio is a "pure number," and there are no units to mention when you name a ratio.

The examples on page 106 show ratios that compare part of a whole to the whole. But ratios often compare two amounts where neither is a part of the other. These comparisons can also be expressed as fractions, decimals, percents, words, or with a colon.

Example A radio sells for $27 at store A. An identical radio sells for $30 at store B. Compare the prices.

In words: The ratio of the price at store A to the price at store B is *27 to 30*.

As a fraction: The fraction $\frac{\text{price at store A}}{\text{price at store B}}$ shows this comparison. The fraction equals $\frac{27}{30}$ after substituting the actual prices.

As a percent: $\frac{27}{30} = \frac{9}{10} = 0.9 = 90\%$. So, the price at store A is 90% of the price at store B.

With a colon: The ratio of price at store A to price at store B is 27:30.

Many statements describe ratios without actually using the word "ratio." Study the examples below.

Examples For each statement, identify the quantities being compared. Then write the ratio as a fraction.

- On an average evening, about $\frac{1}{3}$ of the U.S. population watches TV.

  Quantities compared: the number of people in the U.S. watching TV and the total number of people in the U.S.

  Ratio written as a fraction: $\frac{\text{persons watching TV}}{\text{persons in the U.S}}$, which is said to equal $\frac{1}{3}$.
  This ratio compares part of a whole to the whole.

- By the year 2020, there will be about 5 times as many people who are at least 100 years old as there were in 1990.

  Quantities compared: the number of people at least 100 years old in year 2020 and the number of people at least 100 years old in year 1990

  Ratio written as a fraction: $\frac{\text{persons 100+ in 2020}}{\text{persons 100+ in 9090}}$, which is said to equal $\frac{5}{1}$, or 5.
  This ratio does not compare part of a whole to the whole. It compares the number of people in two separate groups.

Check Your Understanding

Last month, Mark received an allowance of $20. He spent $12 and saved the rest.

1. What is the ratio of the money he spent to his total allowance?

2. What is the ratio of the money he saved to the money he spent?

3. What percent of his allowance did he save?

Check your answers on page 436.

Proportions

A **proportion** is a number sentence stating that two fractions are equal.

Examples
$$\frac{1}{2} = \frac{3}{6} \qquad \frac{2}{3} = \frac{8}{12} \qquad \frac{7}{8} = \frac{14}{16}$$

If you know any three numbers in a proportion, you can find the fourth number. Finding the fourth number is like finding a missing number in a pair of equivalent fractions.

Examples Solve each proportion.

$$\frac{2}{3} = \frac{n}{9} \qquad \frac{3}{4} = \frac{30}{k} \qquad \frac{x}{5} = \frac{6}{15} \qquad \frac{1}{z} = \frac{6}{24}$$
$$n = 6 \qquad\quad k = 40 \qquad\quad x = 2 \qquad\quad z = 4$$

Sample solution: To solve $\frac{1}{z} = \frac{6}{24}$, rename $\frac{6}{24}$ as an equivalent fraction with a numerator of 1.

Since $6 \div 6 = 1$, $\frac{1}{z} = \frac{6}{24} = \frac{6 \div 6}{24 \div 6} = \frac{1}{4}$. So, $z = 4$.

Proportions are useful in problem solving. Writing a proportion can help you organize the numbers in a problem. This can help you decide whether to multiply or divide to find the answer.

Example Gail has 45 baseball cards in her collection. $\frac{3}{5}$ of her cards are for National League players. How many of Gail's cards are for National League players?

The cards for National League players are part of Gail's whole collection of cards.

The ratio of National League cards to the total number of cards can be written as the fraction $\frac{\# \text{ NL cards}}{\text{total } \# \text{ cards}}$.

This fraction is said to be $\frac{3}{5}$, so you can write the proportion $\frac{\# \text{ NL cards}}{\text{total } \# \text{ cards}} = \frac{3}{5}$.

Gail has 45 cards in her collection.
Substitute 45 for "total # cards" in the proportion. $\frac{\# \text{ NL cards}}{45} = \frac{3}{5}$

Rename $\frac{3}{5}$ as an equivalent fraction with a denominator of 45 to solve this proportion.

Since $5 * 9 = 45$, $\frac{\# \text{ NL cards}}{45} = \frac{3}{5} = \frac{3 * 9}{5 * 9} = \frac{27}{45}$

The number of cards for National League players is 27.

Example A calculator sells for $32 at store A. An identical calculator at store B sells for 75% of this price. What is the price of the calculator at store B?

The ratio of the price at store B to the price at store A can be written as a fraction.

$$\frac{\text{price at store B}}{\text{price a-t store A}}$$

But this ratio of prices is 75%.
Since $75\% = \frac{75}{100} = \frac{3}{4}$, you can write the proportion

$$\frac{\text{price at store B}}{\text{price at store A}} = \frac{3}{4}.$$

The calculator sells for $32 at store A.
Substitute 32 for "price at store A" in the proportion.

$$\frac{\text{price at store B}}{32} = \frac{3}{4}$$

Rename $\frac{3}{4}$ as an equivalent fraction with a denominator of 32 to solve this proportion. Since $4 * 8 = 32$,

$$\frac{\text{price at store B}}{32} = \frac{3 * 8}{4 * 8} = \frac{24}{32}.$$

The price of the calculator at store B is $24.

Notice that the ratio of prices compares two separate prices. It does not compare part of a whole to the whole.

Example Ms. Wheeler spends $1,000 a month. This amount is $\frac{4}{5}$ of her monthly earnings. How much does she earn per month?

The ratio of Ms. Wheeler's spending to earnings can be written as a fraction.

$$\frac{\text{spending}}{\text{earnings}}$$

This fraction is $\frac{4}{5}$, so you can write the proportion

$$\frac{\text{spending}}{\text{earnings}} = \frac{4}{5}.$$

Ms. Wheeler's spending is $1,000 each month.
Substitute 1,000 for "spending" in the proportion.

$$\frac{1,000}{\text{earnings}} = \frac{4}{5}$$

Rename $\frac{4}{5}$ as an equivalent fraction with a numerator of 1,000 to solve this proportion. Since $4 * 250 = 1,000$,

$$\frac{1,000}{\text{earnings}} = \frac{4 * 250}{5 * 250} = \frac{1,000}{1,250}.$$

Ms. Wheeler earns $1,250 each month.

Notice that the ratio of spending to earnings compares part of a whole to the whole.

Check Your Understanding

Solve.

1. Francine earned $36 mowing lawns. She spent $\frac{2}{3}$ of her money on CDs. How much did she spend on CDs?

2. $\frac{3}{4}$ of Frank's cousins are girls. Frank has 15 girl cousins. How many cousins does he have in all?

Check your answers on page 436.

Using Ratios to Describe Size Changes

Many situations produce a **size change.** A magnifying glass, a microscope, and an overhead projector all enlarge the original image. Most copying machines can create a variety of size changes—both enlargements and reductions of the original document.

Similar figures are figures that have the same shape but not necessarily the same size. Enlargements or reductions are **similar** to the originals; that is, they have the same shapes as the originals.

The **size-change factor** is a number that tells the amount of enlargement or reduction that takes place. For example, if you use a copy machine to make a 2x change in size, then every length in the copy is twice the size of the original. The size-change factor is 2. If you make a 0.5x change in size, then every length in the copy is half the size of the original. The size-change factor is $\frac{1}{2}$, or 0.5.

You can think of the size-change factor as a ratio. For a 2x size change, the ratio of a length in the copy to the corresponding length in the original is 2 to 1.

size-change factor 2: $\dfrac{\text{copy size}}{\text{original size}} = \dfrac{2}{1}$

For a 0.5x size change, the ratio of a length in the copy to a corresponding length in the original is 0.5 to 1.

size-change factor 0.5: $\dfrac{\text{copy size}}{\text{original size}} = \dfrac{0.5}{1}$

A microscope with 3 levels of size change: 40X, 100X, and 400X

A 3x magnifying glass

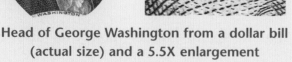

Examples

Head of George Washington from a dollar bill (actual size) and a 5.5X enlargement

Copy-machine reductions using size-change factors of $\frac{1}{5}$ and $\frac{1}{10}$

Scale Models

A model that is a careful copy of an actual object is called a **scale model.** You have probably seen scale models of cars, trains, and airplanes. The size-change factor in scale models is usually called the **scale factor.**

Doll houses often have a scale factor of $\frac{1}{12}$. You can write this scale factor as "$\frac{1}{12}$ of actual size," "scale 1:12," "$\frac{1}{12}$ scale," or as a proportion:

$$\frac{\text{doll house length}}{\text{real house lenght}} = \frac{1 \text{ inch}}{12 \text{ inches}}$$

Every part of the real house is 12 times as long s the corresponding part of the doll house.

Maps

The size-change factor for maps and scale drawings is usually called the **scale.** If a map scale is 1:25,000, then every length on the map is $\frac{1}{25,000}$ of the actual length, and any real distance is 25,000 times the distance shown on the map.

$$\frac{\text{map distance}}{\text{real distance}} = \frac{1}{25,000}$$

Scale Drawings

If an architect's scale drawing shows "scale $\frac{1}{4}$ inch:1 foot," then $\frac{1}{4}$ inch on the drawing represents 1 foot of actual length.

$$\frac{\text{drawing length}}{\text{real length}} = \frac{\frac{1}{4} \text{ inch}}{1 \text{ foot}}$$

Since 1 foot = 12 inches, we can rename $\frac{\frac{1}{4} \text{ inch}}{1 \text{ foot}}$ as $\frac{\frac{1}{4} \text{ inch}}{12 \text{ inches}}$.

Multiply by 4 to change this to an easier fraction.

$$\frac{\frac{1}{4} \text{ inch} * 4}{12 \text{ inches} * 4} = \frac{1 \text{ inch}}{48 \text{ inches}}$$

The drawing is $\frac{1}{48}$ of the actual size.

Did You Know ?

Here are the names and scale factors for some of the most popular model railroads:

Z gauge: $\frac{1}{220}$

N gauge: $\frac{1}{160}$

HO gauge: $\frac{1}{87.1}$

OO gauge: $\frac{1}{76}$

S gauge: $\frac{1}{64}$

O gauge: $\frac{1}{48}$

Gauge 2: $\frac{1}{22.4}$

Gauge 1: $\frac{1}{11.2}$

Note

You may see scales written with an equal sign, such as "$\frac{1}{4}$ inch = 1 foot." But $\frac{1}{4}$ inch is certainly not equal to 1 foot, so this is not mathematically correct. This scale is intended to mean that $\frac{1}{4}$ inch on the map or scale drawing stands for 1 foot in the real world.

Check Your Understanding

Solve.

1. The diameter of a circle is 5 centimeters. A copier is used to make an enlargement of the circle. The size-change factor is 4. What is the diameter of the enlarged circle?

2. Two cities are 6 inches apart on the map. Suppose that $\frac{\text{map distance}}{\text{real distance}} = \frac{1 \text{ inch}}{250 \text{ miles}}$. What is the real distance between the two cities?

Check your answers on page 436.

The Number Pi

Measurements are *always* estimates. But if circles could be measured exactly, the ratio of the circumference to the diameter would be the same for every circle. This ratio is called **pi** and is written as the Greek letter π.

Since ancient times, mathematicians have worked to find the value of π Here are some of the earliest results.

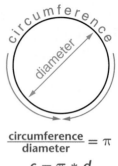

$$\frac{\text{circumference}}{\text{diameter}} = \pi$$

$$c = \pi * d$$

$$d = c / \pi$$

Date	Source	Approximate Value of π
c. 1800–1650 B.C.	Babylonians	$3\frac{1}{8}$
c. 1650 B.C.	Rhind Papyrus (Egypt)	3.16
c. 950 B.C.	Bible (I Kings 7:23)	3
c. 240 B.C.	Archimedes (Greece)	between $3\frac{10}{71}$ and $3\frac{1}{7}$
c. A.D. 470	Tsu Ch'ung Chi (China)	$\frac{355}{113}$ (3.1415929...)
c. A.D. 510	Aryabhata (India)	$\frac{62,832}{20,000}$ (3.1416)
c. A.D. 800	al'Khwarizmi (Persia)	3.1416

NOTE: *c.* stands for *circa*, a Latin word which means "about."

It's not possible to write π exactly with digits, because the decimal for π goes on forever. No repeating pattern has ever been found for the digits in this decimal.

$\pi = 3.14159265358979323846264338327950288419716939937511\ldots$

The number π is so important that most scientific calculators have a π key. If you use the π key on your calculator, be sure to round your results. Results should not be more precise than the original measurements. One or two decimal places in an answer are usually enough.

If you don't have a calculator, you can use an approximation for π. Since few measures are more precise than hundredths, an approximation like 3.14 or $\frac{22}{7}$ is usually close enough.

Check Your Understanding

Use a calculator to find each answer.

1. What is the circumference of a circle with a diameter of 2 inches?

2. What is the diameter of a circle with a circumference of 5 inches?

Check your answers on page 436.

Data and Probability

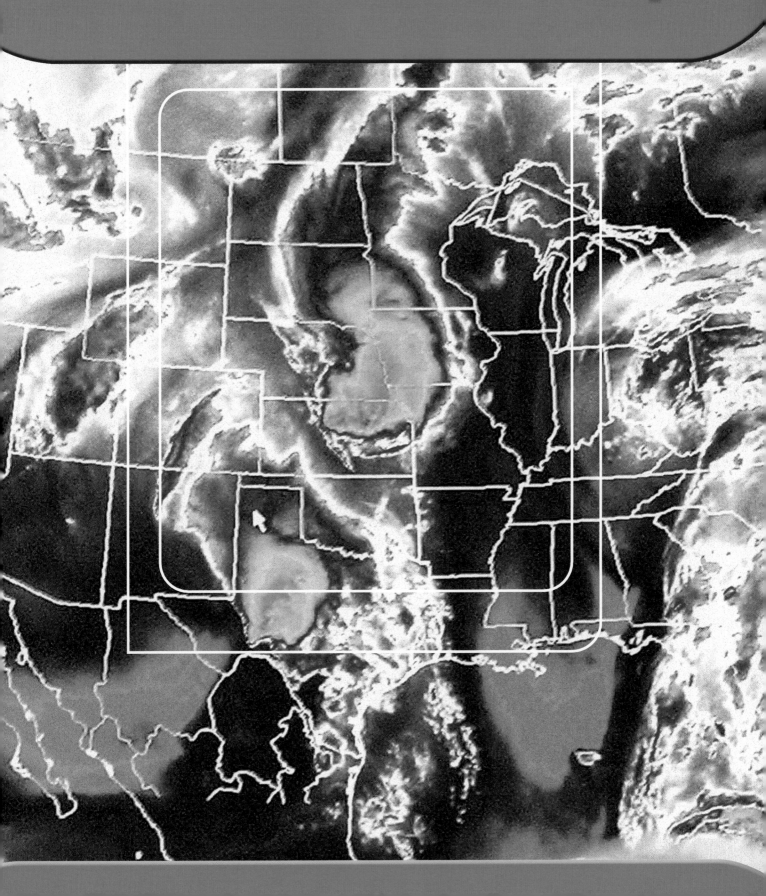

Collecting Data

There are different ways to collect information about something. You can count, measure, ask questions, or observe and describe what you see.

The information you collect is called **data.**

Surveys

A **survey** is a study that collects data. Much of the information used to make decisions comes from surveys. Many surveys collect data about people. Stores survey their customers to find out which products they should carry. Television stations survey viewers to learn what programs are popular. Politicians survey people to learn how they plan to vote in elections.

Survey data are collected in several ways. These include face-to-face interviews, telephone interviews, written questionnaires that are returned by mail, and group discussions (often called *focus groups*).

Not all surveys gather information about people. For example, there are surveys about cars, buildings, and animal groups. These surveys often collect data in other ways than by interviews or questionnaires.

Example Highway engineers sometimes make videotapes of vehicles and drivers along a street or highway. They use the video data to analyze vehicle speeds and driving patterns.

Example A bird survey is conducted during December and January each year in the Chicago area. Bird watchers list the different bird species they see. Then they count the number of each species observed. The lists are combined to create a final data set.

From a recent Chicago bird survey:

Species	Number of Birds Seen
crow	1,335
mallard duck	2,134
mourning dove	213
song sparrow	15

Samples

The **population** for a survey is the group of people or things that is being studied. Because the population may be very large, it may not be possible to collect data from every member of the population. Therefore, data are collected only from a sample group to provide information that is used to describe the entire population. A **sample** is a part of the population that is chosen to represent the whole population.

Large samples usually give more dependable estimates than small ones. For example, if you want to estimate the percent of adults who drive to work, a sample of 100 persons provides a better estimate than a sample of 10.

Example | A survey of teenagers collects data about people aged 13 to 19.

There are about 28 million teenagers in the United States. It would be impossible to collect data from every teenager. Instead, data are collected from a sample of teenagers.

Results from a recent survey of teens:

Reason for Playing Sports	
Love of game	50%
Love to compete and win	24%
Be with friends	14%
Earn college scholarship	6%
Because I'm good at it	4%
Recognition	2%

Source: Reported in the *Chicago Sun-Times*, 1/16/00.

The decennial (every 10 years) **census** is a survey that includes *all* the people in the United States. Every household is required to fill out a census form. But certain questions are asked only in a sample of 1 in 6 households.

A **random sample** is a sample that gives all members of the population the same chance of being selected. Random samples give more dependable information than samples that are not random.

Note

A *census* collects data from every member of the population. In a census, the sample and the population are identical.

Did You Know?

During the 1880s, Herman Hollerith built a set of machines that could count and organize data collected by the 1890 census. Using his system, the census population count was completed within 6 weeks.

Example | Suppose you want to estimate what percentage of the population will vote for Mr. Jones.

If you use a sample of 100 best friends of Mr. Jones, the sample is *not* a random sample. People who do not know Mr. Jones have no chance of being selected. A sample of best friends will not fairly represent the entire population. It will not furnish a dependable estimate of how the entire population will vote.

Student Survey Data

Information was collected from samples of students at Lee Middle School. Three questions were asked.

1. Entertainment Data

Students were asked to select their favorite form of entertainment. They were given four possible choices:

TV: Watch TV/DVDs   **Games:** Play video/computer games
Music: Listen to radio/CDs   **Read:** Read books, magazines

Twenty-four students responded (answered the survey). Here are their data.

TV	TV	Read	TV	Games	TV	Music	Games
Games	TV	Read	Music	TV	TV	Music	TV
Games	Games	Music	TV	TV	Games	TV	Read

2. Favorite Sports Data

Students were asked to select their TWO favorite sports from this list:

Baseball   **Basketball**   **Bicycle riding**
Bowling   **Soccer**   **Swimming**

Twenty students responded. The data below include 40 answers because each student named two sports.

Basketball	Bicycle	Swimming	Soccer	Basketball
Swimming	Baseball	Swimming	Bicycle	Swimming
Bicycle	Swimming	Soccer	Bicycle	Soccer
Bowling	Soccer	Bicycle	Swimming	Bicycle
Bicycle	Swimming	Baseball	Bowling	Bicycle
Baseball	Bowling	Basketball	Basketball	Swimming
Basketball	Swimming	Soccer	Soccer	Baseball
Bicycle	Soccer	Bicycle	Swimming	Bicycle

3. Shower/Bath Time Data

A sample of 40 students was asked to estimate the number of minutes they usually spend taking a shower or bath. Here are the data.

3	20	10	5	8	4	10	7	5	5
25	5	3	25	20	17	5	30	14	35
9	20	15	7	5	10	16	40	10	15
10	5	15	10	15	5	12	22	3	9

Organizing Data

Once the data have been collected, it helps to organize them in order to make them easier to understand. **Line plots** and **tally charts** are two methods of organizing data.

Example Ms. Barton's class got the following scores on a 20-word spelling test. Make a line plot and a tally chart to show the data below.

20	15	18	17	20	12	15	17	19	18	20	16	16
17	14	15	19	18	18	15	10	20	19	18	15	18

Scores on a 20-Word Spelling Test

Number of Students

```
                                    X
                  X           X
                  X           X        X
                  X     X  X  X  X
                  X  X  X  X  X  X
   X     X     X  X  X  X  X  X  X
 ──┼──┼──┼──┼──┼──┼──┼──┼──┼──┼──→
   10 11 12 13 14 15 16 17 18 19 20
              Number Correct
```

Scores on a 20-Word Spelling Test

Number Correct	Number of Students
10	/
11	
12	/
13	
14	/
15	ЖЖ
16	//
17	///
18	ЖЖ /
19	///
20	////

In the line plot, there are 5 Xs above 15.

In the tally chart, there are 5 tallies to the right of 15.

The 5 Xs and the 5 tallies each show that a score of 15 appeared 5 times in the class list of test scores.

Both the line plot and the tally chart help to organize the data. They make it easier to describe the data. For example,

♦ 4 students had 20 correct (a perfect score).

♦ 18 correct is the score that came up most often.

♦ 10, 12, and 14 correct are scores that came up least often.

♦ 0–9, 11, and 13 correct are scores that did not occur at all.

Did You Know?

The Lebombo Bone, which dates from about 35,000 B.C., may be the earliest example of a tally chart. It is a baboon leg bone that clearly shows 29 notches (tally marks). Bushmen clans in Namibia still use "calendar sticks" that resemble the Lebombo Bone.

Check Your Understanding

Here are the numbers of hits made by 15 players in a baseball game. 2 1 0 0 2 3 2 4 2 2 0 3 0 1 0

Organize the data.

1. Make a tally chart.　　**2.** Make a line plot.

Check your answers on page 436.

Sometimes the data are spread over a wide range of numbers. This makes a tally chart and a line plot difficult to draw. In such cases, you can make a tally chart in which the results are **grouped.** Or, you may organize the data by making a **stem-and-leaf plot.**

For a science project, the students in Ms. Beck's class took each other's pulse rates. (A *pulse rate* is the number of heartbeats per minute.) These were the results:

75	86	108	94	75	88	86	99	78	86
90	94	70	94	78	75	90	102	65	94
92	72	90	86	102	78	88	75	72	
70	94	85	88	105	86	78	82		

Tally Chart of Grouped Data

The data have been sorted or grouped into intervals of 10. This is called a tally chart of **grouped data.**

The chart shows that most of the students had a pulse rate from 70 to 99. More students had a pulse rate in the 70s than in any other interval.

Stem-and-Leaf Plot

In a stem-and-leaf plot, the digit or digits in the left column (the **stem**) are combined with a single digit in the right column (a **leaf**) to form a numeral.

Each row has as many entries as there are digits in the right column.

For example, the row with 9 in the left column has ten entries: 94, 99, 90, 94, 94, 90, 94, 92, 90, and 94.

Pulse Rates of Students

Number of Heartbeats	Number of Students
60–69	/
70–79	⊬⊬⊬ ⊬⊬⊬ //
80–89	⊬⊬⊬ ⊬⊬⊬
90–99	⊬⊬⊬ ⊬⊬⊬
100–109	////

Pulse Rates of Students

Stems (10s)	Leaves (1s)
6	5
7	5 5 8 0 8 5 2 8 5 2 0 8
8	6 8 6 6 6 8 5 8 6 2
9	4 9 0 4 4 0 4 2 0 4
10	8 2 2 5

Check Your Understanding

Michael Jordan played in 12 games of the 1996 NBA Playoffs. He scored the following number of points:

35	29	26	44	28	46	27	35	21	35	17	45

Organize the data.

1. Make a tally chart of grouped data.   **2.** Make a stem-and-leaf plot.

Check your answers on page 436.

Statistical Landmarks

The **landmarks** for a set of data are used to describe the data.

♦ The **minimum** is the smallest value.

♦ The **maximum** is the largest value.

♦ The **range** is the difference between the maximum and the minimum.

♦ The **mode** is the value or values that occur most often.

♦ The **median** is the middle value or values.

Example Here is a record of children's absences for one week at Medgar Evers School.

Monday	Tuesday	Wednesday	Thursday	Friday
25	15	10	14	14

Find the landmarks for the data.

Minimum (smallest) number: 10
Maximum (largest) number: 25
Range of numbers: 25 − 10 = 15
Mode (most frequent number): 14

10   14   14   15   25

1̶0̶   14   14   15   2̶5̶

1̶0̶   1̶4̶   [14]   1̶5̶   2̶5̶

↑
median

To find the median (middle value):

• List the numbers in order from smallest to largest or from largest to smallest.
• Cross out one number from each end of the list.
• Continue to cross out one number from each end of the list.
• The median is the number that remains after all others have been crossed out.

There may not be landmarks for some sets of data. For example, if you collect data about hair color, there is no "largest color" or "middle value color." But, you can find the mode. The mode is the hair color that occurs most often.

Check Your Understanding

Here are the math quiz scores (number correct) for 15 students:

4 1 2 4 2 4 3 2 2 0 1 3 2 0 3

Find these landmarks for the data:

1. minimum   **2.** maximum   **3.** range   **4.** mode   **5.** median

Check your answers on page 436.

one hundred nineteen

Example   The **line plot** shows students' scores on a 20-word spelling test. Find the landmarks for the data.

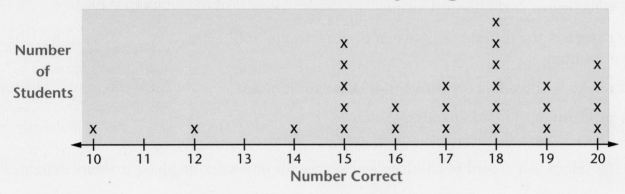

Scores on a 20-Word Spelling Test

Number of Students

Number Correct

Minimum: 10   **Maximum:** 20   **Range:** $20 - 10 = 10$ **Mode:** 18

To find the median (middle value), first list the numbers in order:

10 12 14 15 15 15 15 15 16 16 17 17 17 18 18 18 18 18 18 19 19 19 20 20 20 20

Cross out one number from each end of the list.

Continue to cross out one number from each end until only two numbers are left.

~~10~~ ~~12~~ ~~14~~ ~~15~~ ~~15~~ ~~15~~ ~~15~~ ~~15~~ ~~16~~ ~~16~~ ~~17~~ ~~17~~ 17 18 ~~18~~ ~~18~~ ~~18~~ ~~18~~ ~~18~~ ~~19~~ ~~19~~ ~~19~~ ~~20~~ ~~20~~ ~~20~~ ~~20~~

middle scores

The two numbers remaining are the middle scores.
There are two middle scores, 17 and 18.

The median is 17.5, which is the number halfway between 17 and 18.

Check Your Understanding

1. Here are the math quiz scores (number correct) for 14 students:
   2 0 4 2 2 4 3 4 2 1 3 1 3 3
Find the minimum, maximum, range, mode, and median for this set of data.

2. Find the median for this set of numbers: 15 12 8 21 16 33 16 9 8 12 33 12

Check your answers on page 436.

The Mean (or Average)

The **mean** of a set of numbers is often called the *average*.
To find the mean, do the following:

Step 1: Add the numbers.
Step 2: Then divide the sum by the number of addends.

Example On a 4-day trip, Lisa's family drove 240, 100, 200, and 160 miles. What is the mean number of miles they drove per day?

Step 1: Add the numbers: 240 + 100 + 200 + 160 = 700.
Step 2: Divide by the number of addends: 700 ÷ 4 = 175.

The mean is 175 miles. They drove an average of 175 miles per day.

You can use a calculator:

Add the miles. Key in: 240 [+] 100 [+] 200 [+] 160 [=]
Answer: 700

Divide this sum by 4. 700 [÷] 4 [=]
Answer: 175

Sometimes you will calculate the mean for a set of numbers where many of the numbers are repeated. The shortcut explained below could save you time.

Did You Know?

The average lifespan of a Galapagos tortoise is about 100 years. The oldest known tortoise is now over 160 years old.

Example Calculate the mean for this set of eight numbers:

80   80   80   90   90   90   90   90

You could add the eight numbers, then divide by 8.
80 + 80 + 80 + 90 + 90 + 90 + 90 + 90 = 690; 690 ÷ 8 = 86.25
Or, you could use this shortcut.

- Multiply each data value by the number of times it occurs.
- Add these products.
- Divide by the number of addends.

$$3 * 80 = 240$$
$$\underline{5 * 90 = 450}$$
$$690$$
$$690 ÷ 8 = 86.25$$

The mean is 86.25.

Check Your Understanding

Jason received these scores on math tests: 85 70 80 90 80 80 80 75 85 75 90.
Use your calculator to find Jason's mean score.

Check your answers on page 437.

Bar Graphs

A **bar graph** is a drawing that uses bars to represent numbers. Bar graphs display information in a way that makes it easy to show comparisons.

The title of a bar graph describes the information in the graph. Each bar has a label. Units are given to show how something was counted or measured. When possible, the graph gives the source of the information.

Example   This is a **vertical bar graph.**

• Each bar represents the area (in square miles) of the lake named below the bar.

• It is easy to compare lake areas by comparing the lengths of the bars. Lake Superior is about 5,000 square miles larger than Lake Victoria. Lake Huron and Lake Michigan have about the same area.

Largest Lakes

Area (sq. miles): 0, 10,000, 20,000, 30,000

Superior, Victoria, Huron, Michigan

Source: *The World Almanac*

Example   This is a **horizontal bar graph.**

• Each bar represents the number of grams of fat in the food named to the left of the bar.

• It is easy to compare the fat content of foods by comparing bars. One cup of ice cream contains nearly twice as much fat as one cup of whole milk. One doughnut contains 12 times as much fat as one banana.

Fat Content of Foods

Ice Cream (1 cup), Doughnut, Whole Milk (1 cup), Egg, Banana

Grams of Fat: 0, 5, 10, 15

Source: *The World Almanac*

Check Your Understanding

The table at the right shows the average number of vacation days per year for three countries. Make a bar graph to show this information.

Vacation Days per Year

Country	Average Number of Days
Canada	26
Italy	42
United States	13

Source: World Tourism Organization

Check your answers on page 437.

Side-by-Side and Stacked Bar Graphs

Sometimes there are two or more bar graphs that are related to the same situation. Related bar graphs are often combined into a single graph. The combined graph saves space and makes it easier to compare the data. The examples below show two different ways to draw combined bar graphs.

Did You Know?

The two points farthest apart in the United States are Elliot Key, Florida, and Kure Island, Hawaii. They are 5,859 miles apart.

Example One bar graph shows road miles from Los Angeles to different cities. A second bar graph shows air miles.

The graphs are combined into a **side-by-side bar graph** by drawing the related bars side by side in different colors. It is easy to compare road miles and air miles on the side-by-side graph.

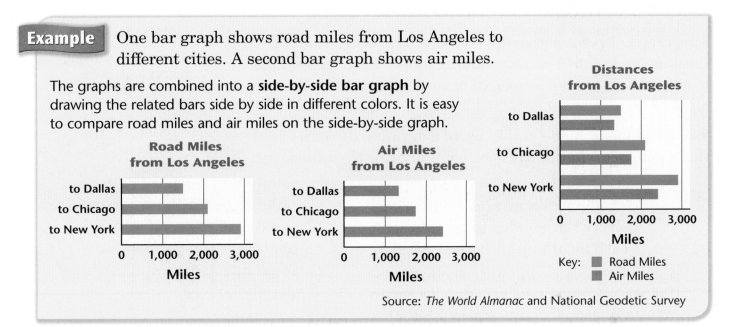

Source: *The World Almanac* and National Geodetic Survey

Example The bar graphs below show the number of sports teams that boys and girls joined during a 1-year period.

The bars within each graph can be stacked on top of one another. The **stacked bar graph** includes each of the stacked bars.

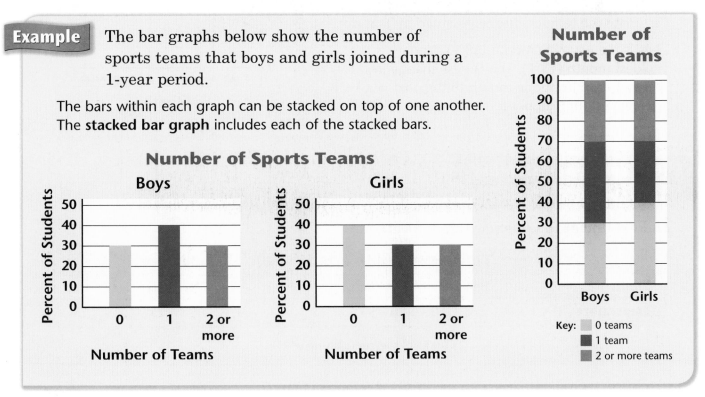

Line Graphs

Line graphs are used to display information that shows trends. They often show how something has changed over a period of time.

Line graphs are often called **broken-line graphs.** Line segments connect the points on the graph. The segments joined end to end look like a broken line.

Line graphs have a horizontal and a vertical scale. Each of these scales is called an **axis** (plural: **axes**). Each axis is labeled to show what is being measured or counted and what the unit of measure or count unit is.

When looking at a line graph, try to determine the purpose of the graph. See what conclusions you can draw from the graph.

Broken-Line Graph

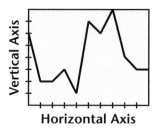

Joined end to end, the segments look like a broken line.

Example The broken-line graph at the right shows the average temperature for each month in Anchorage, Alaska.

The horizontal axis shows each month of the year. The average temperature for a month is shown with a dot above the label for that month. The labels on the vertical axis at the left are used to estimate the temperature represented by that dot.

July is the warmest month (58°F). January is the coldest month (15°F). The largest change in temperature from one month to the next is a 14°F decrease from October to November.

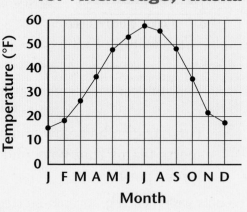

Check Your Understanding

The following table shows average temperatures for Phoenix, Arizona.
Make a line graph to show this information.

Average Temperatures for Phoenix, Arizona											
Month	Jan	Feb	Mar	Apr	May	Jun	Jul	Aug	Sep	Oct	Nov	Dec
Temperature (°F)	54	58	63	70	79	89	93	91	86	75	62	54

Check your answers on page 437.

How to Use the Percent Circle

A **compass** is a device for drawing circles. You can also use your **Geometry Template** to draw circles.

An **arc** is a piece of a circle. If you mark two points on a circle, these points and the part of the circle between them form an arc.

The region inside a circle is called its **interior.**

A **sector** is a wedge-shaped piece of a circle and its interior. A sector consists of two radii (singular: radius), one of the arcs determined by the endpoints of the radii, and the part of the interior of the circle bounded by the radii and the arc.

A **circle graph** is sometimes called a **pie graph** because it looks like a pie that has been cut into several pieces. Each "piece" is a sector of the circle.

You can use the **Percent Circle** on your Geometry Template to find what percent of the circle graph each sector represents. Here are two methods for using the Percent Circle.

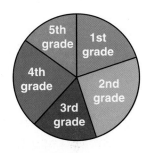

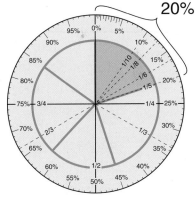

The circle graph shows the distribution of students in grades 1 to 5 at Elm Place School.

Method 1: Direct Measure

♦ Place the center of the Percent Circle over the center of the circle graph.

♦ Rotate the template so that the 0% mark is aligned with one side (line segment) of the sector you are measuring.

♦ Read the percent at the mark on the Percent Circle located over the other side of the sector. This tells what percent the sector represents.

The sector for first grade represents 20%.

Method 2: Difference Comparison

♦ Place the center of the Percent Circle over the center of the circle graph.

♦ Note the percent reading for one side of the sector you are measuring.

♦ Find the percent reading for the other side of the sector.

♦ Find the difference between these readings.

The sector for second grade represents 45% − 20%, or 25%.

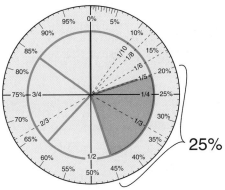

Check Your Understanding

What percents are represented by the other three sectors in the grades circle graph?

Check your answers on page 437.

How to Draw a Circle Graph Using a Percent Circle

Example Draw a circle graph to show the following information. The students in Ms. Ahmad's class were asked to name their favorite colors: 9 students chose blue, 7 students chose green, 4 students chose yellow, and 5 chose red.

Step 1: Find what percent of the total each part represents. The total number of students who voted was $9 + 7 + 4 + 5 = 25$.

- 9 out of 25 chose blue.

  $\frac{9}{25} = \frac{36}{100} = 36\%$, so 36% chose blue.

- 7 out of 25 chose green.

  $\frac{7}{25} = \frac{28}{100} = 28\%$, so 28% chose green.

- 4 out of 25 chose yellow.

  $\frac{4}{25} = \frac{16}{100} = 16\%$, so 16% chose yellow.

- 5 out of 25 chose red.

  $\frac{5}{25} = \frac{20}{100} = 20\%$, so 20% chose red.

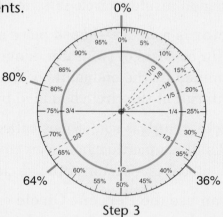

Step 3

Step 2: Check that the sum of the percents is 100%. $36\% + 28\% + 16\% + 20\% = 100\%$

Step 3: Draw the circle. Then use the Percent Circle on the Geometry Template to mark off the sectors.

- To mark off 36%, place the center of the Percent Circle over the center of the circle graph. Make a mark at 0% and 36% on the circle.

- To mark off 28%, make a mark at 64% ($36\% + 28\% = 64\%$), without moving the Percent Circle.

- To mark off 16%, make a mark at 80% ($64\% + 16\% = 80\%$).

- Check that the final sector represents 20%.

Step 4

Step 4: Draw the sector lines (radii). Label each sector of the circle. Color the sectors.

Check Your Understanding

Draw a circle graph to display the following information:

a. The Hot Shots basketball team scored 30 points in one game.

b. Sally scored 15 points.   **c.** Drew scored 6 points.

d. Bonita scored 6 points.   **e.** Damon scored 3 points.

Check your answers on page 437.

How to Draw a Circle Graph Using a Protractor

Example Draw a circle graph to show the following information: In the month of June, there were 19 sunny days, 6 partly-cloudy days, and 5 cloudy days.

Step 1: Find out what fraction or percent of the total each part represents. June has 30 days.

- 5 out of 30 were cloudy days.

    $\frac{5}{30} = \frac{1}{6}$, so $\frac{1}{6}$ of the days were cloudy.

- 6 out of 30 were partly-cloudy days.

    $\frac{6}{30} = \frac{1}{5}$, so $\frac{1}{5}$ of the days were partly cloudy.

- 19 out of 30 were sunny days.

    $\frac{19}{30} = 0.633 \ldots \approx 63.3\%$, so about 63.3% of the days were sunny.

Step 2: Calculate the degree measure of the sector for each piece of data.

- The number of cloudy days in June was $\frac{1}{6}$ of the total number of days. Therefore, the degree measure of the sector for cloudy days is $\frac{1}{6}$ of 360°.

    $\frac{1}{6}$ of 360° $= \frac{1}{6} * 360° = 60°$

- The number of partly-cloudy days in June was $\frac{1}{5}$ of the total number of days. Therefore, the degree measure of the sector for partly-cloudy days is $\frac{1}{5}$ of 360°.

    $\frac{1}{5}$ of 360° $= \frac{1}{5} * 360° = 72°$

- The number of sunny days in June was 63.3% of the total number of days. Therefore, the degree measure of the sector for sunny days is 63.3% of 360°.

    $0.633 * 360° = 228°$, rounded to the nearest degree

Step 3: Check that the sum of the degree measures of the sectors is 360°.

$60° + 72° + 228° = 360°$

Step 4: Draw a circle. Use a protractor to draw 3 sectors with degree measures of 60°, 72°, and 228°. Label each sector.

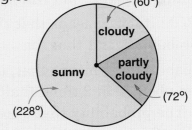

Check Your Understanding

Use your protractor to make a circle graph to display the information in the chart. What is the degree measure of each sector?

Favorite Subjects	
Subject	**Number of Students**
Reading	6
Social Studies	2
Math	4
Music	1
Science	2
Art	5

Check your answers on page 437.

Chance and Probability

Chance

Things that happen are called **events.** There are many events that you can be sure about:

♦ You are **certain** that the sun will set today.

♦ It is **impossible** for you to grow to be 10 feet tall.

There are also many events that you *cannot* be sure about:

♦ You cannot be sure that you will get a letter tomorrow.

♦ You cannot be sure whether it will be sunny next Friday.

You might sometimes talk about the **chance** that something will happen. If Paul is a good chess player, you may say, "Paul has a *good chance* of winning the game." If Paul is a poor player, you may say, "It is *very unlikely* that Paul will win."

Probability

Sometimes a number is used to tell the chance of something happening. This number is called a **probability.** It is a number from 0 to 1. The closer a probability is to 1, the more likely it is that an event will happen.

♦ A probability of 0 means the event is *impossible.* The probability is 0 that you will live to the age of 150.

♦ A probability of 1 means that the event is *certain.* The probability is 1 that the sun will rise tomorrow.

♦ A probability of $\frac{1}{2}$ means that, in the long run, an event will happen about 1 in 2 times (half of the time or 50% of the time). The probability that a tossed coin will land heads up is $\frac{1}{2}$. We often say that the coin has a "50–50 chance" of landing heads up.

A probability can be written as a fraction, a decimal, or a percent. The **Probability Meter** is often used to record probabilities. It is marked to show fractions, decimals, and percents between 0 (or 0%) and 1 (or 100%).

The phrases printed on the bar of the Probability Meter may be used to describe probabilities in words. For example, suppose that the probability of snow tomorrow is 70%. The 70% mark falls within that part of the bar where "LIKELY" is printed. So you might say that "Snow is *likely* tomorrow," instead of stating the probability as 70%.

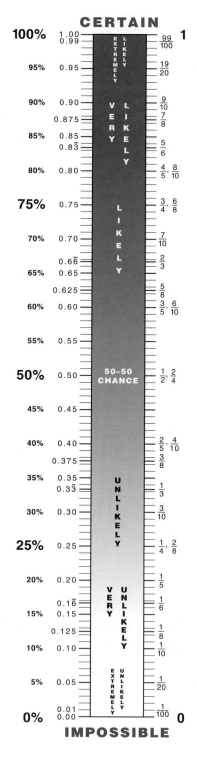

Calculating a Probability

Four common ways for finding probabilities are shown below.

Make a Guess

Vince guesses that he has a 10% chance (a 1 in 10 chance) of returning home by 9 o'clock.

Conduct an Experiment

Elizabeth dropped 80 tacks: 64 landed point up and 16 landed point down. The fraction of tacks that landed point up is $\frac{64}{80}$.

Elizabeth estimates the probability that the next tack she drops will land point up is $\frac{64}{80} = \frac{64 \div 8}{80 \div 8} = \frac{8}{10}$, or 80%.

Use a Data Table

Kenny got 48 hits in his last 100 times at bat. He estimates the probability that he will get a hit the next time at bat is $\frac{48}{100}$, or 48%.

Hits	48
Walks	11
Outs	41
Total	100

Assume That All Possible Results Have the Same Chance

A standard die has 6 faces and is shaped like a cube. You can assume that each face has the same $\frac{1}{6}$ chance of landing up.

An 8-sided die has 8 faces and is shaped like a regular octahedron. You can assume that each face has the same $\frac{1}{8}$ chance of landing up.

A standard deck of playing cards has 52 cards. Suppose the cards are shuffled and one card is drawn. You can assume that each card has the same $\frac{1}{52}$ chance of being drawn.

Suppose that a spinner is divided into 12 equal sections. When you spin the spinner, you can assume that each section has the same $\frac{1}{12}$ chance of being landed on.

Naming a Probability

A spinner is divided into 10 equal sections, numbered 1 through 10. Each of the following statements has the same meaning:

♦ The probability (chance) of landing on 7 is $\frac{1}{10}$.

♦ The probability (chance) of landing on 7 is **0.1**.

♦ The probability (chance) of landing on 7 is **10%**.

♦ The probability (chance) of landing on 7 is **1 out of 10**.

♦ There is a **1 in 10** probability (chance) of landing on 7.

♦ If you spin many times, you can expect to land on 7 **about $\frac{1}{10}$ of the time.**

Equally Likely Outcomes

In solving probability problems, it is often useful to list all the possible results for a situation. Each possible result is called an **outcome.** If all the possible outcomes have the same probability, they are called **equally likely outcomes.**

Example If you roll a 6-sided die, the number of dots on the face that lands up may be 1, 2, 3, 4, 5, or 6. There are six possible outcomes. You can assume that each face of the die has the same $\frac{1}{6}$ chance of landing up. So, the outcomes are equally likely.

Example This spinner is divided into 10 equal parts. If you spin the spinner, it will land on one of the numbers 1 through 10. There are ten possible outcomes. You can assume that each number has a $\frac{1}{10}$ chance of being landed on because the sections are equal parts of the spinner. So, the outcomes are equally likely.

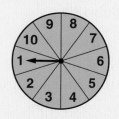

Example This spinner is divided into two sections. If you spin the spinner, it will land either on red or on blue. Red and blue are the two possible outcomes. There is a greater chance of landing on red than on blue because the red section is twice as large as the blue section. So, the outcomes are *not* equally likely.

A Probability Formula

Finding the probability of an event is easy *if all of the outcomes are equally likely.* Follow these steps:

1. List all the possible outcomes.

2. Look for any outcomes that will make the event happen. These outcomes are called **favorable outcomes.** Circle each favorable outcome.

3. Count the number of possible outcomes. Count the number of favorable outcomes. The probability of the event is:

$$\frac{\text{number of favorable outcomes}}{\text{number of possible outcomes}}$$

Example Amy, Beth, Carol, Dave, Eduardo, Frank, and George are on a camping trip. They decide to choose a leader. Each of them writes his or her name on an index card. The cards are put into a paper bag and mixed. One card will be drawn without looking. The child whose name is drawn will become leader. Find the probability that a girl's name will be selected.

What is the *event* for which you want to find the probability? Draw a girl's name.

How many *possible outcomes* are there? Seven. Any of 7 names might be drawn.

Are the outcomes *equally likely*? Yes
Names were written on identical cards and the cards were mixed in the bag. Each name has the same $\frac{1}{7}$ chance of being drawn.

Which of the possible outcomes are *favorable outcomes*? Amy, Beth, and Carol. Drawing any one of these 3 names will make the event happen.

List the possible outcomes, and circle the favorable outcomes.

Ⓐmy Ⓑeth Ⓒarol   Dave   Eduardo   Frank   George

The probability of drawing a girl's name equals $\frac{\text{number of favorable outcomes}}{\text{number of possible outcomes}} = \frac{3}{7}$.

Example Use one each of the number cards 1, 4, 6, 8, and 10. Draw one card without looking. What is the probability that the number drawn is between 5 and 9?

Event: Get a number between 5 and 9.

Possible outcomes: 1, 4, 6, 8, and 10

The card is drawn without looking, so the outcomes are equally likely.

Favorable outcomes: 6 and 8   1   4   ⑥ ⑧   10

The probability of a number between 5 and 9 equals

$$\frac{\text{number of favorable outcomes}}{\text{number of possible outcomes}} = \frac{2}{5}, \text{ or } 0.4, \text{ or } 40\%.$$

Listing all of the possible outcomes is sometimes confusing.
Study the example below.

Example What are the possible outcomes for the spinner at the right?

The spinner is divided into 10 sections. When you spin the spinner, it may land on any one of those 10 sections. So, there are 10 possible outcomes. But how do you list the 10 sections?

If you include both the number and color in your list it will look like this:

1 blue   2 red   3 yellow   4 blue   5 orange   6 yellow   7 red   8 yellow   9 blue   10 orange

If you list only the number for each section, it will look like this:

1   2   3   4   5   6   7   8   9   10

The short list of numbers is good enough. If you know the number, you can always look at the spinner to find the color that goes with that number.

Example What is the probability that the spinner shown above will land on a prime number?

Event: Land on a prime number.

Possible outcomes: 1, 2, 3, 4, 5, 6, 7, 8, 9, 10

The sections are equal parts of the spinner, so the outcomes are equally likely.

Favorable outcomes: the prime numbers 2, 3, 5, and 7

1 ② ③ 4 ⑤ 6 ⑦ 8 9 10

The probability of landing on a prime number equals

$$\frac{\text{number of favorable outcomes}}{\text{number of possible outcomes}} = \frac{4}{10}, \text{ or } 0.4, \text{ or } 40\%.$$

Example What is the probability that the spinner shown above will land on a blue or yellow section that has an even number?

The possible outcomes are 1, 2, 3, 4, 5, 6, 7, 8, 9, and 10.

The favorable outcomes are the sections that have an even number *and* are colored blue or yellow. Only three sections meet these conditions: the sections numbered 4, 6, and 8.

1   2   3 ④ 5 ⑥ 7 ⑧ 9   10

The probability of landing on a section that has an even number and is blue or yellow equals

$$\frac{\text{number of favorable outcomes}}{\text{number of possible outcomes}} = \frac{3}{10}, \text{ or } 0.3, \text{ or } 30\%.$$

In some problems, there may be several outcomes that look exactly the same.

Example Two red blocks and 3 blue blocks are placed in a bag. All of the blocks are cubes of the same size. One block is drawn without looking. What is the probability of drawing a red block?

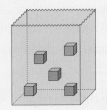

What is the *event* for which you want to find the probability? Draw a red block.

How many *possible outcomes* are there? Five

It is true that the red blocks look the same and the blue blocks look the same. But there are *5 different blocks* in the bag. So when you draw a block from the bag, there are *5 possible results* or outcomes.

When you list the possible outcomes, be sure to include each of the 5 blocks in your list.   red   red   blue   blue   blue

Are the outcomes *equally likely*? Yes. All 5 blocks are cubes of the same size. We can assume that each block has the same $\frac{1}{5}$ chance of being drawn.

Which of the possible outcomes in your list are *favorable outcomes*? The 2 red outcomes. Drawing a red block will make the event happen.

Circle the favorable outcomes.   (red) (red)   blue   blue   blue

The probability of drawing a red block equals $\frac{\text{number of favorable outcomes}}{\text{number of possible outcomes}} = \frac{2}{5}$.

Example A bag contains 1 green, 2 blue, and 3 red counters. The counters are the same, except for color. One counter is drawn. What is the probability of drawing a blue counter?

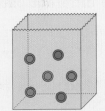

Event: Draw a blue counter.

Possible outcomes: green, blue, blue, red, red, red
The counters are the same, except for color.
So, the 6 outcomes are equally likely.

Favorable outcomes: blue, blue   green (blue) (blue) red   red   red

The probability of drawing a blue counter equals $\frac{\text{number of favorable outcomes}}{\text{number of possible outcomes}} = \frac{2}{6}$, or $\frac{1}{3}$.

Check Your Understanding

Six red, 4 green, and 3 blue blocks are placed in a bag. The blocks are the same, except for color. One block is drawn without looking.

Find the probability of each event.

1. Draw a green block.   2. Draw a block that is *not* green.   3. Draw a blue block.
4. Draw a red block.   5. Draw a block that is *not* red.   6. Draw any block.

Check your answers on page 437.

Tree Diagrams and the Multiplication Counting Principle

Many situations require two or more choices. **Tree diagrams** can be used to count the number of different ways to make those choices.

For example, suppose Vince is buying a new shirt. He must choose among 3 colors—white, blue, and green. He must also decide between long or short sleeves. How many different combinations of color and sleeve length are there?

To count the different combinations and see what they are, make a tree diagram like the one shown at the right. The paths drawn look like the branches of a tree.

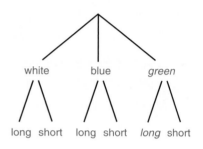

♦ The 3 top branches of the diagram are labeled white, blue, and green to show the color choices.

♦ The 2 branches below each color show the sleeve-length choices that are possible.

Each possible way to choose a shirt is found by following a path from the top to the bottom of the diagram. One possible choice is shown in italics: *green–long.* Counting shows that there are 6 different paths. Six different shirt choices are possible.

Multiplication is used to solve many types of counting problems that involve two or more choices.

Multiplication Counting Principle

Suppose you can make a first choice in m ways and a second choice in n ways. There are $m * n$ ways of making the first choice followed by the second choice.

Note

Cases with 3 or more choices are counted in the same way.

Suppose you can make a first choice in x ways, a second choice in y ways, and a third choice in z ways. Then there are $x * y * z$ different ways to make the 3 choices.

Example | Vince has pants in 4 different colors and shirts in 8 different colors. How many different color combinations for shirts and pants can Vince choose from?

Use the Multiplication Counting Principle: $4 * 8 = 32$. There are 32 different color combinations that Vince could choose from.

Check Your Understanding

Draw a tree diagram that shows all 32 combinations for the example above.

Check your answers on page 437.

Geometry and Constructions

Geometry in Our World

The world is filled with geometry. There are angles, segments, lines, and curves everywhere you look. There are 2-dimensional and 3-dimensional shapes of every type.

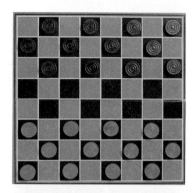

Many wonderful geometric patterns can be seen in nature. You can find patterns in flowers, spider webs, leaves, seashells, even your own face and body.

The ideas of geometry are also found in the things people create. Think of the games you play. Checkers is played with round pieces. The gameboard is covered with squares. Basketball and tennis are played with spheres. They are played on rectangular courts that are painted with straight and curved lines. The next time you play or watch a game, notice how geometry is important to the way the game is played.

The places we live in are built from plans that use geometry. Buildings almost always have rectangular rooms. Outside walls and roofs often include sections that have triangular shapes. Archways are curved and are often shaped like semicircles (half circles). Staircases may be straight or spiral.

Buildings and rooms are often decorated with beautiful patterns. You see these decorations on doors and windows; on walls, floors, and ceilings; and on railings of staircases.

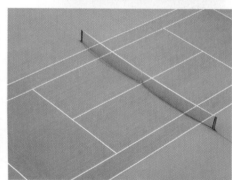

The clothes people wear are often decorated with geometric shapes. So are the things they use every day. All around the world, people create things using geometric patterns. Examples include quilts, pottery, baskets, and tiles. Some patterns are shown here. Which are your favorites?

Make a practice of noticing geometric shapes around you. Pay attention to bridges, buildings, and other structures. Look at the ways in which simple shapes such as triangles, rectangles, and circles are combined. Notice interesting designs. Share these with your classmates and your teacher.

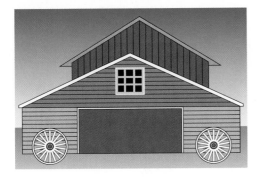

In this section, you will study geometric shapes and learn how to construct them. As you learn, try to create your own beautiful designs.

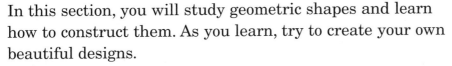

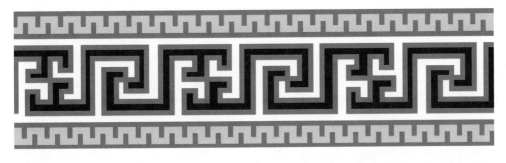

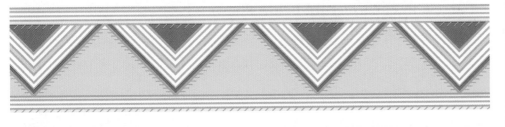

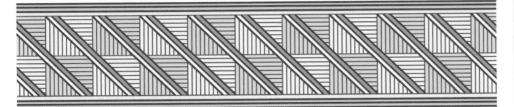

Did You Know?

Many buildings in the Middle East have designs made from geometric shapes. *The Dome of the Rock* in Jerusalem, for example, has an intricate design made from a combination of squares, rectangles, and triangles that can be seen just beneath its golden dome.

Angles

An **angle** is formed by 2 rays or 2 line segments that share the same endpoint.

angle formed by 2 rays   **angle formed by 2 segments**

The endpoint where the rays or segments meet is called the **vertex** of the angle. The rays or segments are called the **sides** of the angle.

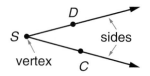

Naming Angles

The symbol for an angle is ∠. An angle can be named in two ways:

1. Name the vertex. The angle shown above is angle *S*. Write this as ∠*S*.

2. Name 3 points: the vertex and one point on each side of the angle. The angle above can be named angle *DSC* (∠*DSC*) or angle *CSD* (∠*CSD*). The vertex must always be listed in the middle, between the points on the sides.

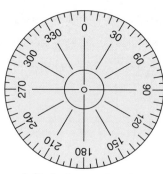

full-circle protractor

Measuring Angles

The **protractor** is a tool used to measure angles. Angles are measured in **degrees.** A degree is the unit of measure for the size of an angle.

The **degree symbol** ° is often used in place of the word *degrees.* The measure of ∠*S* (shown above) is 30 degrees, or 30°.

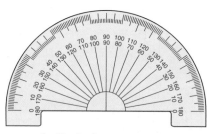

half-circle protractor

Sometimes there is confusion about which angle should be measured. The small curved arrow in each picture below shows which angle opening should be measured.

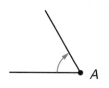

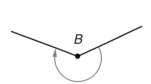

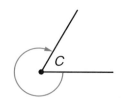

Measure of ∠*A* is 60°.   Measure of ∠*B* is 225°.   Measure of ∠*C* is 300°.

Classifying Angles According to Size

A **right angle**
measures 90°.

An **acute angle**
measures between
0° and 90°.

An **obtuse angle**
measures between
90° and 180°.

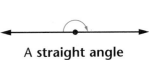

A **straight angle**
measures 180°.

Classifying Pairs of Angles

Vertical angles are angles that are opposite each other when two lines intersect. If two angles are vertical angles, they have the same measure in degrees.

Adjacent angles are angles that are next to each other. They have a common vertex and common side, but no other overlap.

Supplementary angles are two angles whose measures add up to 180°.

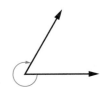

A **reflex angle**
measures between
180° and 360°.

Example   When two lines intersect, the angles formed have special properties.

For the figure at the right, the following statements are true.

Angles 1 and 3 are vertical angles. They have the same measure.
Angles 2 and 4 are vertical angles. They have the same measure.

There are four pairs of adjacent angles:
∠1 and ∠2   ∠2 and ∠3   ∠3 and ∠4   ∠4 and ∠1

There are four pairs of supplementary angles:
∠1 and ∠2   ∠2 and ∠3   ∠3 and ∠4   ∠4 and ∠1

Angles *a* and *b* are adjacent angles. They are *not* supplementary angles because their measures don't add up to 180°.

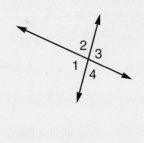

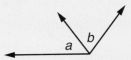

Check Your Understanding

1. Draw and label a right angle named ∠*C*.
2. Draw and label an obtuse angle named ∠*RST*.
3. In the figure at the right, find the measure of each angle.

     a. ∠2   **b.** ∠1   **c.** ∠3

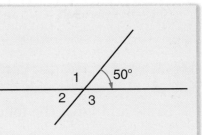

Check your answers on page 437.

Parallel Lines and Segments

Parallel lines are lines on a flat surface that never meet or cross. Think of a railroad track that goes on forever. The two rails are parallel lines. The rails never meet or cross and are always the same distance apart.

Parallel line segments are parts of lines that are parallel. The top and bottom edges of this page are parallel. If each edge were extended forever in both directions, the lines would be parallel.

The symbol for parallel is a pair of vertical lines ||. If $\overline{BF}$ and $\overline{TG}$ are parallel, write $\overline{BF} \parallel \overline{TG}$.

The railroad tracks shown here are parallel line segments.

If lines or segments cross or meet each other, they **intersect.** Lines or segments that intersect and form right angles are called **perpendicular** lines or segments.

The symbol for perpendicular is ⊥, which looks like an upside-down letter T. If $\overleftrightarrow{SU}$ and $\overleftrightarrow{XY}$ are perpendicular, write $\overleftrightarrow{SU} \perp \overleftrightarrow{XY}$.

Examples

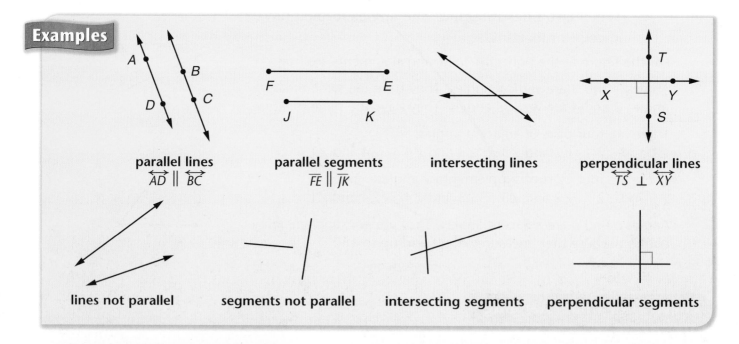

parallel lines
$\overleftrightarrow{AD} \parallel \overleftrightarrow{BC}$

parallel segments
$\overline{FE} \parallel \overline{JK}$

intersecting lines

perpendicular lines
$\overleftrightarrow{TS} \perp \overleftrightarrow{XY}$

lines not parallel

segments not parallel

intersecting segments

perpendicular segments

Check Your Understanding

Draw and label the following.

1. parallel line segments AB and EF

2. a line segment that is perpendicular to both $\overline{AB}$ and $\overline{EF}$

Check your answers on page 438.

Line Segments, Rays, Lines, and Angles

Figure	Symbol	Name and Description
• A	A	**point:** A location in space
C B endpoints	$\overline{BC}$ or $\overline{CB}$	**line segment:** A straight path between two points, called its endpoints
N M endpoint	$\overrightarrow{MN}$	**ray:** A straight path that goes on forever in one direction from an endpoint
T S	$\overleftrightarrow{ST}$ or $\overleftrightarrow{TS}$	**line:** A straight path that goes on forever in both directions
vertex S T P	$\angle T$ or $\angle STP$ or $\angle PTS$	**angle:** Two rays or line segments with a common endpoint, called the vertex
B A D C	$\overleftrightarrow{AB} \parallel \overleftrightarrow{CD}$	**parallel lines:** Lines that never meet or cross and are everywhere the same distance apart
	$\overline{AB} \parallel \overline{CD}$	**parallel line segments:** Segments that are parts of lines that are parallel
R E D S	none	**intersecting lines:** Lines that cross or meet
	none	**intersecting line segments:** Segments that cross or meet
B E F C	$\overleftrightarrow{BC} \perp \overleftrightarrow{EF}$	**perpendicular lines:** Lines that intersect at right angles
	$\overline{BC} \perp \overline{EF}$	**perpendicular line segments:** Segments that intersect at right angles

The white dividing lines illustrate parallel and perpendicular line segments.

Did You Know?

The use of letters to designate points and lines has been traced back to the Greek mathematician Hippocrates of Chios (about 450 B.C.).

Check Your Understanding

Draw and label each of the following.

1. point H 2. $\overleftrightarrow{JK}$ 3. $\angle CAT$ 4. $\overline{TU}$ 5. $\overline{PR} \parallel \overline{JK}$ 6. $\overrightarrow{EF}$

Check your answers on page 438.

Polygons

A **polygon** is a flat, 2-dimensional figure made up of line segments called **sides.** A polygon can have any number of sides, as long as it has at least three. The **interior** (inside) of a polygon is not part of the polygon.

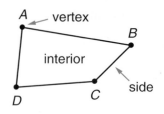

♦ The sides of a polygon are connected end to end and make one closed path.

♦ The sides of a polygon do not cross.

Each endpoint where two sides meet is called a **vertex.** The plural of vertex is **vertices.**

Figures That Are Polygons

4 sides, 4 vertices

3 sides, 3 vertices

7 sides, 7 vertices

Figures That Are NOT Polygons

All sides of a polygon must be line segments. Curved lines are not line segments.

The sides of a polygon must form a closed path.

A polygon must have at least 3 sides.

The sides of a polygon must not cross.

Prefixes	
tri-	3
quad-	4
penta-	5
hexa-	6
hepta-	7
octa-	8
nona-	9
deca-	10
dodeca-	12

Polygons are named after the number of their sides. The prefix in a polygon's name tells the number of sides it has.

Convex Polygons

A **convex** polygon is a polygon in which all the sides are pushed outward. The polygons below are all convex.

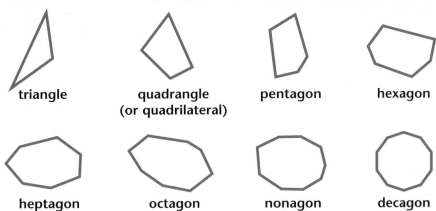

triangle   quadrangle (or quadrilateral)   pentagon   hexagon

heptagon   octagon   nonagon   decagon

Nonconvex (Concave) Polygons

A **nonconvex,** or **concave,** polygon is a polygon in which at least two sides are pushed in. The polygons at the right are all nonconvex.

Regular Polygons

A polygon is a **regular polygon** if (1) the sides all have the same length; and (2) the angles inside the figure are all the same size. A regular polygon is always convex. The polygons below are all regular.

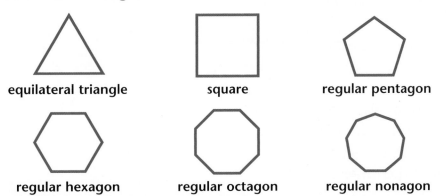

equilateral triangle   square   regular pentagon

regular hexagon   regular octagon   regular nonagon

quadrangle (or quadrilateral)

pentagon

hexagon

octagon

Check Your Understanding

1. What is the name of a polygon that has
   a. 6 sides?   **b.** 4 sides?   **c.** 8 sides?

2. **a.** Draw a convex heptagon.   **b.** Draw a concave decagon.

3. Explain why the cover of your journal is not a regular polygon.

Check your answers on page 438.

Triangles

Triangles are the simplest type of polygon. The prefix
tri- means *three*. All triangles have 3 vertices, 3 sides,
and 3 angles.

For the triangle shown here,

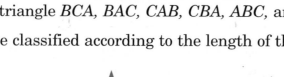

Angle *A* is formed
by sides that meet
at vertex *A*.

♦ the vertices are the points *B, C,* and *A*.

♦ the sides are $\overline{BC}$, $\overline{BA}$, and $\overline{CA}$.

♦ the angles are ∠*B*, ∠*C*, and ∠*A*.

Triangles have 3-letter names. You name a triangle by listing the
letter names for the vertices in order. The triangle above has 6
possible names: triangle *BCA, BAC, CAB, CBA, ABC,* and *ACB*.

Triangles may be classified according to the length of their sides.

A **scalene triangle** is a
triangle whose sides all
have different lengths.

An **isosceles triangle** is
a triangle that has two
sides of the same length.

An **equilateral triangle**
is a triangle whose sides
all have the same length.

A **right triangle** is a triangle with one right angle (square
corner). Right triangles have many different shapes and sizes.

   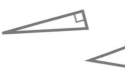

Some right triangles are scalene triangles, and some right
triangles are isosceles triangles. But a right triangle cannot
be an equilateral triangle because the side opposite the right
angle is always longer than each of the other two sides.

Check Your Understanding

1. Draw and label an equilateral triangle named *DEF*.
   Write the five other possible names for this triangle.

2. Draw an isosceles triangle.

3. Draw a right scalene triangle.

Check your answers on page 438.

Quadrangles

A **quadrangle** is a polygon that has 4 sides. Another name for quadrangle is **quadrilateral.** The prefix *quad-* means *four.* All quadrangles have 4 vertices, 4 sides, and 4 angles.

For the quadrangle shown here,

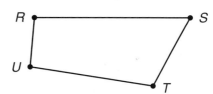

♦ the sides are $\overline{RS}$, $\overline{ST}$, $\overline{TU}$, and $\overline{UR}$.

♦ the vertices are *R, S, T,* and *U.*

♦ the angles are ∠*R,* ∠*S,* ∠*T,* and ∠*U.*

A quadrangle is named by listing the letter names for the vertices in order. The quadrangle above has 8 possible names:

RSTU, RUTS, STUR, SRUT, TURS, TSRU, URST, and *UTSR.*

Some quadrangles have two pairs of parallel sides. These quadrangles are called **parallelograms.**

Reminder: Two sides are parallel if they never meet, no matter how far they are extended.

Figures That Are Parallelograms

Opposite sides are parallel in each figure.

Figures That Are NOT Parallelograms

  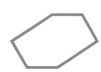

No parallel sides   Only 1 pair of parallel sides   3 pairs of parallel sides

A parallelogram must have exactly 2 pairs of parallel sides.

Check Your Understanding

1. Draw and label a quadrangle named *QUAD* that has exactly one pair of parallel sides.

2. Is *QUAD* a parallelogram?

3. Write the seven other possible names for this quadrangle.

Check your answers on page 438.

Special types of quadrangles have been given names. Some of these are parallelograms, others are not.

The tree diagram below shows how the different types of quadrangles are related. For example, quadrangles are divided into two major groups—"parallelograms" and "not parallelograms." The special types of parallelograms include rectangles, rhombuses, and squares.

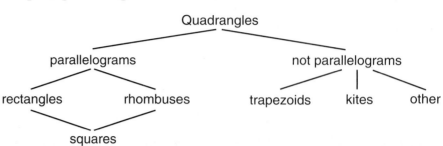

Quadrangles That Are Parallelograms		
rectangle		**Rectangles** are parallelograms. A rectangle has 4 right angles (square corners). The sides do not all have to be the same length.
rhombus		**Rhombuses** are parallelograms. A rhombus has 4 sides that are all the same length. The angles of a rhombus are usually not right angles, but they may be.
square		**Squares** are parallelograms. A square has 4 right angles (square corners). Its 4 sides are all the same length. *All* squares are rectangles. *All* squares are also rhombuses.

Quadrangles That Are NOT Parallelograms		
trapezoid		**Trapezoids** have exactly 1 pair of parallel sides. The 4 sides of a trapezoid can all have different lengths.
kite		A **kite** is a quadrangle with 2 pairs of equal sides. The equal sides are next to each other. The 4 sides cannot all have the same length. (A rhombus is not a kite.)
other		Any polygon with 4 sides that is not a parallelogram, a trapezoid, or a kite.

Check Your Understanding

What is the difference between the quadrangles in each pair below?

1. a square and a rectangle

2. a kite and rhombus

3. a trapezoid and a parallelogram

Check your answers on page 438.

Geometric Solids

Polygons and circles are flat, **2-dimensional** figures. The surfaces they enclose take up a certain amount of area, but they do not have any thickness and do not take up any volume.

Three-dimensional shapes have length, width, *and* thickness. They take up volume. Boxes, chairs, books, cans, and balls are all examples of 3-dimensional shapes.

A **geometric solid** is the surface or surfaces that surround a 3-dimensional shape. The surfaces of a geometric solid may be flat, curved, or both. Despite its name, a geometric solid is hollow; it does not include the points within its interior.

♦ A **flat surface** of a solid is called a **face.**

♦ A **curved surface** of a solid does not have a special name.

Examples Describe the surfaces of each geometric solid.

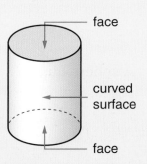

face
face

A cube has 6 square faces that are the same size.

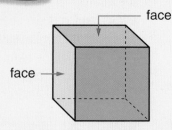

curved surface
face

A cone has 1 circular face and 1 curved surface. The circular face is called its **base**.

face
curved surface
face

A cylinder has 1 curved surface. It has 2 circular faces that are the same size and are parallel. The two faces are called its **bases**.

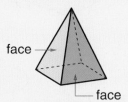

face
face

This pyramid has 4 triangular faces and 1 square face.

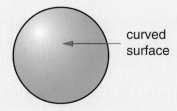

curved surface

A sphere has 1 curved surface.

The **edges** of a geometric solid are the line segments or curves where surfaces meet.

A corner of a geometric solid is called a **vertex** (plural *vertices*).

A vertex is usually a point at which edges meet. The vertex of a cone is an isolated corner completely separated from the edge of the cone.

A cone has 1 edge and 1 vertex. The vertex opposite the circular base is called the **apex.**

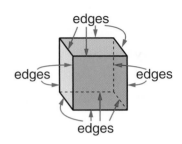

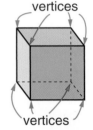

A cube has 12 edges and 8 vertices.

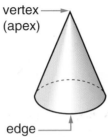

vertex (apex)

edge

The pyramid shown here has 8 edges and 5 vertices. The vertex opposite the rectangular base is called the **apex.**

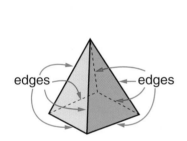

edges — edges

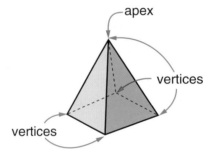

apex

vertices

vertices

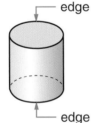

edge

edge

A cylinder has 2 edges. It has no vertices.

A sphere has no edges and no vertices.

Did You Know?

Euler's Theorem is a formula that tells how the number of faces, edges, and vertices in a polyhedron are related. Let *F, E, V* denote the number of faces, edges, and vertices of a polyhedron. Then $F + V - E = 2$. (Polyhedrons are defined on page 149.)

Check Your Understanding

1. **a.** How are cylinders and cones alike?   **b.** How do they differ?
2. **a.** How are pyramids and cones alike?   **b.** How do they differ?

Check your answers on page 439.

Polyhedrons

A **polyhedron** is a geometric solid whose surfaces are all formed by polygons. These surfaces are the faces of the polyhedron. A polyhedron does not have any curved surfaces.

Two important groups of polyhedrons are shown below. These are **pyramids** and **prisms.**

Pyramids

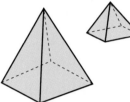

triangular pyramids

rectangular pyramids

pentagonal pyramid

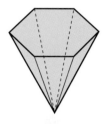

hexagonal pyramid

Prisms

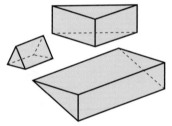

triangular prisms

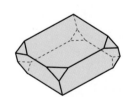

rectangular prisms

hexagonal prism

Many polyhedrons are not pyramids or prisms. Some examples are shown below.

Polyhedrons That Are NOT Pyramids or Prisms

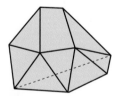
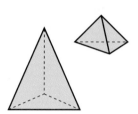

To find out why these are neither pyramids nor prisms, read pages 150 and 151.

Check Your Understanding

1. **a.** How many faces does a rectangular pyramid have?
   b. How many faces have a rectangular shape?
2. **a.** How many faces does a triangular prism have?
   b. How many faces have a triangular shape?

Check your answers on page 439.

Prisms

All of the geometric solids below are **prisms.**

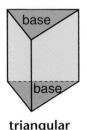

triangular
prism

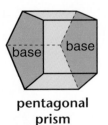

rectangular
prism

pentagonal
prism

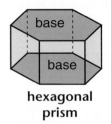

hexagonal
prism

The two shaded faces of each prism are called **bases.**

♦ The bases have the same size and shape.

♦ The bases are parallel. This means that the bases will never meet, no matter how far they are extended.

♦ The other faces connect the bases and are all shaped like parallelograms.

Note

Notice that the edges connecting the bases of a prism are parallel to each other.

The shape of its bases is used to name a prism. If the bases are triangular shapes, it is called a **triangular prism.** If the bases are rectangular shapes, it is called a **rectangular prism.** Rectangular prisms have three possible pairs of bases.

The number of faces, edges, and vertices that a prism has depends on the shape of the base.

Example The triangular prism shown here has 5 faces—3 rectangular faces and 2 triangular bases. It has 9 edges and 6 vertices.

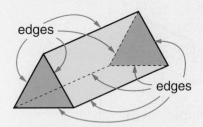

edges

edges

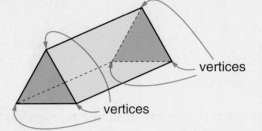

vertices

vertices

Check Your Understanding

1. **a.** How many faces does a hexagonal prism have?

   b. How many edges?

   c. How many vertices?

2. What is the name of a prism that has 10 faces?

   Check your answers on page 439.

Did You Know?

A *rhombohedron* is a six-sided prism whose faces are all parallelograms. Every rectangular prism is also a rhombohedron.

Pyramids

All of the geometric solids below are **pyramids.**

triangular
pyramid

square
pyramid

pentagonal
pyramid

hexagonal
pyramid

The shaded face of each of these pyramids is called the **base** of the pyramid.

- The polygon that forms the base can have any number of sides.
- The faces that are not a base all have a triangular shape.
- The faces that are not a base all meet at the same vertex.

The shape of its base is used to name a pyramid. If the base is a triangle shape, it is called a **triangular pyramid.** If the base is a square shape, it is called a **square pyramid.**

The great pyramids of Giza were built near Cairo, Egypt around 2600 B.C. They have square bases and are square pyramids.

The number of faces, edges, and vertices that a pyramid has depends on the shape of its base.

Did You Know?

The largest of the Giza pyramids measures 230 meters along each side and is 137 meters high.

Example The hexagonal pyramid at the right has 7 faces—6 triangular faces and a hexagonal base.

It has 12 edges. Six edges surround the hexagonal base. The other 6 edges meet at the apex (tip) of the pyramid.

It has 7 vertices. Six vertices are on the hexagonal base. The remaining vertex is the apex of the pyramid.

apex

The apex is the vertex opposite the base.

Check Your Understanding

1. **a.** How many faces does a triangular pyramid have?
   b. How many edges?   **c.** How many vertices?

2. What is the name of a pyramid that has 10 edges?

3. **a.** How are prisms and pyramids alike?   **b.** How are they different?

Check your answers on page 439.

Regular Polyhedrons

A polyhedron is **regular** if:

♦ each face is formed by a regular polygon.

♦ the faces all have the same size and shape.

♦ all of the vertices look exactly the same.

There are only five kinds of regular polyhedrons.

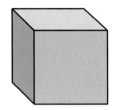

regular tetrahedron

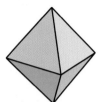

cube

regular octahedron

regular dodecahedron

regular icosahedron

Name	Shape of face	Number of faces
regular tetrahedron	equilateral triangle	4
cube	square	6
regular octahedron	equilateral triangle	8
regular dodecahedron	regular pentagon	12
regular icosahedron	equilateral triangle	20

Did You Know?

The ancient Greeks proved that there are exactly five regular polyhedrons. They were described by Plato (427-327 B.C.) and are often called the *Platonic solids*.

Plato associated the tetrahedron with fire, the cube with earth, the octahedron with air, the dodecahedron with heavenly bodies, and the icosahedron with water.

Check Your Understanding

1. Which regular polyhedrons have faces formed by equilateral triangles?

2. **a.** How many edges does a regular octahedron have?
   b. How many vertices?

3. **a.** How are regular tetrahedrons and regular octahedrons alike?
   b. How are they different?

Check your answers on page 439.

Circles

A **circle** is a curved line that forms a closed path on a flat surface. All of the points on a circle are the same distance from the **center of the circle.**

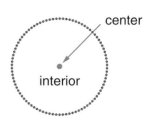

center

interior

The center is not part of the circle. The interior is not part of the circle.

The **compass** is a tool used to draw circles.

♦ The point of a compass, called the **anchor**, is placed at the center of the circle.

♦ The pencil in a compass traces out a circle. Every point on the circle is the same distance from the anchor.

The **radius** of a circle is any line segment that connects the center of the circle with any point on the circle. The word *radius* can also refer to the length of this segment.

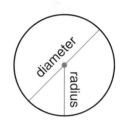

diameter

radius

The **diameter** of a circle is any line segment that passes through the center of the circle and has both of its endpoints on the circle. The word *diameter* can also refer to the length of this segment.

An **arc** is part of a circle, going from one point on the circle to another. For example, a **semicircle** is an arc: its endpoints are the endpoints of a diameter of the circle.

All circles are similar because they have the same shape, but circles do not all have the same size.

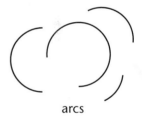

arcs

Examples   Many pizzas have a circular shape. You can order a pizza by saying the diameter that you want.

A "12-inch pizza" means a pizza with a 12-inch diameter.

A "16-inch pizza" means a pizza with a 16-inch diameter.

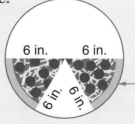

6 in.   6 in.

6 in.   6 in.

A 12-inch pizza

This pizza is 12 inches across. The diameter is 12 inches.

Each slice is a wedge that has 6-inch long sides.

Spheres

A **sphere** is a geometric solid that has a single curved surface. It is shaped like a ball, a marble, or a globe. All of the points on the surface of a sphere are the same distance from the **center of the sphere.**

All spheres have the same shape, but all spheres do not have the same size. The size of a sphere is the distance across its center.

◆ The line segment *RS* passes through the center of the sphere. This line segment is called a **diameter of the sphere.**

◆ The length of line segment *RS* is also called the diameter of the sphere.

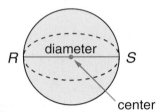

Globes and basketballs are examples of spheres that are hollow. The interior of each is empty. The hollow interior is not part of the sphere. The sphere includes only the points on the curved surface.

Marbles and baseballs are examples of spheres that have solid interiors. In cases like these, think of the solid interior as part of the sphere.

Example   The Earth is shaped very nearly like a sphere.

The diameter of the Earth is about 8,000 miles.

The distance from the Earth's surface to the center of the Earth is about 4,000 miles.

Every point on the Earth's surface is about 4,000 miles from the center of the Earth.

Layers Inside the Earth

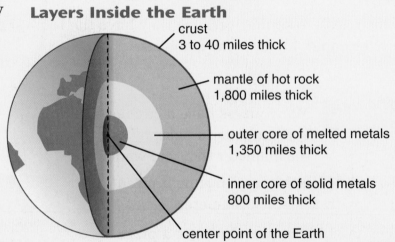

crust
3 to 40 miles thick

mantle of hot rock
1,800 miles thick

outer core of melted metals
1,350 miles thick

inner core of solid metals
800 miles thick

center point of the Earth

Did You Know?

The diameter of Pluto is about 1,500 miles, and the diameter of Neptune is about 31,000 miles.

Congruent Figures

It sometimes happens that figures have the same shape and size. They are **congruent.** Figures are congruent if they match exactly when one figure is placed on top of the other.

Example   Line segments are congruent if they have the same length.

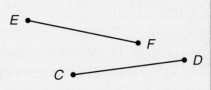

$\overline{EF}$ and $\overline{CD}$ are both 3 centimeters long. They have the same shape and the same length. These line segments are congruent.

Example   Angles are congruent if they have the same degree measure.

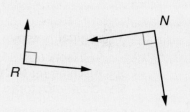

Angle *R* and angle *N* are both right angles. They have the same shape, and they each measure 90°. The angle openings match exactly when one angle is placed on top of the other.

Example   Circles are congruent if their diameters are the same length.

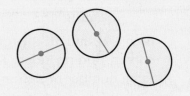

The circles here each have $\frac{1}{2}$-inch diameters. They have the same shape and the same size. The three circles are congruent.

Example   A copy machine was used to copy the pentagon *ABCDE*.

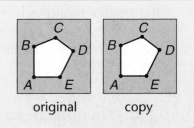

original   copy

If you cut out the copy, it will match exactly when placed on top of the original. The sides will match exactly. All the angles will match exactly. The original figure and the copy are congruent.

Check Your Understanding

Which of these methods could you use to make a congruent copy of this square?
a. Use a copy machine to copy the square.
b. Use tracing paper and trace the square.
c. Cut out the square and trace around it.
d. Measure the sides with a ruler, then draw the sides at right angles to each other using a protractor.

Check your answers on page 439.

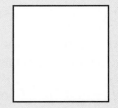

Similar Figures

Figures that have exactly the same shape are called **similar figures.** Usually, one figure is an enlargement or a reduction of the other. The **size-change factor** tells the amount of enlargement or reduction. Congruent figures are similar because they have the same shape.

> **Note**
>
> The size-change factor for congruent figures is 1X because they have the same size.

Examples If a copy machine is used to copy a drawing or picture, the copy will be similar to the original.

original copy	original copy	original copy
Exact copy	Enlargement	Reduction
Copy machine set to 100%.	Copy machine set to 200%.	Copy machine set to 50%.
Size-change factor is 1X.	Size-change factor is 2X.	Size-change factor is $\frac{1}{2}$X.

Example The triangles *CAT* and *DOG* are similar. The larger triangle is an enlargement of the smaller triangle.

Each side and its enlargement form a pair of sides called **corresponding sides.** These sides are marked with the same number of slash marks.

The size-change factor is 2X. Each side in the larger triangle is twice the size of the corresponding side in the smaller triangle. The size of the angles is the same for both triangles. For example, ∠*T* and ∠*G* have the same degree measure.

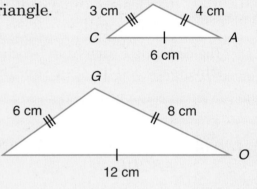

Example Quadrangles *ABCD* and *MNOP* are similar. How long is side *MN*? How long is side *AD*?

$\overline{NO}$ is $\frac{1}{3}$ as long as $\overline{BC}$. $\overline{OP}$ is $\frac{1}{3}$ as long as side *CD*.

So, the size-change factor is $\frac{1}{3}$X.
$\overline{AB}$ and $\overline{MN}$ are corresponding sides. $\overline{AB}$ is 15 feet long.

So, $\overline{MN}$ must be $\frac{1}{3} * 15 = 5$ feet long.

$\overline{AD}$ and $\overline{MP}$ are corresponding sides.
$\overline{MP}$ is $\frac{1}{3}$ as long as $\overline{AD}$ and equals 7 feet.

So, $\overline{AD}$ must be 21 feet long.

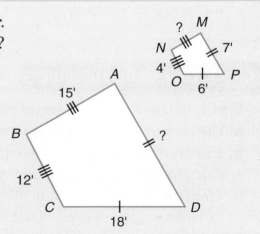

Reflections, Translations, and Rotations

In geometry, a figure can be moved from one place to another. Three different ways to move a figure are shown below.

♦ A **reflection** moves a figure by "flipping" it over a line.

♦ A **translation** moves a figure by "sliding" it to a new location.

♦ A **rotation** moves a figure by "turning" it around a point.

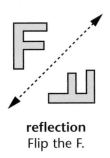

reflection
Flip the F.

translation
Slide the F.

rotation
Turn the F.

An approximate reflection is shown here. The line of reflection is the water's edge, along the bank.

The original figure, before it has been moved, is called the **preimage.** The new figure produced by the move is called the **image.**

Each point of the preimage is moved to a new point of the image called its **matching point.** A point and its matching point are also called **corresponding points.**

For each of the moves shown above, the image has the same size and shape as the preimage. The image and preimage are congruent shapes.

Reflections

A reflection is a "flipping" motion of a figure. The line that the figure is flipped over is called the **line of reflection.** The preimage and the image are on opposite sides of the line of reflection.

For any reflection:

♦ The preimage and the image have the same size and shape.

♦ The preimage and the image are reversed.

♦ Each point and its matching point are the same distance from the line of reflection.

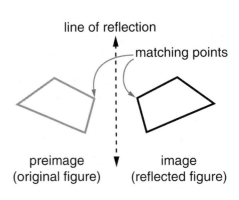

line of reflection

matching points

preimage
(original figure)

image
(reflected figure)

Translations

A translation is a "sliding" motion of a figure. Each point of the figure slides the same distance in the same direction. Imagine the letter T drawn on grid paper.

♦ If each point of the letter T slides 6 grid squares to the right, the result is a *horizontal translation*.

♦ If each point of the letter T slides 8 grid squares upward, the result is a *vertical translation*.

♦ Suppose that each point of the letter T slides 6 grid squares to the right, then 8 grid squares upward. The result is the same as a *diagonal translation*.

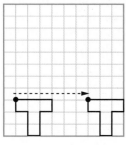
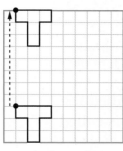
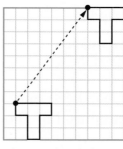

| horizontal translation | vertical translation | diagonal translation |

Rotations

When a figure is rotated, it is turned a certain number of degrees around a particular point.

A figure can be rotated *clockwise* (the direction in which clock hands move). A figure can also be rotated *counterclockwise* (the opposite direction to the way clock hands move).

preimage
(original figure)

270° clockwise
90° counterclockwise

90° clockwise
270° counterclockwise

P

180° clockwise
180° counterclockwise

Check Your Understanding

1. Copy the figure and reflect it over $\overleftrightarrow{AB}$.

CHIN

A   *B*

2. Which figure is a 90° clockwise rotation of 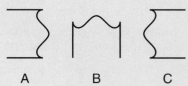 ?

A   B   C

Check your answers on page 439.

Line Symmetry

A dashed line is drawn through the figure at the right. The line divides the figure into two parts. Both parts look alike, but are facing in opposite directions.

The figure is **symmetric about a line.** The dashed line is called a **line of symmetry** for the figure.

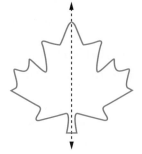

line of symmetry

You can use a reflection to get the figure shown above.

◆ Think of the line of symmetry as a line of reflection.

◆ Reflect the left side of the figure over the line.

◆ The left side and its reflection (the right side) together form the figure.

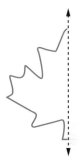

Reflect the left side to get the figure above.

An easy way to check whether a figure has *line symmetry* is to fold it in half. If the two halves match exactly, then the figure is symmetric. The fold line is the line of symmetry.

Examples   The letters M, A, C, and H are symmetric. The lines of symmetry are drawn for each letter.

The letter H has two lines of symmetry. If you fold along either line, the two halves match exactly.

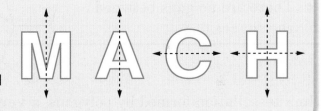

The figures below are all symmetric. The line of symmetry is drawn for each figure. If there is more than one line of symmetry, they are all drawn.

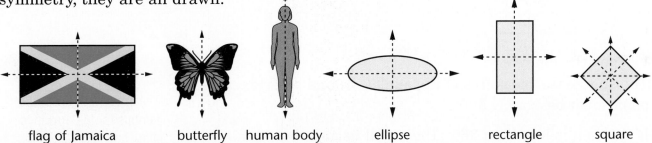

flag of Jamaica   butterfly   human body   ellipse   rectangle   square

Check Your Understanding

1. Trace each pattern-block (PB) shape on your Geometry Template onto a sheet of paper. Draw the lines of symmetry for each shape.

2. How many lines of symmetry does a circle have?

Check your answers on page 439.

Tessellations

A **tessellation** is a pattern formed by repeated use of polygons or other figures that completely cover a surface.

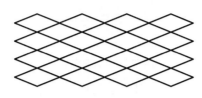 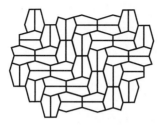 

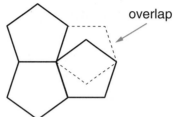

♦ The shapes in a tessellation do not overlap.

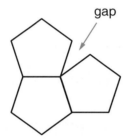

overlap

♦ There are no gaps between the shapes.

gap

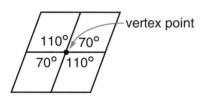

In a tessellation formed by polygons, a **vertex point** is a point where vertices of the polygons meet.

vertex point

110° 70°
70° 110°

110° + 70° + 110° + 70° = 360°

♦ The sum of the measures of the angles around a vertex point must be exactly 360°.

♦ If the sum is less than 360°, there will be gaps between the shapes. The pattern is not a tessellation.

♦ If the sum is greater than 360°, the shapes will overlap. The pattern is not a tessellation.

Did You Know ?

The artist M.C. Escher is famous for his drawings that show tessellations.

Regular Tessellations

A tessellation made by repeating congruent copies of one kind of regular polygon is called a **regular tessellation.**

In a regular tessellation:

♦ All of the polygons are *regular* and *congruent*. This means that only one kind of regular polygon is used, and all copies of this polygon are the same size.

♦ If the vertex of one polygon meets another polygon, the meeting point must be a vertex of both polygons.

The figure to the right uses congruent copies of a square to cover the surface. The pattern forms a tessellation, but it is *not* a regular tessellation: squares 1 and 2 each have a vertex that meets square 3, but the meeting point is not a vertex of square 3.

There are exactly three possible regular tessellations. These are shown to the right.

Semiregular Tessellations

Tessellations may involve more than one type of shape.

A tessellation is called a **semiregular tessellation** if it satisfies these conditions:

♦ It uses at least two different regular polygon shapes.

♦ For each polygon shape, all copies of that polygon are congruent.

♦ If the vertex of one polygon meets another polygon, the meeting point must be a vertex of both polygons.

♦ The same polygons, in the same order, surround each polygon vertex.

The example at the right is a semiregular tessellation made up of squares and equilateral triangles. As you move around any vertex point, in order, there are 2 triangles, 1 square, 1 triangle, and then 1 square.

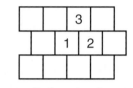

a tessellation that is *not* a regular tessellation

the three possible regular tessellations

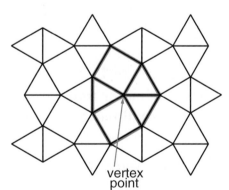

vertex point

The Geometry Template

The **Geometry Template** has many uses.

The template has two rulers. The inch scale measures in inches and fractions of an inch. The centimeter scale measures in centimeters or millimeters. Use either edge of the template as a straightedge for drawing line segments.

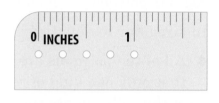

There are 17 different geometric figures on the template. The figures labeled "PB" are **pattern-block shapes.** These are half the size of real pattern blocks. There is a hexagon, a trapezoid, two different rhombuses, an equilateral triangle, and a square. These will come in handy for some of the activities you do this year.

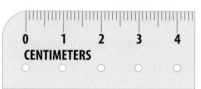

Each triangle on the template is labeled with a T and a number. Triangle "T1" is an equilateral triangle whose sides all have the same length. Triangles "T2" and "T5" are right triangles. Triangle "T3" has sides that all have different lengths. Triangle "T4" has two sides of the same length.

The remaining shapes are circles, squares, a regular octagon, a regular pentagon, a kite, a rectangle, a parallelogram, and an ellipse.

The two circles near the inch scale can be used as ring-binder holes so you can store your template in your notebook.

Use the **half-circle** and **full-circle protractors** at the bottom of the template to measure and draw angles. Use the **Percent Circle** at the top of the template to construct and measure circle graphs. The Percent Circle is divided into 1% intervals and some common fractions of the circle are marked.

Notice the tiny holes near the 0-, $\frac{1}{4}$-, $\frac{2}{4}$-, and $\frac{3}{4}$-inch marks of the inch scale and at each inch mark from 1 to 7. On the centimeter side, the holes are placed at each centimeter mark from 0 to 10. These holes can be used to draw circles of various sizes.

Example   Draw a circle with a 3-inch radius.

Place one pencil point in the hole at 0. Place another pencil point in the hole at 3 inches. Hold the pencil at 0 inches steady while rotating the pencil at 3 inches (along with the template) to draw the circle.

Hold this pencil steady.

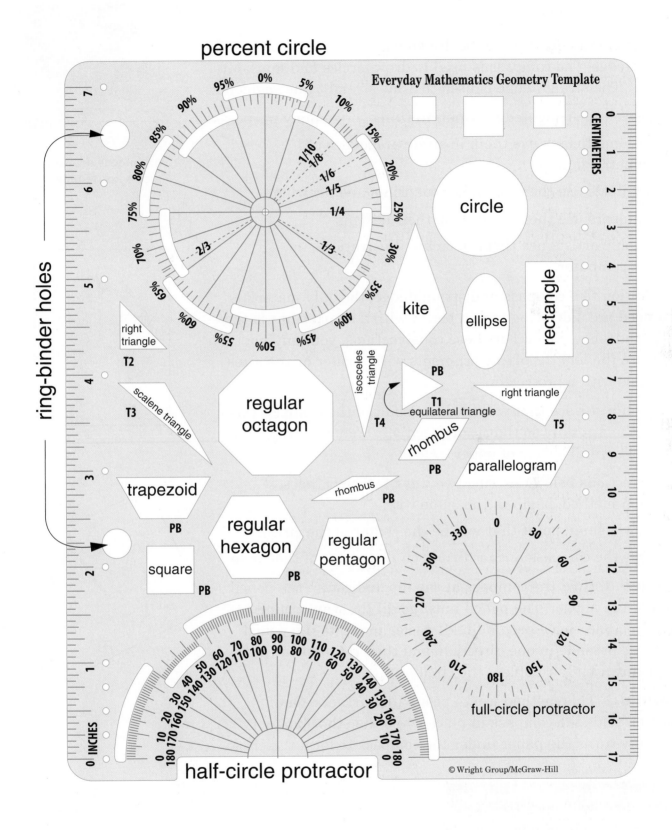

percent circle

ring-binder holes

Everyday Mathematics Geometry Template

circle

rectangle

kite

ellipse

right triangle

T2

isosceles triangle

PB

T1

right triangle

scalene triangle

T3

regular octagon

equilateral triangle

T4

T5

rhombus

PB

parallelogram

trapezoid

rhombus

PB

PB

regular hexagon

regular pentagon

PB

square

PB

full-circle protractor

half-circle protractor

CENTIMETERS

INCHES

© Wright Group/McGraw-Hill

Compass-and-Straightedge Constructions

Many geometric figures can be drawn using only a compass and straightedge. The compass is used to draw circles and mark off lengths. The straightedge is used to draw straight line segments.

Compass-and-straightedge **constructions** serve many purposes:

♦ Mathematicians use them to study properties of geometric figures.

♦ Architects use them to make blueprints and drawings.

♦ Engineers use them to develop their designs.

♦ Graphic artists use them to create illustrations on a computer.

Architect's drawing of a house plan

In addition to a compass and straightedge, the only materials you need are a drawing tool (a pencil with a sharp point is the best) and some paper. For these constructions, you may not measure the lengths of line segments with a ruler or the sizes of angles with a protractor.

Draw on a surface that will hold the point of the compass (also called the **anchor**) so that it does not slip. You can draw on a stack of several sheets of paper.

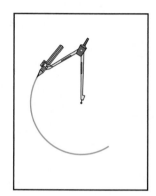

Method 1

The directions below describe two ways to draw circles. For each method, begin in the same way:

♦ Draw a small point that will be the center of the circle.

♦ Press the compass anchor firmly on the center of the circle.

Method 1 Hold the compass at the top and rotate the pencil around the anchor. The pencil must go all the way around to make a circle. Some people find it easier to rotate the pencil as far as possible in one direction, and then rotate it in the other direction to complete the circle.

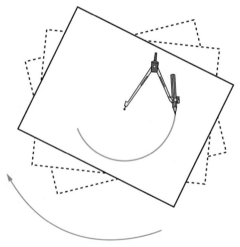

Method 2 This method works best with partners. One partner holds the compass in place while the other partner carefully turns the paper under the compass to form the circle.

Method 2

Check Your Understanding

Concentric circles are circles that have the same center. Use a compass to draw 3 concentric circles.

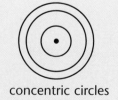

concentric circles

Copying a Line Segment

Follow each step carefully. Use a clean sheet of paper.

Step 1: Draw line segment *AB*.

Step 2: Draw a second line segment. It should be longer than segment *AB*. Label one of its endpoints as *A'* (read "*A* prime").

Step 3: Place the compass anchor at *A* and the pencil point at *B*.

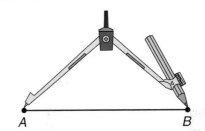

Step 4: Without changing the compass opening, place the compass anchor on *A'* and draw a small arc that crosses the line segment. Label the point where the arc crosses the line segment as *B'*.

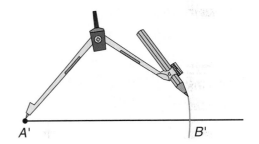

The segments *A'B'* and *AB* have the same length.

Line segment *A'B'* is **congruent** to line segment *AB*.

Check Your Understanding

Draw a line segment. Using a compass and straightedge only, copy the line segment. After you make your copy, measure the segments with a ruler to see how accurately you copied the original line segment.

Copying a Triangle

Follow each step carefully. Use a clean sheet of paper.

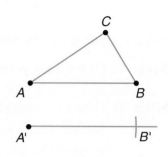

Step 1: Draw a triangle *ABC*. Draw a line segment that is longer than line segment *AB*. Copy line segment *AB* onto the segment you just drew. (See page 165.) Label the end points of the copy as *A'* and *B'* (read as "A prime" and "B prime").

Step 2: Place the compass anchor at *A* and the pencil point at *C*. Without changing the compass opening, place the compass anchor on *A'* and draw an arc.

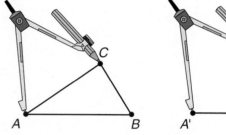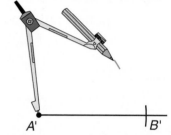

Step 3: Place the compass anchor at *B* and the pencil point at *C*. Without changing the compass opening, place the compass anchor on *B'* and draw another arc. Label the point where the arcs intersect as *C'*.

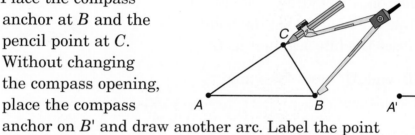

Step 4: Draw line segments *A'C'* and *B'C'*.

Triangles *ABC* and *A'B'C'* are congruent. That is, they are the same size and shape.

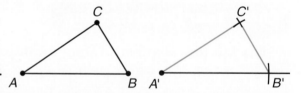

Check Your Understanding

Draw a triangle. Using a compass and straightedge, copy the triangle. Cut out the copy and place it on top of the original triangle to check that the triangles are congruent.

Constructing a Parallelogram

Follow each step carefully. Use a clean sheet of paper.

Step 1: Draw an angle *ABC*.

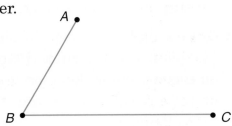

Step 2: Place the compass anchor at *B* and the pencil point at *C*. Without changing the compass opening, place the compass anchor on *A* and draw an arc.

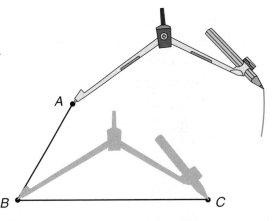

Step 3: Place the compass anchor at *B* and the pencil point at *A*. Without changing the compass opening, place the compass anchor on *C* and draw another arc that crosses the first arc. Label the point where the two arcs cross as *D*.

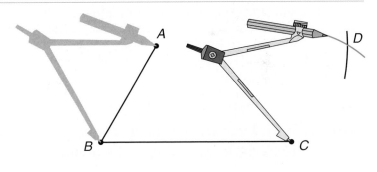

Step 4: Draw line segments *AD* and *CD*.

Check Your Understanding

1. Use a compass and straightedge to construct a parallelogram.
2. Use a compass and straightedge to construct a rhombus.

(*Hint:* A rhombus is a parallelogram whose sides are all the same length.)

Constructing a Regular Inscribed Hexagon

Follow each step carefully. Use a clean sheet of paper.

Step 1: Draw a circle and keep the same compass opening. Make a dot on the circle. Place the compass anchor on the dot and make a mark with the pencil point on the circle. Keep the same compass opening for Steps 2 and 3.

Step 2: Place the compass anchor on the mark you just made. Make another mark with the pencil point on the circle.

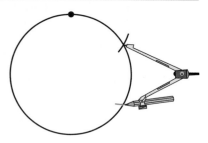

Step 3: Do this four more times to divide the circle into 6 equal parts. The sixth mark should be on the dot you started with or very close to it.

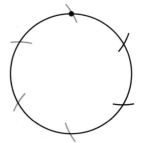

Step 4: With your straightedge, connect the 6 marks on the circle to form a regular hexagon.

Use your compass to check that the sides of the hexagon are all the same length.

The hexagon is **inscribed** in the circle because each vertex of the hexagon is on the circle.

Check Your Understanding

1. Draw a circle. Using a compass and straightedge, construct a regular hexagon that is inscribed in the circle.
2. Draw a line segment from the center of the circle to each vertex of the hexagon to form 6 triangles. Use your compass to check that the sides of each triangle are the same length.

Constructing an Inscribed Square

Follow each step carefully. Use a clean sheet of paper.

Step 1: Draw a circle with a compass.

Step 2: Draw a line segment through the center of the circle with endpoints on the circle. Label the endpoints as *A* and *B*.

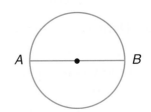

Step 3: Increase the compass opening. Place the compass anchor on *A*. Draw an arc below the center of the circle and another arc above the center of the circle.

Step 4: Without changing the compass opening, place the compass anchor on *B*. Draw arcs that cross the arcs you drew in Step 3. Label the points where the arcs intersect as *C* and *D*.

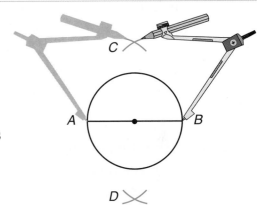

Step 5: Draw a line through points *C* and *D*.

Label the points where line *CD* intersects the circle as *E* and *F*.

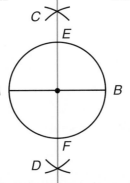

Step 6: Draw line segments *AE*, *EB*, *BF*, and *FA*.

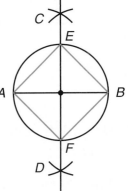

Check with your compass that all four line segments are the same length. Check with the corner of your straightedge or another square corner that all four angles are right angles.

The square is **inscribed** in the circle because all vertices of the square are on the circle.

Check Your Understanding

Use a compass and a straightedge to construct an inscribed square.

Did You Know?

Using only a straightedge and a compass, it is possible to construct a regular polygon with 3, 4, 5, 6, 8, 10, 12, 15, 16, or 17 sides—but not with 7, 9, 11, 13, 14 or 18 sides.

Bisecting a Line Segment

Follow each step carefully. Use a clean sheet of paper.

Step 1: Draw line segment *AB*.

A ●————————● B

Step 2: Open your compass so that the compass opening is greater than half the distance between point *A* and point *B*. Place the compass anchor on *A*. Draw a small arc below $\overline{AB}$ and another small arc above $\overline{AB}$.

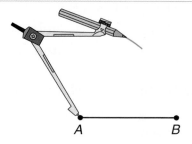

A ●————————● B

Step 3: Without changing the compass opening, place the compass anchor on *B*. Draw an arc below $\overline{AB}$ and another arc above $\overline{AB}$, so that the arcs cross the first arcs you drew. Label the points where the pairs of arcs intersect as *M* and *N*.

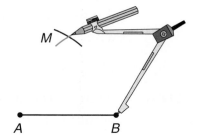

M ✕

A ●————————● B

N ✕

Step 4: Draw $\overline{MN}$. Label the point where $\overline{MN}$ intersects $\overline{AB}$ as *O*.

We say that line segment *MN* **bisects** line segment *AB* at point *O*. The distance from *A* to *O* is the same as the distance from *B* to *O*.

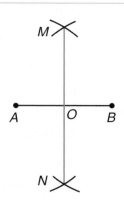

M ✕

A ●———|O——● B

N ✕

Check Your Understanding

Draw a line segment. Use a compass and a straightedge to bisect it. Measure to check that the line segment has been divided into two equal parts.

Constructing a Perpendicular Line Segment (Part 1)

Let P be a point *on* the line segment AB. You can construct a line segment that is perpendicular to $\overline{AB}$ at the point P.

Follow each step carefully. Use a clean sheet of paper.

Step 1: Draw line segment AB. Make a dot on $\overline{AB}$ and label it as P.

Step 2: Place the compass anchor on P and draw an arc that crosses $\overline{AB}$. Label the point where the arc crosses the segment as C.

Keeping the compass anchor on point P and keeping the same compass opening, draw another arc that crosses $\overline{AB}$. Label the point where the arc crosses the segment as D.

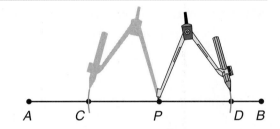

Step 3: Make sure the compass opening is greater than the length of $\overline{CP}$. Place the compass anchor on C and draw an arc above $\overline{AB}$.

Keeping the same compass opening, place the compass anchor on D and draw another arc above $\overline{AB}$ that crosses the first arc.

Label the point where the two arcs cross as Q.

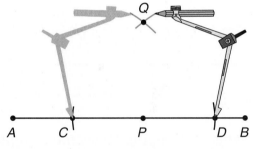

Step 4: Draw $\overline{QP}$.

$\overline{QP}$ is **perpendicular** to $\overline{AB}$.

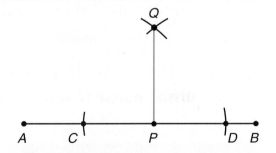

Check Your Understanding

Draw a line segment. Draw a point on the segment and label it as R.

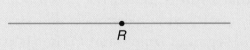

Use a compass and straightedge. Construct a line segment through point R that is perpendicular to the segment you drew.

Use a protractor to check that the segments are perpendicular.

Constructing a Perpendicular Line Segment (Part 2)

Let M be a point that is *not on* the line segment PQ. You can construct a line segment with one endpoint at M that is perpendicular to $\overline{PQ}$.

Follow each step carefully. Use a clean sheet of paper.

Step 1: Draw line segment PQ. Draw a point M not on $\overline{PQ}$.

Step 2: Place the compass anchor on M and draw an arc that crosses $\overline{PQ}$ at two points.

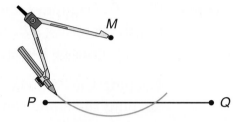

Step 3: Place the compass anchor on one of the points and draw an arc below $\overline{PQ}$.

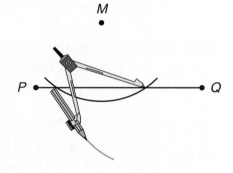

Step 4: Keeping the same compass opening, place the compass anchor on the other point and draw another arc that crosses the first arc.

Label the point where the two arcs cross as N.

Then draw the line segment MN.

$\overline{MN}$ is **perpendicular** to $\overline{PQ}$.

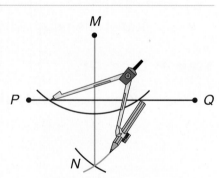

Check Your Understanding

1. Draw a line segment AT and a point C above the line segment. Using a compass and straightedge, construct a line segment from point C that is perpendicular to $\overline{AT}$.

2. Use the Geometry Template to draw a parallelogram. Then construct a line segment to show the height of the parallelogram.

Copying an Angle

Follow each step carefully. Use a clean sheet of paper.

Step 1: Draw an angle *B*.

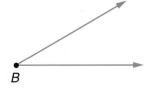

Step 2: To start copying the angle, draw a ray. Label the endpoint of the ray as *B'*.

Step 3: Place the compass anchor on *B*. Draw an arc that crosses both sides of angle *B*. Label the point where the arc crosses one side as *A*. Label the point where the arc crosses the other side as *C*.

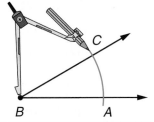

Step 4: Without changing the compass opening, place the compass anchor on *B'*. Draw an arc about the same size as the one you drew in Step 3. Label the point where the arc crosses the ray as *A'*.

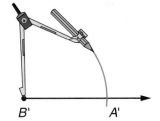

Step 5: Place the compass anchor on *A* and the pencil point on *C*.

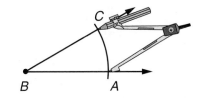

Step 6: Without changing the compass opening, place the compass anchor on *A'*. Draw a small arc where the pencil point crosses the larger arc and label the crossing point as *C'*.

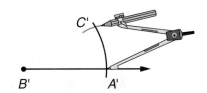

Step 7: Draw a ray from point *B'* through point *C'*. ∠*A'B'C'* is **congruent** to ∠*ABC*. That is, the two angles have the same degree measure.

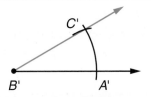

Check Your Understanding

Draw an angle. Use a compass and straightedge to copy the angle. Then measure the two angles with a protractor to check that they are the same size.

Copying a Quadrangle

Before you can copy a quadrangle with a compass and straightedge, you need to know how to copy line segments and angles. Those constructions are described on pages 165 and 173.

Follow each step carefully. Use a clean sheet of paper.

Step 1: Draw a quadrangle *ABCD*. Copy ∠*BAD*. Label the vertex of the new angle as *A*'. The sides of your new angle should be longer than $\overline{AB}$ and $\overline{AD}$.

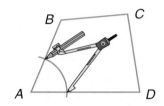 

Step 2: Mark off the distance from *A* to *D* on the horizontal side of your new angle. Label the endpoint as *D*'.

Mark off the distance from *A* to *B* on the other side of your new angle. Label the endpoint as *B*'.

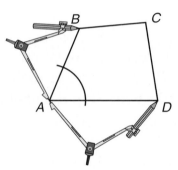 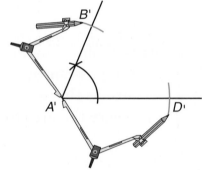

Step 3: Place the compass anchor on *B* and the pencil point on *C*. Without changing the compass opening, place the compass anchor on *B*' and make an arc.

Step 4: Place the compass anchor on *D* and the pencil point on *C*. Without changing the compass opening, place the compass anchor on *D*' and make an arc that crosses the arc you made in Step 3. Label the point where the two arcs meet as *C*'.

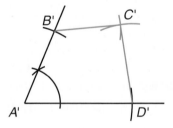

Step 5: Draw $\overline{B'C'}$ and $\overline{D'C'}$.
Quadrangle *A*'*B*'*C*'*D*' is **congruent** to quadrangle *ABCD*. The two quadrangles are the same size and shape.

Check Your Understanding

Draw a quadrangle. Use a compass and straightedge to copy the quadrangle.

Mathematics and Architecture

Architecture is the planning of structures that have function and beauty. Architecture showcases the mathematical and technological achievements of the human race. Throughout history, architects and builders have often competed to build the biggest, the most beautiful, or the most unusual structures.

◄ During the economic boom of the 1920s, the "race" to design the tallest skyscrapers on earth began. In 1933, the world's four tallest buildings could be found in New York City. The Empire State Building, at 381 meters (1,250 ft), held the height record for more than 40 years.

◄

The Empire State Building was built using huge steel beams. Engineers had to calculate whether the beams were sturdy enough to support the immense weight of the structure.

In the late 1990s, the title of tallest building in the world belonged to a building on another continent. The Petronas Towers in Malaysia measure 452 meters (1,483 ft). ►

The Petronas Towers rely on steel-reinforced concrete construction for support. Here are the towers under construction. ►

Ancient Architecture

We cannot be sure when architecture began, but there are many structures still standing from the ancient world that show us about the mathematical abilities of the people who designed them.

◀ Many ancient mathematicians were also astronomers. When Stonehenge was built nearly 5,000 years ago, it was oriented to track the movements of celestial bodies. Stonehenge may have even been used to forecast eclipses of the sun and moon.

Mayan astronomers of Central America designed their observatories for mathematical studies of stars and planets. The Spaniards called this structure El Caracol. ➤

◀ The pyramids of Egypt were incredible engineering feats. The Great Pyramid of Khufu was built in about 20 years. Every year, 100,000 stones weighing about 2.5 tons each had to be placed in precise locations, so the four faces would meet at the vertex.

Throughout his studies of music, the Greek mathematician Pythagoras developed theories that had a great influence on Western architecture. Pythagoras observed how notes of plucked strings created harmonious sounds when the string lengths were ratios of small integers. To create visual harmony, architects adopted the use of similar ratios in their building designs.

◀ The Parthenon in Athens, Greece was built around the time of Pythagoras. The ratio 2:3 and its square, 4:9 can be seen in its construction.

▼ The Greeks developed three architectural styles—the Doric, the Ionic, and the Corinthian.

▲ The Doric style, like that of the Parthenon, is sturdy and plain.

▲ The Ionic style is ornate and elegant.

▲ The Corinthian style is elaborately decorated.

Architecture of Eastern and Western Europe

Great architecture is both beautiful and functional. Architects incorporate aesthetically pleasing patterns, symmetry, and proportion as they design buildings. They also make sure the buildings are useful and structurally sound.

▲ In 530 A.D., the Byzantine emperor Justinian wanted a building to surpass anything ever built. He asked two architects to design the Hagia Sofia. One major design challenge was to figure out how to support a round dome over a square building.

▲ Inside the Hagia Sofia

◄ King's College at Cambridge University, England was built in the 15th century.

Note the patterns and symmetry inside King's College Chapel. This fan vaulting pattern is unique to English architecture of this period. The narrow radiating vault lines add delicacy to the ceiling and help to distribute its weight. ➤

Architecture of East Asia

Beauty means different things to different people, and ideas of beauty change over time. Architects in Asia and other parts of the Eastern hemisphere use many of the same mathematical concepts as their Western counterparts. However, the resulting structures can be very different.

◄ This is a gate in the Forbidden Purple City in Vietnam. Note the symmetry, proportion, and aesthetic elements as you scan the building from ground level to the top.

▲ Detailed patterns and gold artwork on the Grand Palace

The Grand Palace in Bangkok, Thailand has a number of complex geometrical forms. ▼

What Will They Think of Next?

Architects often break from tradition as they experiment with new ideas, technologies, and materials. Here are some unusual buildings built within the last 100 years.

The Opera house in Sydney, Australia was designed by Jorn Utzon and completed in 1973. ➤

▲ Casa Battlo in Barcelona, Spain is one of many buildings designed by Antoni Gaudi in the early 20th century.

The Guggenheim Museum in Bilbao, Spain was designed by Frank Gehry in the 1990s. ▼

What do you think the buildings of the next 100 years will look like?

Measurement

Natural Measures and Standard Units

Systems of weights and measures have been used in many parts of the world since ancient times. People measured lengths and weights long before they had rulers and scales.

Ancient Measures of Weight

Shells and grains such as wheat or rice were often used as units of weight. For example, a small item might be said to weigh 300 grains of rice. Large weights were often compared to the load that could be carried by a man or a pack animal.

Ancient Measures of Length

People used **natural measures** based on the human body to measure length and distance. Some of these units are shown below.

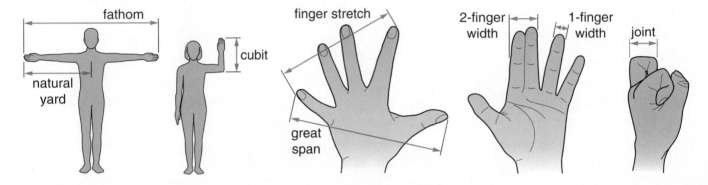

Standard Units of Length and Weight

Using shells and grains to measure weight is not exact. Even if the shells and grains are of the same type, they vary in size and weight.

Using body lengths to measure length is also not exact. Body measures depend upon the person who is doing the measuring. The problem is that different persons have hands and arms of different lengths.

One way to solve this problem is to make **standard units** of length and weight. Most rulers are marked off with inches and centimeters as standard units. Bath scales are marked off using pounds and kilograms as standard units. The standard units never change and are the same for everyone. If two people measure the same object using standard units, their measurements will be the same or almost the same.

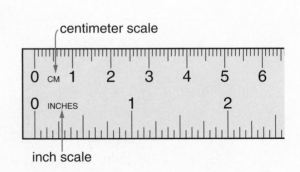

The Metric System and the U.S. Customary System

About 200 years ago, a system of weights and measures called the **metric system** was developed. It uses standard units for length, weight, and temperature. In the metric system:

♦ The **meter** is the standard unit for length. The symbol for a meter is **m.** A meter is about the width of a front door.

♦ The **gram** is the standard unit for weight. The symbol for a gram is **g.** A paper clip weighs about $\frac{1}{2}$ gram.

♦ The **Celsius degree** is the standard unit for temperature. The symbol for degrees Celsius is **°C.** Water freezes at 0°C and boils at 100°C. Normal room temperature is about 20°C.

Scientists almost always use the metric system for measurement. The metric system is easy to use because it is a base-ten system. Larger and smaller units are defined by multiplying or dividing the units given above by powers of ten: 10, 100, 1,000, and so on.

about 1 meter

Examples All metric units of length are based on the meter. Each unit is defined by multiplying or dividing the meter by a power of 10.

Units of Length Based on the Meter	Prefix	Meaning
1 decimeter (dm) = $\frac{1}{10}$ meter	deci-	$\frac{1}{10}$
1 centimeter (cm) = $\frac{1}{100}$ meter	centi-	$\frac{1}{100}$
1 millimeter (mm) = $\frac{1}{1,000}$ meter	milli-	$\frac{1}{1,000}$
1 kilometer (km) = 1,000 meters	kilo-	1,000

Note

The U.S. customary system is not based on powers of 10. This makes it more difficult to use than the metric system. For example, in order to change inches to yards, you must know that 36 inches equals 1 yard.

The metric system is used in most countries around the world. In the United States, the **U.S. customary system** is used for everyday purposes. The U.S. customary system uses standard units like the **inch, foot, yard, mile, ounce, pound,** and **ton.**

Check Your Understanding

1. Which of these units are in the metric system?
   foot   millimeter   pound   inch   gram   meter   centimeter   yard

2. What does the prefix "milli-" mean?

3. 2 grams = ? milligrams

Check your answers on page 439.

Converting Units of Length

The table below shows how different units of length in the metric system compare. You can use this table to rewrite a length using a different unit.

Comparing Metric Units of Length				Symbols for Units of Length	
1 cm = 10 mm	1 m = 1,000 mm	1 m = 100 cm	1 km = 1,000 m	mm = millimeter	cm = centimeter
1 mm = $\frac{1}{10}$ cm	1 mm = $\frac{1}{1,000}$ m	1 cm = $\frac{1}{100}$ m	1 m = $\frac{1}{1,000}$ km	m = meter	km = kilometer

Examples Use the table to rewrite each length using a different unit. Replace the unit given first with an equal length that uses the new unit.

Problem	Solution
56 centimeters = ? millimeters	56 cm = 56 * 10 mm = 560 mm
56 centimeters = ? meters	56 cm = 56 * $\frac{1}{100}$ m = $\frac{56}{100}$ m = 0.56 m
9.3 kilometers = ? meters	9.3 km = 9.3 * 1,000 m = 9,300 m
6.9 meters = ? centimeters	6.9 m = 6.9 * 100 cm = 690 cm

The table below shows how different units of length in the U.S. customary system compare. You can use this table to rewrite a length using a different unit.

Comparing U.S. Customary Units of Length				Symbols for Units of Length	
1 ft = 12 in.	1 yd = 36 in.	1 yd = 3 ft	1 mi = 5,280 ft	in. = inch	ft = foot
1 in. = $\frac{1}{12}$ ft	1 in. = $\frac{3}{16}$ yd	1 ft = $\frac{1}{3}$ yd	1 ft = $\frac{1}{5,280}$ mi	yd = yard	mi = mile

Examples Use the table to rewrite each length using a different unit. Replace the unit given first with an equal length that uses the new unit.

Problem	Solution
14 ft = ? inches	14 ft = 14 * 12 in. = 168 in.
21 ft = ? yards	21 ft = 21 * $\frac{1}{3}$ yd = $\frac{21}{3}$ yd = 7 yd
7 miles = ? feet	7 mi = 7 * 5,280 ft = 36,960 ft
180 inches = ? yards	180 in. = 180 * $\frac{1}{36}$ yd = $\frac{180}{36}$ yd = 5 yd

Personal References for Units of Length

Sometimes it is hard to remember just how long a centimeter or a yard is, or how a kilometer and a mile compare. You may not have a ruler, yardstick, or tape measure handy. When this happens, you can estimate lengths by using the lengths of common objects and distances that you know well.

Some examples of personal references for length are given below. A good personal reference is something that you see or use often, so you don't forget it. A good personal reference also doesn't change size. For example, a wooden pencil is not a good personal reference for length because it gets shorter as it is sharpened.

The thickness of a dime is about 1 mm.

The diameter of a quarter is about 1 in.

Personal References for Metric Units of Length	
About 1 millimeter	**About 1 centimeter**
Thickness of a dime	Thickness of a crayon
Thickness of the point of a thumbtack	Width of the head of a thumbtack
Thickness of the thin edge of a paper match	Thickness of a pattern block
About 1 meter	**About 1 kilometer**
One big step (for an adult)	1,000 big steps (for an adult)
Width of a front door	Length of 10 football fields (including the end zones)
Tip of the nose to tip of the thumb, with arm extended (for an adult)	

Personal References for U.S. Customary Units of Length	
About 1 inch	**About 1 foot**
Length of a paper clip	A man's shoe length
Width (diameter) of a quarter	Length of a license plate
Width of a man's thumb	Length of this book
About 1 yard	**About 1 mile**
One big step (for an adult)	2,000 average-size steps (for an adult)
Width of a front door	Length of 15 football fields (including the end zones)
Tip of the nose to tip of the thumb, with arm extended (for an adult)	

Note

The personal references for 1 meter can also be used for 1 yard. 1 yard equals 36 inches, while 1 meter is about 39.37 inches. One meter is often called a "fat yard," which means 1 yard plus 1 hand width.

Did You Know?

The longest wall in the world is the Great Wall of China. It is 2,150 miles long, the length of 32,000 football fields. It would take about 3,800,000 big steps for an adult to walk its length.

Perimeter

Sometimes we want to know the **distance around** a shape, which is called the **perimeter** of the shape. To measure perimeter, use units of length such as inches, meters, or miles.

To find the perimeter of a polygon, add the lengths of all its sides. Remember to name the unit of length used to measure the shape.

Example Find the perimeter of polygon *ABCDE*.

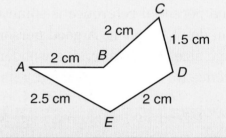

2 cm + 2 cm + 1.5 cm + 2 cm + 2.5 cm = 10 cm

The perimeter is 10 centimeters.

Perimeter Formulas	
Rectangles	**Squares**
p = 2 * (l + w)	**p = 4 * s**
p is the perimeter, l is the length, w is the width of the rectangle.	p is the perimeter, s is the length of one of the sides of the square.

Examples Find the perimeter of each polygon.

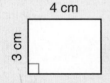

Rectangle

Use the formula p = 2 * (l + w).
- length (l) = 4 cm
- width (w) = 3 cm
- perimeter (p) = 2 * (4 cm + 3 cm)
             = 2 * 7 cm = 14 cm

The perimeter is 14 centimeters.

Square

Use the formula p = 4 * s.
- length of side (s) = 9 ft
- perimeter (p) = 4 * 9 ft
             = 36 ft

The perimeter is 36 feet.

Check Your Understanding

1. Find the perimeter of a rectangle whose dimensions are 4 feet 2 inches and 9 feet 5 inches.

2. Measure the sides of this book to the nearest half-inch. What is the perimeter of the book?

Check your answers on page 439.

Circumference

The perimeter of a circle is the **distance around** the circle.

The perimeter of a circle has a special name. It is called the **circumference** of the circle.

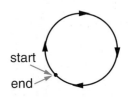

start
end

> **Example** Most food cans are cylinders. Their tops and bottoms have circular shapes. The circumference of a circular can top is how far the can turns when opened by a can opener.

The **diameter** of a circle is any line segment that passes through the center of the circle and has both endpoints on the circle.

The length of a diameter segment is also called the diameter.

If you know the diameter, there is a simple formula for finding the circumference of a circle.

> **circumference = pi * diameter**   or $c = \pi * d$
> (c is the circumference and d is the diameter)

The Greek letter π is called **pi.** It is approximately equal to 3.14. In working with the number π, you can either use 3.14 or $3\frac{1}{7}$ as approximate values for π, or a calculator with a π key.

> **Example**   Find the circumference of the circle.
>
> Use the formula $c = \pi * d$.
> - diameter (d) = 8 cm
> - circumference (c) = $\pi * 8$ cm
>
>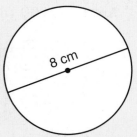
>
> 8 cm
>
> Use either the π key on the calculator or use 3.14 as an approximate value for π.
>
> Circumference (c) = 25.1 cm, rounded to the nearest tenth of a centimeter.
>
> The circumference of the circle is 25.1 cm.

Check Your Understanding

1. Measure the diameter of the quarter in millimeters.
2. Find the circumference of the quarter in millimeters.
3. What is the circumference of a pizza with a 14-inch diameter?

Check your answers on page 439.

Area

Area is a measure of the amount of surface inside a closed boundary. You can find the area by counting the number of squares of a certain size that cover the region inside the boundary. The squares must cover the entire region. They must not overlap, have any gaps, or extend outside the boundary.

Sometimes a region cannot be covered by an exact number of squares. In that case, count the number of whole squares and fractions of squares that cover the region.

Area is reported in square units. Units of area for small regions are square inches (in.^{2}), square feet (ft^2), square yards (yd^2), square centimeters (cm^2), and square meters (m^2). For large regions, square miles (mi^2) are used in the United States, while square kilometers (km^2) are used in other countries.

You may report area using any of the square units. But you should choose a square unit that makes sense for the region being measured.

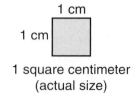

1 square centimeter
(actual size)

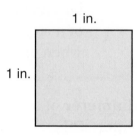

1 square inch
(actual size)

Examples The area of a field-hockey field is reported below in three different ways.

Area of the field is 6,000 square yards.	Area of the field is 54,000 square feet.	Area of the field is 7,776,000 square inches.
Area = 6,000 yd^2	Area = 54,000 ft^2	Area = 7,776,000 in.^{2}

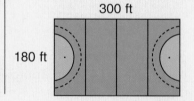

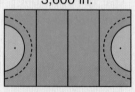

Although each of the measurements above is correct, reporting the area in square inches really doesn't give a good idea about the size of the field. It is hard to imagine 7,776,000 of anything!

Did You Know?

The International Space Station (ISS) orbits the Earth at an altitude of 250 miles. It is 356 feet wide and 290 feet long, and has an area of over 100,000 square feet.

Area of a Rectangle

When you cover a rectangular shape with unit squares, the squares can be arranged into rows. Each row will contain the same number of squares and fractions of squares.

Example Find the area of the rectangle.

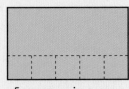

5 squares in a row

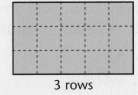

3 rows

3 rows with 5 squares in each row for a total of 15 squares

Area = 15 square units

height base

Either pair of parallel sides in a rectangle can be chosen as its **bases**. The **height** of a rectangle is the shortest distance between its bases.

To find the area of a rectangle, use either of these formulas:

Area = (the number of squares in 1 row) * (the number of rows)
Area = length of a base * height

Area Formulas	
Rectangles	**Squares**
$A = b * h$	$A = s^2$
A is the area, b is the length of a base, h is the height of the rectangle.	A is the area, s is the length of a side of the square.

Examples Find the area of the rectangle.

Use the formula $A = b * h$.
• length of base (b) = 4 in.
• height (h) = 3 in.
• area (A) = 4 in. * 3 in.
  = 12 in.^{2}

The area of the rectangle is 12 in.^{2}.

Find the area of the square.

Use the formula $A = s^2$.
• length of a side (s) = 9 ft
• area (A) = 9 ft * 9 ft
  = 81 ft^2

The area of the square is 81 ft^2.

Check Your Understanding

Find the area of these figures. Include the unit in each answer.

1. 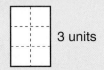 3 units

   2 units

2. 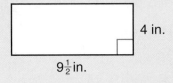 4 in.

   $9\frac{1}{2}$ in.

3. 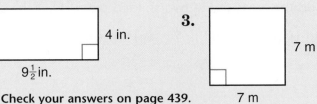 7 m

   7 m

Check your answers on page 439.

The Rectangle Method of Finding Area

Many times you will need to find the area of a polygon that is not a rectangle. Unit squares will not fit neatly inside the figure, and you won't be able to use the formula for the area of a rectangle.

One approach that works well in cases such as these is called the **rectangle method.** Rectangles are used to surround the figure or parts of the figure. Then the only areas that you must calculate are for rectangles and triangular halves of rectangles.

Example What is the area of triangle *JKL*?

Draw a rectangle around the triangle. Rectangle *JKLM* surrounds the triangle.

The area of rectangle *JKLM* is 10 square units. The segment *JL* divides the rectangle into two congruent triangles that have the same area.

The area of triangle *JKL* is 5 square units.

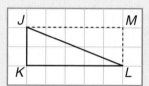

Example What is the area of triangle *ABC*?

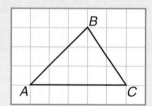

Step 1: Divide triangle *ABC* into two parts.

Step 2: Draw a rectangle around the left shaded part. The area of the rectangle is 9 square units. The shaded area is $4\frac{1}{2}$ square units.

Step 3: Draw a rectangle around the right shaded part. The area of the rectangle is 6 square units. The shaded area is 3 square units.

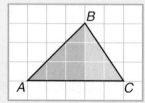

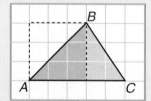

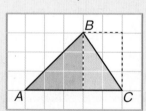

Step 4: Add the areas of the two shaded parts: $4\frac{1}{2} + 3 = 7\frac{1}{2}$ square units.

The area of triangle *ABC* is $7\frac{1}{2}$ square units.

Example What is the area of triangle *XYZ*?

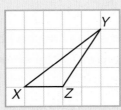

Step 1: Draw a rectangle around the triangle.

Step 2: The area of rectangle *XRYS* is 12 square units. So, the area of triangle *XRY* is 6 square units.

Step 3: Draw a rectangle around triangle *ZSY*. The area of the rectangle is 6 square units, so the area of triangle *ZSY* is 3 square units.

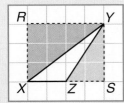

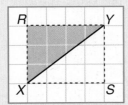

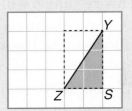

Step 4: Subtract the areas of the two shaded triangles from the area of rectangle *XRYS*: 12 − 6 − 3 = 3 square units.

The area of triangle *XYZ* is 3 square units.

Check Your Understanding

Use the rectangle method to find the area of each figure below.

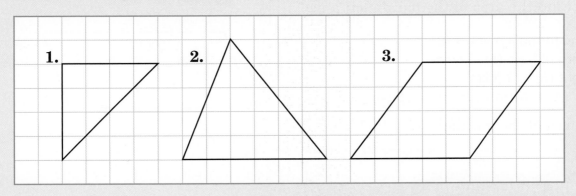

1. **2.** **3.**

Check your answers on page 439.

Area of a Parallelogram

In a parallelogram, either pair of opposite sides can be chosen as its **bases.** The **height** of the parallelogram is the shortest distance between the two bases.

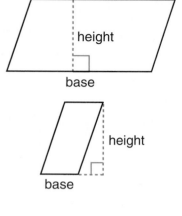

height

base

In the parallelograms at the right, the height is shown by a dashed line that is **perpendicular** (at a right angle) to the base. In the second parallelogram, the base has been extended and the dashed height line falls outside the parallelogram.

height

base

Any parallelogram can be cut into two pieces that will form a rectangle. This rectangle will have the same base length and height as the parallelogram. The rectangle will also have the same area as the parallelogram. So you can find the area of the parallelogram in the same way you find the area of the rectangle—by multiplying the length of the base by the height.

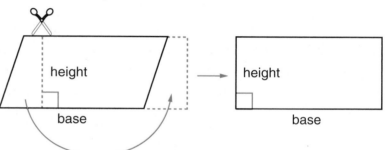

height

base

height

base

Formula for the Area of a Parallelogram

$$A = b * h$$

A is the area, *b* is the length of the base, *h* is the height of the parallelogram.

Example Find the area of the parallelogram.

Use the formula $A = b * h$.
- length of base (b) = 6 cm
- height (h) = 2.5 cm
- area (A) = 6 cm $*$ 2.5 cm = 15 cm^2

The area of the parallelogram is 15 cm^2.

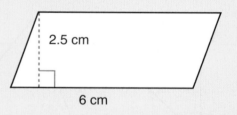

2.5 cm

6 cm

Check Your Understanding

Find the area of each parallelogram. Include the unit in each answer.

1.

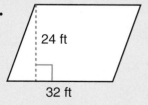

24 ft

32 ft

2.

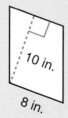

10 in.

8 in.

3.

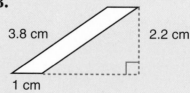

3.8 cm

2.2 cm

1 cm

Check your answers on page 439.

Area of a Triangle

Any of the sides of a triangle can be chosen as its **base.**
The **height** of the triangle is the shortest distance
between the chosen base and the vertex opposite this
base. The height is shown by a dashed line that is
perpendicular (at a right angle) to the base. In
some triangles, the base is extended and the dashed
height line falls outside the triangle. In the right
triangle shown, the height line is one of the sides of
the triangle.

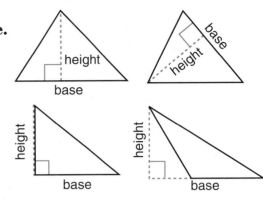

Any triangle can be combined with a second triangle of the
same size and shape to form a parallelogram. Each triangle
at the right has the same size base and height as the
parallelogram. The area of each triangle is half the area of
the parallelogram. Therefore, the area of a triangle is half
the product of the base length multiplied by the height.

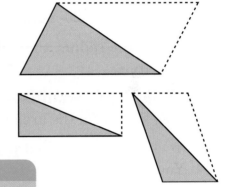

Area Formulas	
Parallelograms	**Triangles**
$A = b * h$	$A = \frac{1}{2} * (b * h)$
A is the area, b is the length of a base, h is the height.	A is the area, b is the length of the base, h is the height.

Example   Find the area of the triangle.

Use the formula $A = \frac{1}{2} * (b * h)$.
- length of base $(b) = 7$ in.
- height $(h) = 4$ in.
- area $(A) = \frac{1}{2} * (7 \text{ in.} * 4 \text{ in.}) = \frac{1}{2} * 28 \text{ in.}^2 = \frac{28}{2} \text{ in.}^2 = 14 \text{ in.}^2$

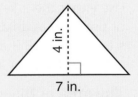

The area of the triangle is 14 in.^2.

Check Your Understanding

Find the area of each triangle. Include the unit in each answer.

1.
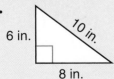

2.

3.
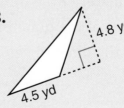

Check your answers on page 440.

Area of a Circle

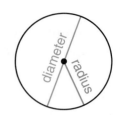

The **radius** of a circle is any line segment that connects the center of the circle with any point on the circle. The length of a radius segment is also called the radius.

The **diameter** of a circle is any segment that passes through the center of the circle and has both endpoints on the circle. The length of a diameter segment is also called the diameter.

If you know either the radius or the diameter, you can find the other length too. Use the following formulas:

$$\textbf{diameter} = 2 * \textbf{radius} \qquad \textbf{radius} = \tfrac{1}{2} * \textbf{diameter}$$

If you know the radius, there is a simple formula for finding the area of a circle:

$$\textbf{Area} = \textbf{pi} * \textbf{(radius squared)} \qquad \text{or} \qquad A = \pi * r^2$$
(*A* is the area and *r* is the radius.)

The Greek letter π is called **pi,** and it is approximately equal to 3.14. You can either use 3.14 or $3\tfrac{1}{7}$ as approximate values for π, or a calculator with a π key.

Example Find the area of the circle.

Use the formula $A = \pi * r^2$.
- radius (*r*) = 3 in.
- area (*A*) = π * 3 in. * 3 in.

Use either the π key on the calculator or use 3.14 as an approximate value for π.
- area (*A*) = 28.3 in.^{2}, rounded to the nearest tenth of a square inch.

The area of the circle is 28.3 in.^{2}.

Check Your Understanding

1. Measure the diameter of the dime in millimeters.
2. What is the radius of the dime in millimeters?
3. Find the area of the dime in square millimeters.

Check your answers on page 440.

Volume and Capacity

Volume

The **volume** of a solid object such as a brick or a ball is a measure of *how much space the object takes up*. The volume of a container such as a freezer is a measure of *how much the container will hold*.

Volume is measured in **cubic units,** such as cubic inches (in.³), cubic feet (ft³), and cubic centimeters (cm³). It is easy to find the volume of an object that is shaped like a cube or other rectangular prism. For example, picture a container in the shape of a 10-centimeter cube (that is, a cube that is 10 cm by 10 cm by 10 cm). It can be filled with exactly 1,000 centimeter cubes. Therefore, the volume of a 10-centimeter cube is 1,000 cubic centimeters (1,000 cm³).

To find the volume of a rectangular prism, all you need to know are the length and width of its base and its height. The length, width, and height are called the **dimensions** of the prism.

You can also find the volume of another solid, such as a triangular prism, pyramid, cone, or sphere, by measuring its dimensions. It is even possible to find the volume of an irregular object such as a rock or your own body.

Capacity

We often measure things that can be poured into or out of containers such as liquids, grains, salt, and so on. The volume of a container that is filled with a liquid or a solid that can be poured is often called its **capacity.**

Capacity is usually measured in units such as **gallons, quarts, pints, cups, fluid ounces, liters,** and **milliliters.**

The tables at the right compare different units of capacity. These units of capacity are not cubic units, but liters and milliliters are easily converted to cubic units:

$$1 \text{ milliliter} = 1 \text{ cm}^3 \qquad 1 \text{ liter} = 1,000 \text{ cm}^3$$

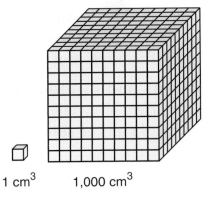

1 cm³   1,000 cm³

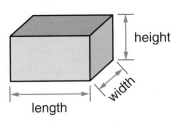

The dimensions of a rectangular prism

U.S. Customary Units
1 gallon (gal) = 4 quarts (qt)
1 gallon = 2 half-gallons
1 half-gallon = 2 quarts
1 quart = 2 pints (pt)
1 pint = 2 cups (c)
1 cup = 8 fluid ounces (fl oz)
1 pint = 16 fluid ounces
1 quart = 32 fluid ounces
1 half-gallon = 64 fluid ounces
1 gallon = 128 fluid ounces

Metric Units
1 liter (L) = 1,000 milliliters (mL)
1 milliliter = $\frac{1}{1,000}$ liter
1 liter = 1,000 cubic centimeters
1 milliliter = 1 cubic centimeter

Volume of a Geometric Solid

You can think of the volume of a geometric solid as the total number of whole unit cubes and fractions of unit cubes needed to fill the interior of the solid without any gaps or overlaps.

Prisms and Cylinders

In a prism or cylinder, the cubes can be arranged in layers that each contain the same number of cubes or fractions of cubes.

> **Example**   Find the volume of the prism.
>
> 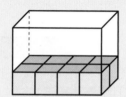   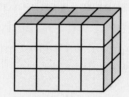
>
>   8 cubes in 1 layer   3 layers
>
> 3 layers with 8 cubes in each layer makes a total of 24 cubes.
>
> Volume = 24 cubic units

Note

For a *right rectangular prism* with side lengths l, w, and h units, the volume V can be found using the formula $V = l * w * h$.

The **height** of a prism or cylinder is the shortest distance between its **bases.** The volume of a prism or cylinder is the product of the area of the base (the number of cubes in one layer) multiplied by its height (the number of layers).

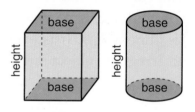

Pyramids and Cones

The height of a pyramid or cone is the shortest distance between its base and the vertex opposite its base.

If a prism and a pyramid have the same size base and height, then the volume of the pyramid is one-third the volume of the prism. If a cylinder and a cone have the same size base and height, then the volume of the cone is one-third the volume of the cylinder.

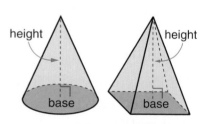

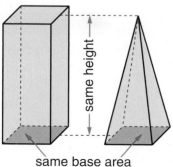

same base area

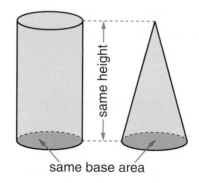

same base area

Volume of a Rectangular or Triangular Prism

Volume of a Prism	Area of a Rectangle	Area of a Triangle
$V = B * h$	$A = b * h$	$A = \frac{1}{2} * (b * h)$
V is the volume, B is the area of the base, h is the height of the prism.	A is the area, b is the length of the base, h is the height of the rectangle.	A is the area, b is the length of the base, h is the height of the triangle.

Example   Find the volume of the rectangular prism.

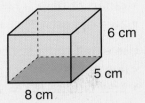

Step 1: Find the area of the base (B). Use the formula $A = b * h$.
- length of the rectangular base (b) = 8 cm
- height of the rectangular base (h) = 5 cm
- area of base (B) = 8 cm * 5 cm = 40 cm²

Step 2: Multiply the area of the base by the height of the rectangular prism. Use the formula $V = B * h$.
- area of base (B) = 40 cm²
- height of prism (h) = 6 cm
- volume (V) = 40 cm² * 6 cm = 240 cm³

The volume of the rectangular prism is 240 cm³.

Example   Find the volume of the triangular prism.

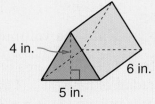

Step 1: Find the area of the base (B). Use the formula $A = \frac{1}{2}(b * h)$.
- length of the triangular base (b) = 5 in.
- height of the triangular base (h) = 4 in.
- area of base (B) = $\frac{1}{2}$ * (5 in. * 4 in.) = 10 in.²

Step 2: Multiply the area of the base by the height of the triangular prism. Use the formula $V = B * h$.
- area of base (B) = 10 in.²
- height of prism (h) = 6 in.
- volume (V) = 10 in.² * 6 in. = 60 in.³

The volume of the triangular prism is 60 in.³.

Check Your Understanding

Find the volume of each prism. Include the unit in each answer.

1. 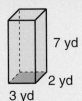 7 yd, 2 yd, 3 yd

2. 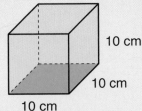 10 cm, 10 cm, 10 cm

3. 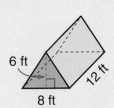 6 ft, 12 ft, 8 ft

Check your answers on page 440.

Volume of a Cylinder or Cone

Volume of a Cylinder	Volume of a Cone	Area of a Circle
$V = B * h$	$V = \frac{1}{3} * (B * h)$	$A = \pi * r^2$
V is the volume, B is the area of the base, h is the height of the cylinder.	V is the volume, B is the area of the base, h is the height of the cone.	A is the area, r is the radius of the circle.

Example   Find the volume of the cylinder.

Step 1: Find the area of the base (B). Use the formula $A = \pi * r^2$.
- radius of base (r) = 5 cm
- area of base (B) = $\pi * 5$ cm $* 5$ cm

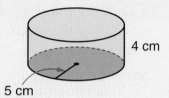

Use either the π key on a calculator or 3.14 as an approximate value for π.
- area of base (B) = 78.5 cm^2, rounded to the nearest tenth of a square centimeter.

Step 2: Multiply the area of the base by the height of the cylinder. Use the formula $V = B * h$.
- area of base (B) = 78.5 cm^2
- height of cylinder (h) = 4 cm
- volume (V) = 78.5 cm$^2 * 4$ cm = 314.0 cm^3

The volume of the cylinder is 314.0 cm^3.

Example   Find the volume of the cone.

Step 1: Find the area of the base (B). Use the formula $A = \pi * r^2$.
- radius of base (r) = 3 in.
- area of base (B) = $\pi * 3$ in. $* 3$ in.

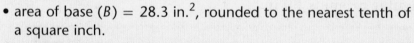

Use either the π key on a calculator or 3.14 as an approximate value for π.
- area of base (B) = 28.3 in.^{2}, rounded to the nearest tenth of a square inch.

Step 2: Find $\frac{1}{3}$ of the product of the area of the base multiplied by the height of the cone.

Use the formula $V = \frac{1}{3} * (B * h)$.
- area of base (B) = 28.3 in.^{2}
- height of cone (h) = 6 in.
- volume (V) = $\frac{1}{3} * (28.3$ in.$^2 * 6$ in.$) = 56.6$ in.^{3}

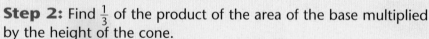

The volume of the cone is 56.6 in.^{3}.

Volume of a Rectangular or Triangular Pyramid

Volume of a Pyramid	Area of a Rectangle	Area of a Triangle
$V = \frac{1}{3} * (B * h)$	$A = b * h$	$A = \frac{1}{2} * (b * h)$
V is the volume, B is the area of the base, h is the height of the pyramid.	A is the area, b is the length of the base, h is the height of the rectangle.	A is the area, b is the length of the base, h is the height of the triangle.

Example   Find the volume of the rectangular pyramid.

Step 1: Find the area of the base (B). Use the formula $A = b * h$.
- length of base (b) = 4 cm
- height of base (h) = 2.5 cm
- area of base (B) = 4 cm * 2.5 cm = 10 cm^2

Step 2: Find $\frac{1}{3}$ of the product of the area of the base multiplied by the height of the pyramid. Use the formula $V = \frac{1}{3} * (B * h)$.
- area of base (B) = 10 cm^2
- height of pyramid (h) = 9 cm
- volume (V) = $\frac{1}{3} * (10$ cm$^2 * 9$ cm$) = 30$ cm^3

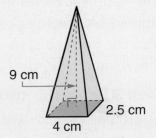
9 cm
2.5 cm
4 cm

The volume of the rectangular pyramid is 30 cm^3.

Example   Find the volume of the triangular pyramid.

Step 1: Find the area of the base (B). Use the formula $A = \frac{1}{2} * (b * h)$
- length of base (b) = 10 in.
- height of base (h) = 6 in.
- area of base (B) = $\frac{1}{2} * (10$ in. * 6 in.$) = 30$ in.^{2}

Step 2: Find $\frac{1}{3}$ of the product of the area of the base multiplied by the height of the pyramid. Use the formula $V = \frac{1}{3} * (B * h)$.
- area of base (B) = 30 in.^{2}
- height of pyramid (h) = $4\frac{1}{2}$ in.
- volume (V) = $\frac{1}{3} * (30$ in.$^2 * 4\frac{1}{2}$ in.$) = 45$ in.^{3}

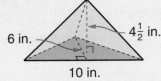
6 in.
$4\frac{1}{2}$ in.
10 in.

The volume of the triangular pyramid is 45 in.^{3}.

Check Your Understanding

Find the volume of each pyramid. Include the unit in each answer.

1.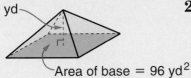
4 yd
Area of base = 96 yd^2

2.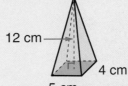
12 cm
4 cm
5 cm

3.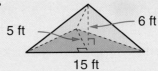
5 ft
6 ft
15 ft

Check your answers on page 440.

Surface Area of a Rectangular Prism

A rectangular prism has six flat surfaces called **faces.** The **surface area** of a rectangular prism is the sum of the areas of all six of its faces. Think of the six faces as three pairs of opposite, parallel faces. Since opposite faces have the same area, you find the surface area of one face in each pair of opposite faces. Then find the sum of these three areas and double the result.

The simplest rectangular prisms have all six of their faces shaped like rectangles. These prisms look like boxes. You can find the surface area of a box-like prism if you know its dimensions: length (l), width (w), and height (h).

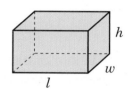

A box-like prism has all 6 faces shaped like rectangles.

Step 1: Find the area of one face in each pair of opposite faces.

area of base $= l * w$ area of front face $= l * h$ area of side face $= w * h$

Step 2: Find the sum of the areas of the three faces.

sum of areas $= (l * w) + (l * h) + (w * h)$

Step 3: Multiply the sum of the three areas by 2.

surface area of prism $= 2 * ((l * w) + (l * h) + (w * h))$

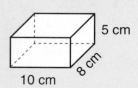

Surface Area of a Box-Like Rectangular Prism

$$S = 2 * ((l * w) + (l * h) + (w * h))$$

S is the surface area, l the length of the base, w the width of the base, h the height of the prism.

Example   Find the surface area of the box-like rectangular prism.

Use the formula $S = 2 * ((l * w) + (l * h) + (w * h))$.
- length (l) = 4 in.   width (w) = 3 in.   height (h) = 2 in.
- surface area (S) $= 2 * ((4 \text{ in.} * 3 \text{ in.}) + (4 \text{ in.} * 2 \text{ in.}) + (3 \text{ in.} * 2 \text{ in.}))$
   $= 2 * (12 \text{ in.}^2 + 8 \text{ in.}^2 + 6 \text{ in.}^2)$
   $= 2 * 26 \text{ in.}^2 = 52 \text{ in.}^2$

The surface area of the rectangular prism is 52 in.^{2}.

Check Your Understanding

Find the surface area of the box-like prism. Include the unit in your answer.

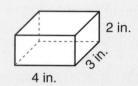

Check your answers on page 440.

Surface Area of a Cylinder

The simplest cylinders look like food cans and are called **right cylinders.** Their bases are perpendicular to the line joining the centers of the bases.

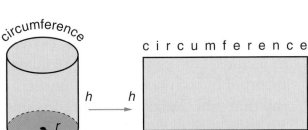

cylinder

To find the area of the curved surface of a right cylinder, imagine a soup can with a label. If you cut the label perpendicular to the top and bottom of the can, peel it off, and lay it flat on a surface, you will get a rectangle. The length of the rectangle is the same as the circumference of a base of the cylinder. The width of the rectangle is the same as the height of the can. Therefore, the area of the curved surface is the product of the circumference of the base and the height of the can.

circumference of base $= 2 * \pi * r$

area of curved surface $= (2 * \pi * r) * h$

The surface area of a cylinder is the sum of the areas of the two bases ($2 * \pi * r^2$) and the curved surface.

Surface Area of a Right Cylinder

$$S = (2 * \pi * r^2) + ((2 * \pi * r) * h)$$

S is the surface area, r is the radius of the base, h is the height of the cylinder.

Example   Find the surface area of the right cylinder.

Use the formula $S = (2 * \pi * r^2) + ((2 * \pi * r) * h)$.
- radius of base (r) = 3 cm
- height (h) = 5 cm

5 cm   3 cm

Use either the π key on a calculator or 3.14 as an approximate value for π.
- surface area (S) = $(2 * \pi * 3$ cm $* 3$ cm$) + ((2 * \pi * 3$ cm$) * 5$ cm$)$
  $= (\pi * 18$ cm$^2) + (\pi * 30$ cm$^2)$
  $= 150.8$ cm^2, rounded to the nearest tenth of a square centimeter

The surface area of the cylinder is 150.8 cm^2.

Check Your Understanding

Find the surface area of the right cylinder to the nearest tenth of a square inch. Include the unit in your answer.

2 in.   3 in.

Check your answers on page 440.

Weight

Today, in the United States, two different sets of standard units are used to measure weight.

♦ The standard unit for weight in the metric system is the **gram.** A small, plastic base-10 cube weighs about 1 gram. Heavier weights are measured in **kilograms.** One kilogram equals 1,000 grams.

♦ Two standard units for weight in the U.S. customary system are the **ounce** and the **pound.** Heavier weights are measured in pounds. One pound equals 16 ounces. Some weights are reported in both pounds and ounces. For example, we might say that "the suitcase weighs 14 pounds, 6 ounces."

Metric Units	U.S. Customary Units
1 gram (g) = 1,000 milligrams (mg)	1 pound (lb) = 16 ounces (oz)
1 milligram = $\frac{1}{1,000}$ gram	1 ounce = $\frac{1}{16}$ pound
1 kilogram (kg) = 1,000 grams	1 ton (t) = 2,000 pounds
1 gram = $\frac{1}{1,000}$ kilogram	1 pound = $\frac{1}{2,000}$ ton
1 metric ton (t) = 1,000 kilograms	
1 kilogram = $\frac{1}{1,000}$ metric ton	

Rules of Thumb	Exact Equivalents
1 ounce equals about 30 grams.	1 ounce = 28.35 grams
1 kilogram equals about 2 pounds.	1 kilogram = 2.205 pounds

Note

The Rules of Thumb table shows how units of weight in the metric system relate to units in the U.S. customary system. You can use this table to convert between ounces and grams, and between pounds and kilograms. For most everyday purposes, you need only remember the simple Rules of Thumb.

Example A bicycle weighs 17 kilograms. How many pounds is that?

Rough Solution: Use the Rule of Thumb. Since 1 kg equals about 2 lb, 17 kg equals about 17 * 2 lb = 34 lb.

Exact Solution: Use the exact equivalent. Since 1 kg = 2.205 lb, 17 kg = 17 * 2.205 lb = 37.485 lb.

Check Your Understanding

1. A softball weighs 6 ounces. How many grams is that? Use both a Rule of Thumb and an exact equivalent.

2. Andy's brother weighs 58 pounds, 9 ounces. How many ounces is that?

Check your answers on page 440.

Temperature

Temperature is a measure of the hotness or coldness of something. To read a temperature in degrees, you need a reference frame that begins with a zero point and has a number-line scale. The two most commonly used temperature scales, Fahrenheit and Celsius, have different zero points.

Fahrenheit

This scale was invented in the early 1700s by the German physicist G.D. Fahrenheit. On the Fahrenheit scale, pure water freezes at 32°F and boils at 212°F. A salt-water solution freezes at 0°F (the zero point) at sea level. The normal temperature for the human body is 98.6°F. The Fahrenheit scale is used primarily in the United States.

Celsius

This scale was developed in 1742 by the Swedish astronomer Anders Celsius. On the Celsius scale, the zero point (0 degrees Celsius or 0°C) is the freezing point of pure water. Pure water boils at 100°C. The Celsius scale divides the interval between these two points into 100 equal parts. For this reason, it is sometimes called the *centigrade* scale. The normal temperature for the human body is 37°C. The Celsius scale is the standard for most people outside of the United States and for scientists everywhere.

A **thermometer** measures temperature. The common thermometer is a glass tube that contains a liquid. When the temperature goes up, the liquid expands and moves up the tube. When the temperature goes down, the liquid shrinks and moves down the tube.

Here are two formulas for converting from degrees Fahrenheit (°F) to degrees Celsius (°C) and vice versa:

$$F = \frac{9}{5} * C + 32 \quad \text{and} \quad C = \frac{5}{9} * (F - 32).$$

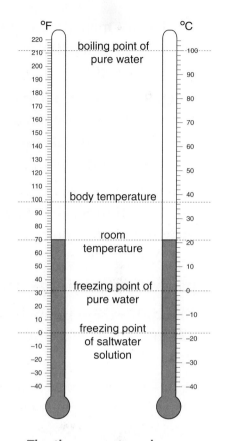

The thermometers show both the Fahrenheit and Celsius scales. Key reference temperatures, such as the boiling and freezing points of water, are indicated. A thermometer reading of 70°F (or about 21°C) is normal room temperature.

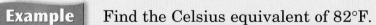

Example   Find the Celsius equivalent of 82°F.

Use the formula $C = \frac{5}{9} * (F - 32)$ and replace F with 82:

$C = \frac{5}{9} * (82 - 32) = \frac{5}{9} * (50) = 27.77$

The Celsius equivalent of 82°F is about 28°C.

Measuring and Drawing Angles

Angles are measured in **degrees.** When writing the measure of an angle, a small raised circle (°) is used as a symbol for the word *degree*.

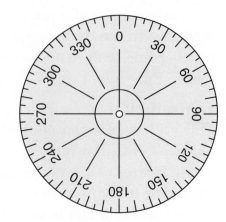

Angles are measured with a tool called a **protractor.** You will find both a full-circle and a half-circle protractor on your Geometry Template. Since there are 360 degrees in a circle, a 1° angle marks off $\frac{1}{360}$ of a circle.

The **full-circle protractor** on the Geometry Template is marked off in 5° intervals from 0° to 360°. It can be used to measure angles, but it cannot be used to draw angles.

Sometimes you will use a full-circle protractor that is a paper cutout. This *can* be used to draw angles.

The **half-circle protractor** on the Geometry Template is marked off in 1° intervals from 0° to 180°.

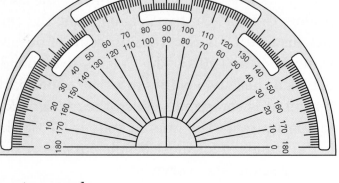

It has two scales, each of which starts at 0°. One scale is read clockwise, the other is read counterclockwise.

The half-circle protractor can be used both to measure angles and to draw angles.

Two rays starting from the same endpoint form two angles. The smaller angle measures between 0° and 180°. The larger angle is called a **reflex angle.** It measures between 180° and 360°. The sum of the measures of the smaller angle and the reflex angle is 360°.

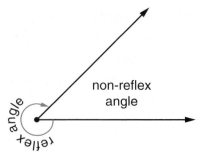

non-reflex angle

reflex angle

Measuring an Angle with a Full-Circle Protractor

Think of the angle as a rotation of the minute hand of a clock. One side of the angle represents the minute hand at the beginning of a time interval. The other side of the angle represents the minute hand some time later.

Example To measure angle *ABC* with a full-circle protractor:

Step 1: Place the center of the protractor over the vertex of the angle, point *B*.

Step 2: Line up the 0° mark on the protractor with $\overrightarrow{BA}$.

Step 3: Read the degree measure where $\overrightarrow{BC}$ crosses the edge of the protractor.

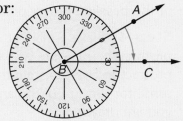

The measure of angle *ABC* = 30°.

Example To measure reflex angle *EFG:*

Step 1: Place the center of the protractor over point *F*.

Step 2: Line up the 0° mark on the protractor with $\overrightarrow{FG}$.

Step 3: Read the degree measure where $\overrightarrow{FE}$ crosses the edge of the protractor.

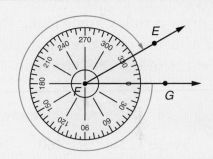

The measure of angle *EFG* = 330°.

Measuring an Angle with a Half-Circle Protractor

Example To measure angle *PQR* with a half-circle protractor:

Step 1: Lay the baseline of the protractor on $\overrightarrow{QR}$.

Step 2: Slide the protractor so that the center of the baseline is over the vertex of the angle, point *Q*.

Step 3: Read the degree measure where $\overrightarrow{QP}$ crosses the edge of the protractor. There are two scales on the protractor. Use the scale that makes sense for the size of the angle that you are measuring.

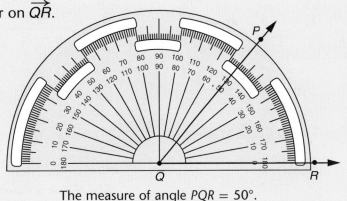

The measure of angle *PQR* = 50°.

Drawing an Angle with a Half-Circle Protractor

Example   Draw a 40° angle.

Step 1: Draw a ray from point *A*.

Step 2: Lay the baseline of the protractor on the ray.

Step 3: Slide the protractor so that the center of the baseline is over point *A*.

Step 4: Make a mark at 40° near the protractor. There are two scales on the protractor. Use the scale that makes sense for the size of the angle that you are drawing.

Step 5: Draw a ray from point *A* through the mark.

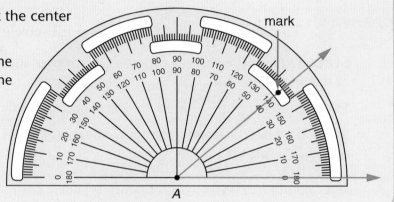

To draw a reflex angle using the half-circle protractor, subtract the measure of the reflex angle from 360°. Use this as the measure of the smaller angle.

Example   Draw a 240° angle.

Step 1: Subtract: 360° − 240° = 120°.

Step 2: Draw a 120° angle.

The larger angle is the reflex angle. It measures 240°.

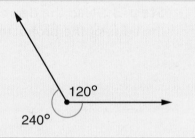

Check Your Understanding

Measure each angle to the nearest degree.

1.   **2.**   **3.**

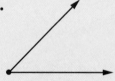  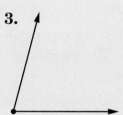

Draw each angle.

4. 70° angle   **5.** 280° angle   **6.** 55° angle

Check your answers on page 440.

The Measures of the Angles of Polygons

Any polygon can be divided into triangles.

♦ The measures of the three angles of each triangle add up to 180°.

♦ To find the sum of the measures of all the angles inside a polygon, multiply:
(the number of triangles inside the polygon) * 180°.

Example What is the sum of the measures of the angles of a hexagon?

Step 1: Draw any hexagon; then divide it into triangles. The hexagon can be divided into four triangles.

Step 2: Multiply the number of triangles by 180°.

4 * 180° = 720°

The sum of the measures of all the angles inside a hexagon equals 720°.

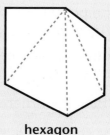

hexagon

Finding the Measure of an Angle of a Regular Polygon

All the angles of a regular polygon have the same measure. So the measure of one angle is equal to the sum of the measures of the angles of the polygon divided by the number of angles.

Example What is the measure of one angle of a regular hexagon?

The sum of the measures of the angles of any hexagon is 720°. A regular hexagon has 6 congruent angles.

Therefore, the measure of one angle of a regular hexagon is $\frac{720°}{6} = 120°$.

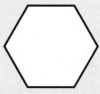

regular hexagon
(6 congruent sides and 6 congruent angles)

Check Your Understanding

1. Into how many triangles can you divide

    a. a quadrilateral?   **b.** a pentagon?   **c.** an octagon?   **d.** a 12-sided polygon?

2. What is the sum of the measures of the angles of a pentagon?

3. What is the measure of an angle of a regular octagon?

4. Suppose that you know the number of sides of a polygon. Without drawing a picture, how can you calculate the number of triangles into which it can be divided?

Check your answers on page 440.

Plotting Ordered Number Pairs

A **rectangular coordinate grid** is used to name points in the plane. It is made up of two number lines, called **axes,** that meet at right angles at their zero points. The point where the two lines meet is called the **origin.**

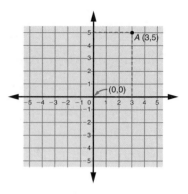

The ordered pair (0,0) names the origin.

Every point on a rectangular coordinate grid can be named by an **ordered number pair.** The two numbers that make up an ordered number pair are called the **coordinates** of the point. The first coordinate is always the *horizontal* distance of the point from the vertical axis. The second coordinate is always the *vertical* distance of the point from the horizontal axis. For example, the ordered pair (3,5) names point *A* on the grid at the right. The numbers 3 and 5 are the coordinates of point *A*.

Example Plot the ordered pair (5,3).

Step 1: Locate 5 on the horizontal axis. Draw a vertical line.

Step 2: Locate 3 on the vertical axis. Draw a horizontal line.

Step 3: The point (5,3) is located at the intersection of the two lines.

The order of the numbers in an ordered pair is important. The pair (5,3) does not name the same point as the pair (3,5).

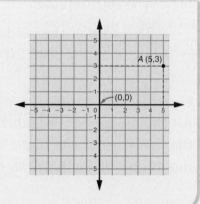

Example Locate $(-2,3)$, $(-4,-1)$, and $(3\frac{1}{2},0)$.

For each ordered pair:

Locate the first coordinate on the horizontal axis and draw a vertical line.

Locate the second coordinate on the vertical axis and draw a horizontal line.

The two lines intersect at the point named by the ordered pair.

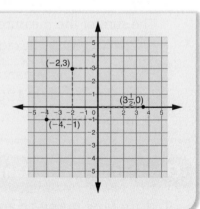

Check Your Understanding

Draw a coordinate grid on graph paper and plot the following points.

1. (2,4) **2.** $(-1,-3)$ **3.** (0,5) **4.** $(-2,2)$

Check your answers on page 440.

Latitude and Longitude

The Earth is almost a perfect **sphere.** All points on Earth are about the same distance from its center. The Earth rotates on an **axis,** which is an imaginary line through the center of the Earth connecting the **North Pole** and the **South Pole.**

Reference lines are drawn on globes and maps to make places easier to find. Lines that go east and west around the Earth are called **lines of latitude.** The **equator** is a special line of latitude. Every point on the equator is the same distance from the North Pole and the South Pole. The lines of latitude are often called **parallels** because each one is a circle that is parallel to the equator.

The **latitude** of a place is measured in **degrees.** The symbol for degrees is (°). Lines north of the equator are labeled °N (degrees north), lines south of the equator are labeled °S (degrees south). The number of degrees tells how far north or south of the equator a place is. The area north of the equator is called the **Northern Hemisphere.** The area south of the equator is called the **Southern Hemisphere.**

> **Did You Know?**
>
> In order to locate places more accurately, each degree is divided into 60 *minutes*. One minute equals $\frac{1}{60}$ degree. The symbol for minutes is (′). For example, 31°23′N means $31\frac{23}{60}$ degrees north.

Examples   The latitude of the North Pole is 90°N.
The latitude of the South Pole is 90°S.

The poles are the points farthest north and farthest south on Earth.

The latitude of Cairo, Egypt, is 30°N. We say that Cairo is 30 degrees north of the equator.

The latitude of Durban, South Africa, is 30°S. Durban is in the Southern Hemisphere.

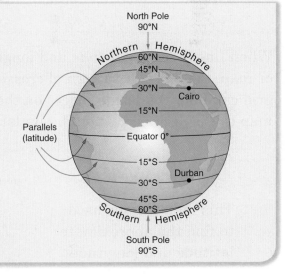

A second set of lines runs from north to south. These are semicircles (half-circles) that connect the poles. They are called **lines of longitude** or **meridians.** The meridians are not parallel, since they meet at the poles.

The **prime meridian** is the special meridian labeled 0°. The prime meridian passes through Greenwich, England (near London). Another special meridian is the **international date line.** This meridian is labeled 180° and is exactly opposite the prime meridian on the other side of the world.

The **longitude** of a place is measured in degrees. Lines west of the prime meridian are labeled °W. Lines east of the prime meridian are labeled °E. The number of degrees tells how far west or east of the prime meridian a place is. The area west of the prime meridian is called the **Western Hemisphere.** The area east of the prime meridian is called the **Eastern Hemisphere.**

Examples

The longitude of Greenwich, England is 0° because it lies on the prime meridian.

The longitude of Durban, South Africa, is 30°E. Durban is in the Eastern Hemisphere.

The longitude of Gambia (a small country in Africa) is about 15°W. We say that Gambia is 15 degrees west of the prime meridian.

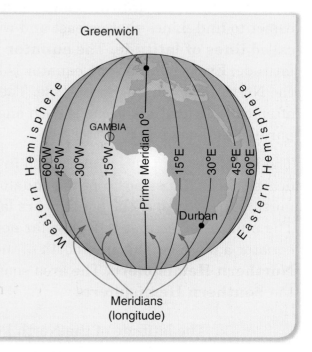

When lines of both latitude and longitude are shown on a globe or map, they form a pattern of crossing lines called a **grid.** The grid can help you locate places on the map. Any place on the map can be located by naming its latitude and longitude.

Examples

This map may be used to find the approximate latitude and longitude for the cities shown. For example, Denver, Colorado, is about 40° North and 105° West.

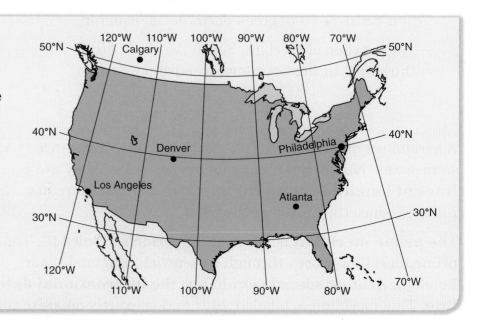

Map Scales and Distances

Map Scales

Mapmakers show large areas of land and water on small pieces of paper. Places that are actually thousands of miles apart may be only inches apart on a map. When you use a map, you can estimate real distances by using a **map scale.**

Different maps use different scales. On one map, 1 inch may represent 10 miles in the real world. On another map, 1 inch may represent 100 miles.

On this map scale, the bar is 2 inches long. Two inches on the map represent 2,000 real miles. One inch on the map represents 1,000 real miles.

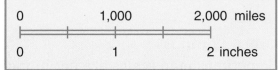

You may see a map scale written as "2 inches = 2,000 miles." This statement is not mathematically correct because 2 inches is not equal to 2,000 miles. What is meant is that a 2-inch distance on the map represents 2,000 miles in the real world.

Measuring Distances on a Map

There are many ways to measure distances on a map. Here are several.

Use a Ruler

Sometimes the distance you want to measure is along a straight line. Measure the straight-line distance with a ruler. Then use the map scale to change the map distance to the real distance.

Example Use this map and scale to find the air distance from Denver to Chicago. The air distance is the straight-line distance between the two cities.

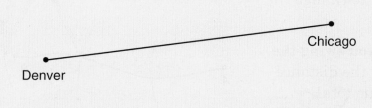
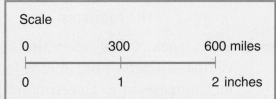

The line segment connecting Denver and Chicago is 3 inches long. The map scale shows that 1 inch represents 300 miles. So 3 inches must represent 3 * 300 miles, or 900 miles. The air distance from Denver to Chicago is 900 miles.

Use String and a Ruler

Sometimes you may need to find the length of a curved path, such as a road or river. You can use a piece of string, a ruler, and the map scale to find the length.

◆ Lay a piece of string along the path you want to measure. Mark the beginning and ending points on the string.

◆ Straighten out the string. Be careful not to stretch it. Use a ruler to measure between the beginning and ending points.

◆ Use the map scale to change the map distance into the real distance.

Use a Compass

Sometimes map scales are not in inches or centimeters, so a ruler is not much help. In these cases you can use a compass to find distances. Using a compass can also be easier than using a ruler, especially if you are measuring a curved path and you do not have any string.

Step 1: Adjust the compass so that the distance between the anchor point and the pencil point is the same as a distance on the map scale.

Step 2: Imagine a path connecting the starting point and ending point of the distance you want to measure. Put the anchor point of the compass at the starting point. Use the pencil point to make an arc on the path. Move the anchor point to the spot where the arc and the path meet. Continue moving the compass along the path and making arcs until you reach or pass the ending point. Be careful not to change the opening of the compass.

Step 3: Keep track of how many times you swing the compass. Each swing stands for the distance on the map scale. To estimate the total distance, multiply the number of swings by the distance each swing stands for.

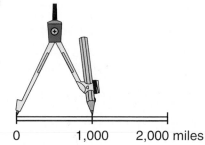

0   1,000   2,000 miles

The compass opening is set
to represent 1,000 miles.

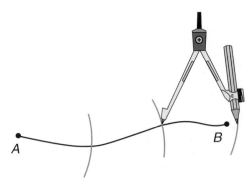

A

B

The real length of the curve
is about 3,000 miles.

If you use a compass to measure distance along a curved path, your estimate will be less than the actual distance. The distance along a straight line between two points is less than the distance along a curved path between the same two points.

Perpetual Calendar

The **perpetual calendar** consists of 14 different one-year calendars. It shows all the possible one-year calendars. The calendar for a year is determined by which day is January 1. There are 7 calendars for years with 365 days. There are another 7 calendars for years with 366 days.

Years that have 366 days are called **leap years.** They occur every four years. The extra day is added to February. Years that are divisible by 4 are leap years, except for years that are multiples of 100. Those years (1600, 1700, 1800, 1900, 2000, and so on) are leap years only if they are divisible by 400. The years 1600 and 2000 were leap years, but the years 1700, 1800, and 1900 were not leap years.

Calendar number to use for the years 1899 to 2028

Year	Cal	Year	Cal	Year	Cal	Year	Cal	Year	Cal
1899	1	1925	5	1951	2	1977	7	2003	4
1900	2	1926	6	1952	10	1978	1	2004	12
1901	3	1927	7	1953	5	1979	2	2005	7
1902	4	1928	8	1954	6	1980	10	2006	1
1903	5	1929	3	1955	7	1981	5	2007	2
1904	15	1930	4	1956	8	1982	6	2008	10
1905	1	1931	5	1957	3	1983	7	2009	5
1906	2	1932	13	1958	4	1984	8	2010	6
1907	3	1933	1	1959	5	1985	3	2011	7
1908	11	1934	2	1960	13	1986	4	2012	8
1909	6	1935	3	1961	1	1987	5	2013	3
1910	7	1936	11	1962	2	1988	13	2014	4
1911	1	1937	6	1963	3	1989	1	2015	5
1912	9	1938	7	1964	11	1990	2	2016	13
1913	4	1939	1	1965	6	1991	3	2017	1
1914	5	1940	9	1966	7	1992	11	2018	2
1915	6	1941	4	1967	1	1993	6	2019	3
1916	14	1942	5	1968	9	1994	7	2020	11
1917	2	1943	6	1969	4	1995	1	2021	6
1918	3	1944	14	1970	5	1996	9	2022	7
1919	4	1945	2	1971	6	1997	4	2023	1
1920	12	1946	3	1972	14	1998	5	2024	9
1921	7	1947	4	1973	2	1999	8	2025	4
1922	1	1948	12	1974	3	2000	14	2026	5
1923	2	1949	7	1975	4	2001	2	2027	6
1924	10	1950	1	1976	12	2002	3	2028	14

Calendar 1

JANUARY: 1 2 3 4 5 6 / 7 8 9 10 11 12 13 / 14 15 16 17 18 19 20 / 21 22 23 24 25 26 27 / 28 29 30 31
FEBRUARY: 1 2 3 / 4 5 6 7 8 9 10 / 11 12 13 14 15 16 17 / 18 19 20 21 22 23 24 / 25 26 27 28
MARCH: 1 2 3 4 / 5 6 7 8 9 10 11 / 12 13 14 15 16 17 18 / 19 20 21 22 23 24 25 / 26 27 28 29 30 31
APRIL: 1 / 2 3 4 5 6 7 8 / 9 10 11 12 13 14 15 / 16 17 18 19 20 21 22 / 23 24 25 26 27 28 29 / 30
MAY: 1 2 3 4 / 5 6 7 8 9 10 11 / 12 13 14 15 16 17 18 / 19 20 21 22 23 24 25 / 26 27 28 29 30 31
JUNE: 1 / 2 3 4 5 6 7 8 / 9 10 11 12 13 14 15 / 16 17 18 19 20 21 22 / 23 24 25 26 27 28 29 / 30
JULY: 1 / 2 3 4 5 6 7 8 / 9 10 11 12 13 14 15 / 16 17 18 19 20 21 22 / 23 24 25 26 27 28 29 / 30 31
AUGUST: 1 2 3 4 5 / 6 7 8 9 10 11 12 / 13 14 15 16 17 18 19 / 20 21 22 23 24 25 26 / 27 28 29 30 31
SEPTEMBER: 1 2 / 3 4 5 6 7 8 9 / 10 11 12 13 14 15 16 / 17 18 19 20 21 22 23 / 24 25 26 27 28 29 30
OCTOBER: 1 2 3 4 5 6 7 / 8 9 10 11 12 13 14 / 15 16 17 18 19 20 21 / 22 23 24 25 26 27 28 / 29 30 31
NOVEMBER: 1 2 3 4 / 5 6 7 8 9 10 11 / 12 13 14 15 16 17 18 / 19 20 21 22 23 24 25 / 26 27 28 29 30
DECEMBER: 1 2 / 3 4 5 6 7 8 9 / 10 11 12 13 14 15 16 / 17 18 19 20 21 22 23 / 24 25 26 27 28 29 30 / 31

Calendar 2

JANUARY: 1 2 / 3 4 5 6 7 8 9 / 10 11 12 13 14 15 16 / 17 18 19 20 21 22 23 / 24 25 26 27 28 29 30 / 31
FEBRUARY: 1 2 3 / 4 5 6 7 8 9 10 / 11 12 13 14 15 16 17 / 18 19 20 21 22 23 24 / 25 26 27 28
MARCH: 1 2 3 / 4 5 6 7 8 9 10 / 11 12 13 14 15 16 17 / 18 19 20 21 22 23 24 / 25 26 27 28 29 30 31
APRIL: 1 2 3 4 5 6 7 / 8 9 10 11 12 13 14 / 15 16 17 18 19 20 21 / 22 23 24 25 26 27 28 / 29 30
MAY: 1 2 3 4 5 / 6 7 8 9 10 11 12 / 13 14 15 16 17 18 19 / 20 21 22 23 24 25 26 / 27 28 29 30 31
JUNE: 1 2 / 3 4 5 6 7 8 9 / 10 11 12 13 14 15 16 / 17 18 19 20 21 22 23 / 24 25 26 27 28 29 30
JULY: 1 2 3 4 5 6 7 / 8 9 10 11 12 13 14 / 15 16 17 18 19 20 21 / 22 23 24 25 26 27 28 / 29 30 31
AUGUST: 1 2 3 4 / 5 6 7 8 9 10 11 / 12 13 14 15 16 17 18 / 19 20 21 22 23 24 25 / 26 27 28 29 30 31
SEPTEMBER: 1 / 2 3 4 5 6 7 8 / 9 10 11 12 13 14 15 / 16 17 18 19 20 21 22 / 23 24 25 26 27 28 29 / 30
OCTOBER: 1 2 3 4 5 6 / 7 8 9 10 11 12 13 / 14 15 16 17 18 19 20 / 21 22 23 24 25 26 27 / 28 29 30 31
NOVEMBER: 1 2 3 / 4 5 6 7 8 9 10 / 11 12 13 14 15 16 17 / 18 19 20 21 22 23 24 / 25 26 27 28 29 30
DECEMBER: 1 / 2 3 4 5 6 7 8 / 9 10 11 12 13 14 15 / 16 17 18 19 20 21 22 / 23 24 25 26 27 28 29 / 30 31

Calendar 3

JANUARY: 1 2 3 4 5 / 6 7 8 9 10 11 12 / 13 14 15 16 17 18 19 / 20 21 22 23 24 25 26 / 27 28 29 30 31
FEBRUARY: 1 2 / 3 4 5 6 7 8 9 / 10 11 12 13 14 15 16 / 17 18 19 20 21 22 23 / 24 25 26 27 28
MARCH: 1 2 / 3 4 5 6 7 8 9 / 10 11 12 13 14 15 16 / 17 18 19 20 21 22 23 / 24 25 26 27 28 29 30 / 31
APRIL: 1 2 3 4 5 6 / 7 8 9 10 11 12 13 / 14 15 16 17 18 19 20 / 21 22 23 24 25 26 27 / 28 29 30
MAY: 1 2 3 4 / 5 6 7 8 9 10 11 / 12 13 14 15 16 17 18 / 19 20 21 22 23 24 25 / 26 27 28 29 30 31
JUNE: 1 / 2 3 4 5 6 7 8 / 9 10 11 12 13 14 15 / 16 17 18 19 20 21 22 / 23 24 25 26 27 28 29 / 30
JULY: 1 2 3 4 5 6 / 7 8 9 10 11 12 13 / 14 15 16 17 18 19 20 / 21 22 23 24 25 26 27 / 28 29 30 31
AUGUST: 1 2 3 / 4 5 6 7 8 9 10 / 11 12 13 14 15 16 17 / 18 19 20 21 22 23 24 / 25 26 27 28 29 30 31
SEPTEMBER: 1 2 3 4 5 6 7 / 8 9 10 11 12 13 14 / 15 16 17 18 19 20 21 / 22 23 24 25 26 27 28 / 29 30
OCTOBER: 1 2 3 4 5 / 6 7 8 9 10 11 12 / 13 14 15 16 17 18 19 / 20 21 22 23 24 25 26 / 27 28 29 30 31
NOVEMBER: 1 2 / 3 4 5 6 7 8 9 / 10 11 12 13 14 15 16 / 17 18 19 20 21 22 23 / 24 25 26 27 28 29 30
DECEMBER: 1 2 3 4 5 6 7 / 8 9 10 11 12 13 14 / 15 16 17 18 19 20 21 / 22 23 24 25 26 27 28 / 29 30 31

Calendar 4

JANUARY: 1 2 3 4 / 5 6 7 8 9 10 11 / 12 13 14 15 16 17 18 / 19 20 21 22 23 24 25 / 26 27 28 29 30 31
FEBRUARY: 1 / 2 3 4 5 6 7 8 / 9 10 11 12 13 14 15 / 16 17 18 19 20 21 22 / 23 24 25 26 27 28
MARCH: 1 / 2 3 4 5 6 7 8 / 9 10 11 12 13 14 15 / 16 17 18 19 20 21 22 / 23 24 25 26 27 28 29 / 30 31
APRIL: 1 2 3 4 5 / 6 7 8 9 10 11 12 / 13 14 15 16 17 18 19 / 20 21 22 23 24 25 26 / 27 28 29 30
MAY: 1 2 3 / 4 5 6 7 8 9 10 / 11 12 13 14 15 16 17 / 18 19 20 21 22 23 24 / 25 26 27 28 29 30 31
JUNE: 1 2 3 4 5 6 7 / 8 9 10 11 12 13 14 / 15 16 17 18 19 20 21 / 22 23 24 25 26 27 28 / 29 30
JULY: 1 2 3 4 5 / 6 7 8 9 10 11 12 / 13 14 15 16 17 18 19 / 20 21 22 23 24 25 26 / 27 28 29 30 31
AUGUST: 1 2 / 3 4 5 6 7 8 9 / 10 11 12 13 14 15 16 / 17 18 19 20 21 22 23 / 24 25 26 27 28 29 30 / 31
SEPTEMBER: 1 2 3 4 5 6 / 7 8 9 10 11 12 13 / 14 15 16 17 18 19 20 / 21 22 23 24 25 26 27 / 28 29 30
OCTOBER: 1 2 3 4 / 5 6 7 8 9 10 11 / 12 13 14 15 16 17 18 / 19 20 21 22 23 24 25 / 26 27 28 29 30 31
NOVEMBER: 1 / 2 3 4 5 6 7 8 / 9 10 11 12 13 14 15 / 16 17 18 19 20 21 22 / 23 24 25 26 27 28 29 / 30
DECEMBER: 1 2 3 4 5 6 / 7 8 9 10 11 12 13 / 14 15 16 17 18 19 20 / 21 22 23 24 25 26 27 / 28 29 30 31

Calendar 5

JANUARY: 1 2 3 / 4 5 6 7 8 9 10 / 11 12 13 14 15 16 17 / 18 19 20 21 22 23 24 / 25 26 27 28 29 30 31
FEBRUARY: 1 2 3 4 5 6 7 / 8 9 10 11 12 13 14 / 15 16 17 18 19 20 21 / 22 23 24 25 26 27 28
MARCH: 1 2 3 4 5 6 7 / 8 9 10 11 12 13 14 / 15 16 17 18 19 20 21 / 22 23 24 25 26 27 28 / 29 30 31
APRIL: 1 2 3 4 / 5 6 7 8 9 10 11 / 12 13 14 15 16 17 18 / 19 20 21 22 23 24 25 / 26 27 28 29 30
MAY: 1 2 / 3 4 5 6 7 8 9 / 10 11 12 13 14 15 16 / 17 18 19 20 21 22 23 / 24 25 26 27 28 29 30 / 31
JUNE: 1 2 3 4 5 6 / 7 8 9 10 11 12 13 / 14 15 16 17 18 19 20 / 21 22 23 24 25 26 27 / 28 29 30
JULY: 1 2 3 4 / 5 6 7 8 9 10 11 / 12 13 14 15 16 17 18 / 19 20 21 22 23 24 25 / 26 27 28 29 30 31
AUGUST: 1 / 2 3 4 5 6 7 8 / 9 10 11 12 13 14 15 / 16 17 18 19 20 21 22 / 23 24 25 26 27 28 29 / 30 31
SEPTEMBER: 1 2 3 4 5 / 6 7 8 9 10 11 12 / 13 14 15 16 17 18 19 / 20 21 22 23 24 25 26 / 27 28 29 30
OCTOBER: 1 2 3 / 4 5 6 7 8 9 10 / 11 12 13 14 15 16 17 / 18 19 20 21 22 23 24 / 25 26 27 28 29 30 31
NOVEMBER: 1 2 3 4 5 6 7 / 8 9 10 11 12 13 14 / 15 16 17 18 19 20 21 / 22 23 24 25 26 27 28 / 29 30
DECEMBER: 1 2 3 4 5 / 6 7 8 9 10 11 12 / 13 14 15 16 17 18 19 / 20 21 22 23 24 25 26 / 27 28 29 30 31

6

JANUARY	MAY	SEPTEMBER
FEBRUARY	JUNE	OCTOBER
MARCH	JULY	NOVEMBER
APRIL	AUGUST	DECEMBER

7

JANUARY	MAY	SEPTEMBER
FEBRUARY	JUNE	OCTOBER
MARCH	JULY	NOVEMBER
APRIL	AUGUST	DECEMBER

8

JANUARY	MAY	SEPTEMBER
FEBRUARY	JUNE	OCTOBER
MARCH	JULY	NOVEMBER
APRIL	AUGUST	DECEMBER

9

JANUARY	MAY	SEPTEMBER
FEBRUARY	JUNE	OCTOBER
MARCH	JULY	NOVEMBER
APRIL	AUGUST	DECEMBER

10

JANUARY	MAY	SEPTEMBER
FEBRUARY	JUNE	OCTOBER
MARCH	JULY	NOVEMBER
APRIL	AUGUST	DECEMBER

11

JANUARY	MAY	SEPTEMBER
FEBRUARY	JUNE	OCTOBER
MARCH	JULY	NOVEMBER
APRIL	AUGUST	DECEMBER

12

JANUARY	MAY	SEPTEMBER
FEBRUARY	JUNE	OCTOBER
MARCH	JULY	NOVEMBER
APRIL	AUGUST	DECEMBER

13

JANUARY	MAY	SEPTEMBER
FEBRUARY	JUNE	OCTOBER
MARCH	JULY	NOVEMBER
APRIL	AUGUST	DECEMBER

14

JANUARY	MAY	SEPTEMBER
FEBRUARY	JUNE	OCTOBER
MARCH	JULY	NOVEMBER
APRIL	AUGUST	DECEMBER

Algebra

Algebra

Algebra is a type of arithmetic that uses letters (or other symbols such as blanks or question marks) as well as numbers. You use algebra when you write and solve number sentences such as $8 + n = 13$ or $y = x + 3$.

In early times, algebra involved solving number problems for which one or more of the numbers was not known. The objective was to find these missing numbers, called the "unknowns." Words were used for the unknowns, as in "Five plus some number equals eight." Then, in the late 1500s, François Viète began using letters, as in $5 + x = 8$, to stand for unknown quantities. Viète's invention made solving number problems much easier and led to many discoveries in mathematics and science.

Variables and Unknowns

The letters that you sometimes see in number sentences are called **variables.**

Variables Can Be Used to Stand for Unknown Numbers.

In the number sentence $5 + x = 8$, the variable x stands for an unknown number. To make the sentence true, the correct number for x must be found. Finding the correct number is called "solving the number sentence." Sometimes symbols such as question marks or blanks are used for unknown numbers.

Variables Can Be Used to State Properties of the Number System.

Properties of the number system are things that are true for all numbers. For example, any number multiplied by 1 is equal to itself. Variables are often used in statements that describe properties.

A portion of the Rhind Papyrus

Examples

Property	Number sentence example of the property
$1 * a = a$	$1 * 3.5 = 3.5$
$a + b = b + a$	$5 + 8 = 8 + 5$
$a * b = b * a$	$5 * 2 = 2 * 5$
$a = a$	$47.5 = 47.5$
$0 + a = a$	$0 + 5 = 5$

Variables Can Be Used in Formulas.

Formulas are used in everyday life, in science, in business, and in many other situations as an easy way to describe relationships. The formula for the area of a circle, for example, is $A = \pi * r^2$, where A is the area, r is the radius, and π is the number 3.1415.... The formula $A = \pi * r^2$ can also be written without a multiplication symbol: $A = \pi r^2$. Putting numbers and variables next to each other like this means they are to be multiplied. The formula for the circumference of a circle is $c = 2 * \pi * r$, or $c = 2\pi r$.

Note

π is not a variable; it is a number so important that it has been named after a letter in the Greek alphabet, *pi* or π.

Variables Can Be Used to Express Rules or Functions.

Function machines and "What's My Rule?" tables have rules that tell you how to get the "out" numbers from the "in" numbers. These rules can be written using variables. For example, a "What's My Rule?" table might have the rule, "Triple the 'in' number." This rule can be written as $y = 3 * x$ by using variables.

Rule

$y = 3 * x$

in	out
x	y
0	0
1	3
2	6
3	9
...	...

Variables Can Be Used in Computers and Calculators.

Variables are used in computer spreadsheets, which makes it possible to evaluate formulas quickly and efficiently. Computer programs are made up of a series of "commands" that contain variables. These commands look very much like number sentences that contain variables.

Certain calculators, especially graphing calculators, use variables to name calculator key functions.

Check Your Understanding

For each problem, write a number sentence using a letter for the unknown.

1. Half of some number equals 28.

2. Some number equals 15 times 4.

Find the circumference of the circles below. Use the formula $c = \pi * d$, where c is the circumference and d is the diameter. Use 3.14 for π.

3.
$d = 2$ cm

4.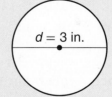
$d = 3$ in.

Check your answers on page 440.

Algebraic Expressions

Variables can be used to express relationships between quantities.

Example Claude earns $6 an hour. Use a variable to express the relationship between Claude's earnings and the amount of time worked.

If you use the variable H to stand for the number of hours Claude worked, you can write his pay as $H * 6$.

$H * 6$ is an example of an **algebraic expression.** An algebraic expression uses operation symbols $(+, -, *, /,$ and so on) to combine variables and numbers.

Example Write the statement as an algebraic expression.

Statement	Algebraic Expression
Marshall is 5 years older than Carol.	If Carol is C years old, then Marshall's age in years is $C + 5$.

Some algebraic expressions:

$2 - x$
$m * m$
$C + 5$
$6 * H$
$(C + 5) / (6 * H)$

Other expressions that are *not* algebraic:

$7 + 5$
$6 * 11$
$(7 + 5) / (6 * 11)$

Evaluating Expressions

To **evaluate** something is to find out what it is worth. To evaluate an algebraic expression, first replace each variable with its value.

Examples Evaluate each algebraic expression.

$6 * H$

If $H = 3$, then $6 * H$ is $6 * 3$, or 18.

$x * x * x$

If $x = 3$, then $x * x * x$ is $3 * 3 * 3$, or 27.

Check Your Understanding

Write an algebraic expression for each situation using the suggested variable.

1. Alan is A inches tall. If Barbara is 3 inches shorter than Alan, what is Barbara's height in inches?

2. Toni runs 2 miles every day. How many miles will she run in D days?

What is the value of each expression when $k = 4$?

3. $k + 2$ 　　　　4. $k * k$ 　　　　5. $k / 2$ 　　　　6. $k^2 + k - 2$

Check your answers on page 440.

Number Sentences

Number sentences are made up of **mathematical symbols.**

Mathematical Symbols				
Digits	**Variables**	**Operation Symbols**	**Relation Symbols**	**Grouping Symbols**
0, 1, 2, 3, 4, 5, 6, 7, 8, 9	$n\ x\ y\ z$ $a\ b\ c\ d$ $C\ M\ P\ ?\ \square$	$+\ -$ $\times\ *$ $/\ \div$	$=\ \neq$ $<\ >$ $\leq\ \geq$	$(\ )$ $[\ ]$

A number sentence must contain numbers (or variables) and a **relation symbol.** Number sentences that contain the $=$ symbol are **equations.** Number sentences that contain any one of the symbols $\neq$, $<$, $>$, $\leq$, or $\geq$ are **inequalities.**

A number sentence that does not contain a variable is either **true** or **false.** For example, the number sentence $10 + 3 = 13$ is true; the number sentence $8 < 5$ is false.

Some equations:

$4 - 3 = 1$

$5 * x = x$

$N = -7$

Some inequalities:

$C > 3.1$

$8 \neq 5$

Open Sentences

In some number sentences, one or more of the numbers may be missing. In place of each missing number there is a letter, a question mark, or some other symbol. These number sentences are called **open sentences.** A symbol used in place of a missing number is called a **variable.**

For most open sentences you can't tell whether the sentence is true or false until you know which number replaces the variable. For example, $9 + x = 15$ is an open sentence in which x stands for some number.

♦ If you replace x with 10, you get $9 + 10 = 15$, which is false.
♦ If you replace x with 6, you get $9 + 6 = 15$, which is true.

If a number used in place of a variable makes the number sentence true, that number is called a **solution** of the open sentence. For example, the number 6 is a solution of the open sentence $9 + x = 15$, because the number sentence $9 + 6 = 15$ is true. When you are being asked to **solve** a number sentence, you are being asked to find its solution(s).

Note

Some open sentences are always true. $5 + x = x + 5$ is true if you replace x with any number.

Some open sentences are always false. $N < (N - 1)$ is false if you replace N with any number.

Check Your Understanding

Solve.

1. $4 + y = 20$

2. $\frac{3}{8} = \frac{z}{16}$

3. $14 - m = 3$

Check your answers on page 440.

Relations

A **relation** tells how two things compare. The table below shows the most common relations that compare numbers and lists their symbols.

Symbol	Relation
=	is equal to
≠	is not equal to
<	is less than
>	is greater than
≤	is less than or equal to
≥	is greater than or equal to

Note

Reminder:
When writing > or <, be sure the arrow tip points to the smaller number.

Equations

A number has many different names. For example, the expressions 16, $4 * 4$, 4^2, and $1.6 * 10$ are different names for the same number (sixteen). Expressions that name the same number are called **equivalent names.** Expressions that name the same number are equal.

One way to state that two things are equal is to write a number sentence using the = symbol. Any two of the expressions 16, $4 * 4$, 4^2, and $1.6 * 10$ are equal because they all name the same number. So we can write many true number sentences using the = symbol: $4 * 4 = 1.6 * 10$, $4^2 = 16$, and so on.

Number sentences that contain the = symbol are called **equations.** An equation that does not contain a variable is either true or false.

Note

An equation that contains a variable is an open sentence. Any number that makes this sentence true when it is used in place of the variable is called a **solution** of the equation.

Examples Here are some equations:

$$5 + 8 = 13 \qquad (32 - 6) * 4 = 68 \qquad 58 = 58$$

The first and third equations above are true. The second equation is false.

Check Your Understanding

True or false?

1. $17 + 4 = 21$ **2.** $96 = 7 * 12$ **3.** $50 - 23 = 3 * 9$

4. $1{,}492 = 1{,}492$ **5.** $60 / 5 = 2 * 7$ **6.** $24 = 15 + 12 - 3$

Check your answers on page 440.

Inequalities

Number sentences that do not contain the = symbol are called **inequalities.** Like equations, inequalities may be true or false.

Note

Most inequalities that contain a variable (like $x > 7$ and $3 * N < 12$) are neither true nor false. You usually can't tell whether an inequality like this is true or false until you know which number replaces the variable.

Examples Here are some inequalities:

$$5 + 6 < 15 \quad | \quad 25 > 12 * 3 \quad | \quad 36 \neq 7 * 6$$

The first and third inequalities above are true; the second inequality is false.

The symbols $\leq$ and $\geq$ combine two meanings. $\leq$ means "is less than or equal to"; $\geq$ means "is greater than or equal to."

Examples Here are some other inequalities:

$5 \leq 5$ True	$300 \geq 350$ False	$5 + 8 \geq 10$ True
$35 \geq 40 + 5$ False	$60 \leq 100 - 25$ True	$40 - 5 \leq 35$ True

Inequalities on the Number Line

For any pair of numbers on the number line, the number to the left is less than the number to the right.

Examples Use the number line to complete each statement. $-5 \;\square\; 2 \quad 3 \;\square\; -4$

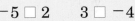

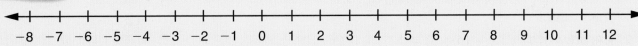

-5 is to the left of 2. So, -5 is less than 2. $(-5 < 2)$

3 is to the right of -4. So, 3 is greater than -4. $(3 > -4)$

Check Your Understanding

True or false?

1. $-10 \geq 0$ **2.** $4 \leq 2 * 2 * 2$ **3.** $-20 \leq -50$

Compare. Use =, <, or > to make each number sentence true.

4. $-30 \;\square\; 10$ **5.** $\frac{1}{8} \;\square\; 0.125$ **6.** $-9 \;\square\; -10$

Check your answers on page 440.

Algebra

Parentheses

The meaning of a number sentence is not always clear. You may not know which operation to do first. For example, solving $17 - 4 * 3 = n$ requires subtraction and multiplication.

◆ Should you subtract 4 from 17 first, and then multiply the result by 3? This will give 39 as the answer.

◆ Or should you multiply 4 and 3 first, and then subtract the result from 17? This will give 5 as the answer.

You can use parentheses in a number sentence to make the meaning clear. When there are parentheses in a number sentence or expression, **the operations inside the parentheses are always done first.**

Example   Solve. $(17 - 4) * 3 = n$

The parentheses tell you to subtract $17 - 4$ first.	$(17 - 4) * 3$
Then multiply by 3.	$13 * 3$
The answer is 39. $(17 - 4) * 3 = 39$	39

So $n = 39$.

Example   Solve. $17 - (4 * 3) = n$

The parentheses tell you to multiply $4 * 3$ first.	$17 - (4 * 3)$
Then subtract.	$17 - 12$
The answer is 5. $17 - (4 * 3) = 5$	5

So $n = 5$.

Check Your Understanding

Solve.

1. $(5 * 5) + 20 = x$ 2. $(100 - 70) * 20 = y$
3. $w = (17 - 12) + (5 * 6)$ 4. $n = (12 - 4) * 7$

Insert parentheses to make each number sentence true.

5. $25 - 15 + 10 = 0$ 6. $100 = 10 * 9 + 1$
7. $5 = 3 + 6 * 3 / 3 * 3$ 8. $26 = 7 + 6 * 2$

Check your answers on page 440.

Order of Operations

In arithmetic and algebra, there are rules that tell you what to do first and what to do next. Without these rules, it may be hard to tell what the solution of a problem should be. For example, what is the answer to $8 + 4 * 3$? Is the answer 36 or 20? You must know whether to multiply first or to add first.

Rules for the Order of Operations

1. Do operations inside **parentheses** first.
   Follow rules 2–4 when you are computing inside parentheses.
2. Calculate all expressions with **exponents.**
3. **Multiply** and **divide** in order, from left to right.
4. **Add** and **subtract** in order, from left to right.

Did You Know?

The rules for order of operations described here have been widely used since the late 1500s, when letters were first used to stand for unknown numbers.

Some people find it's easier to remember the order of operations by memorizing this sentence:

Please Excuse My Dear Aunt Sally.

Parentheses Exponents Multiplication Division Addition Subtraction

Example Evaluate. $17 - 4 * 3 = ?$

Multiply first.	$17 - 4 * 3$
Then subtract.	$17 - 12$
The answer is 5.	5

$17 - 4 * 3 = 5$

Example Evaluate $5^2 + (3 * 4 - 2) / 5$.

Clear parentheses first.	$5^2 + (3 * 4 - 2) / 5$
Calculate exponents next.	$5^2 + 10 / 5$
Divide.	$25 + 10 / 5$
Then add.	$25 + 2$
The answer is 27.	27

$5^2 + (3 * 4 - 2) / 5 = 27$

Check Your Understanding

Evaluate each expression.

1. $15 - 6 / 2 + 1$
2. $22 + (10 + 5) / 5$
3. $5 * (3 / 3 - 2 / 2) / 8 + 1$

Check your answers on page 440.

Some Properties of Arithmetic

Certain facts are true of all numbers. Some of them are obvious—"every number equals itself," for example—but others are less obvious. Since you have been working with numbers for years, you probably already know most of these facts, or properties. But you probably don't know their mathematical names.

The Identity Properties

The sum of any number and 0 is that number. For example, $15 + 0 = 15$. The **identity for addition** is 0. Using variables, you write this as $a + 0 = a$, where a is any number.

$$a + 0 = a$$
$$0 + a = a$$

The product of any number and 1 is that number. For example, $75 * 1 = 75$. The **identity for multiplication** is 1. Using variables, you write this as $a * 1 = a$, where a is any number.

$$a * 1 = a$$
$$1 * a = a$$

The Commutative Properties

When two numbers are added, the order of the numbers makes no difference. For example, $8 + 5 = 5 + 8$. This is known as the **commutative property of addition.** Using variables, you write this as $a + b = b + a$, where a and b are any numbers.

$$a + b = b + a$$

When two numbers are multiplied, the order of the numbers also makes no difference. For example, $7 * 2 = 2 * 7$. This is known as the **commutative property of multiplication.** Using variables, you write this as $a * b = b * a$, where a and b are any numbers.

$$a * b = b * a$$

The Associative Properties

When three numbers are added, it makes no difference which two are added first. For example, $(3 + 4) + 5 = 3 + (4 + 5)$. This is known as the **associative property of addition.** Using variables, you write this as $(a + b) + c = a + (b + c)$, where a, b, and c are any numbers.

$$(a + b) + c = a + (b + c)$$

When three numbers are multiplied, it makes no difference which two are multiplied first. For example, $(3 * 4) * 5 = 3 * (4 * 5)$. This is known as the **associative property of multiplication.** Using variables, you write this as $(a * b) * c = a * (b * c)$, where a, b, and c are any numbers.

$$(a * b) * c = a * (b * c)$$

The Distributive Property

When you play *Multiplication Wrestling* or multiply with the partial-products method, you use the **distributive property.**

$$a * (b + c) = (a * b) + (a * c)$$

For example, when you solve $6 * 58$ with partial products, you think of 58 as $50 + 8$ and multiply each part by 6.

The distributive property says: $6 * (50 + 8) = (6 * 50) + (6 * 8)$.

$$
\begin{array}{r}
58 \\
* \quad 6 \\
\hline
\end{array}
$$

$6 * 50 = \quad 300$
$6 * 8 = + \; 48$
$\overline{6 * 58 = \quad 348}$

Example   Show how the distributive property works by finding the area of Rectangle A in two different ways.

Method 1 Find the total width $(3 + 4)$ of the rectangle, and multiply that by the height (5).

$5 * (3 + 4) = 5 * 7$
$= 35$

Method 2 Find the area of each smaller rectangle, and then add these areas.

$(5 * 3) + (5 * 4) = 15 + 20$
$= 35$

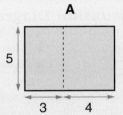

Both methods show that the area of Rectangle A is 35 square units.
$5 * (3 + 4) = (5 * 3) + (5 * 4)$

The distributive property works with subtraction too.

$$a * (b - c) = (a * b) - (a * c)$$

Example   Find the area of the shaded rectangle.

$5 * (7 - 3) = 5 * 4 = 20$, or
$(5 * 7) - (5 * 3) = 35 - 15 = 20$

Both methods show that the area of the shaded rectangle is 20 square units.

$5 * (7 - 3) = (5 * 7) - (5 * 3)$

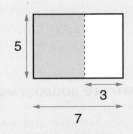

Check Your Understanding

Use the distributive property to fill in the blanks.

1. $8 * (15 + 6) = (8 * \underline{\quad}) + (8 * \underline{\quad})$ **2.** $(5 * 41) + (5 * 11) = 5 * (\underline{\quad} + \underline{\quad})$

3. $16 * (\underline{\quad} - \underline{\quad}) = (16 * 10) - (16 * 8)$

Check your answers on page 440.

Mathematical Models

A good way to learn about something is to work with a model of it. For example, a scale model of the human body can help you understand how the different systems in your own body work together.

Models are important in mathematics, too. A mathematical model can be as simple as acting out a problem with chips or blocks. Other mathematical models use drawings or diagrams. Mathematical models can help you understand and solve problems.

Situation Diagrams

Examples Here are some examples of how you can use diagrams to model simple problems.

Problem	Diagram
Parts-and-Total Situation Kaitlin's class has 19 girls and 12 boys. How many students are there in all?	Total ? Part 19 — Part 12
Change Situation Jonathan had $20 and spent $12.89 on a CD. How much money did he have left?	Change Start $20.00 — −$12.89 — End ?
Comparison Situation The average summer high temperature in Cairo, Egypt is 95°F. The average summer high temperature in Reykjavik, Iceland is 56°F. How much warmer is it in Cairo than in Reykjavik?	Quantity 95°F Quantity 56°F ? Difference
Rate Situation Mitch bought 4 packages of pencils. There were 12 pencils in each package. How many pencils did Mitch buy?	packages: 4 | pencils per package: 12 | pencils in all: ?

The diagrams on page 226 work well for some problems, but for others you need to use more powerful tools such as graphs, tables, and number models.

Number Models

Number sentences and expressions provide another way to model situations. A number sentence or an expression that describes some situation is called a **number model.** Often, two or more number models can fit a given situation.

Note

Every number sentence contains two expressions.

$17 + 13 = (2 * C) + 5$ contains the expressions $17 + 13$ and $(2 * C) + 5$.

Examples   Write number models that describe each problem situation.

Problem	Number Models	
	number sentence	expression
Kaitlin's class has 19 girls and 12 boys. How many students are there in all?	$19 + 12 = n$	$19 + 12$
Jonathan had $20 and spent $12.89 on a CD. How much money did he have left?	$r = \$20 - \$12.89$ or $\$20 = \$12.89 + r$	$\$20 - \$12.89$
The average summer high temperature in Cairo, Egypt is 95°F. The average summer high temperature in Reykjavik, Iceland is 56°F. How much warmer is it in Cairo than in Reykjavik?	$d = 95°F - 56°F$ or $95°F = 56°F + d$	$95°F - 56°F$
Mitch bought 4 packages of pencils. There were 12 pencils in each package. How many pencils did Mitch buy?	$4 * 12 = n$	$4 * 12$

Number models can help you solve problems. For example, the number sentence $\$20 = \$12.89 + r$ suggests counting up to find Jonathan's change from buying a $12.89 CD with a $20 bill.

Number models can also help you show the answer after you have solved the problem: $\$20 = \$12.89 + \$7.11$.

Check Your Understanding

Draw a diagram and write a number model for each problem. Then solve each problem.

1. Becky had $9.50. She wanted to buy a CD that cost $12.95. How much more did she need?

2. Dr. O'Malley's class is going on a field trip. The cost is $4.50 for each of 26 students. What is the total cost?

Check your answers on page 441.

Pan-Balance Problems and Equations

If two different kinds of objects are placed in the pans of a balance so that the pans balance, then you can find the weight of one kind of object in terms of the other kind of object.

Pan-Balance Problems

When you solve a pan-balance problem, the pans must balance after each step. If you *always do the same thing to the objects in both pans,* then the pans will remain balanced. For example, you can remove the same number of the same kind of object from each pan. If the pans balanced before you removed the objects, they will remain balanced after you remove them.

Did You Know ?

An *equation* is a number sentence that contains an = symbol. The = symbol was first used by Robert Recorde in 1557. He justified using 2 parallel line segments "…because noe 2 thynges can be moare equalle."

Example How many paper clips balance 1 pen?

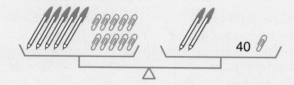

Step 1: Removing 10 paper clips from each pan will keep the pans balanced.

Step 2: Removing 2 pens from each pan will keep the pans balanced.

Step 3: Removing $\frac{2}{3}$ of the objects from each pan will keep the pans balanced. ($\frac{2}{3}$ of 3 pens is 2 pens; $\frac{2}{3}$ of 30 paper clips is 20 paper clips.)

1 pen weighs the same as 10 paper clips.

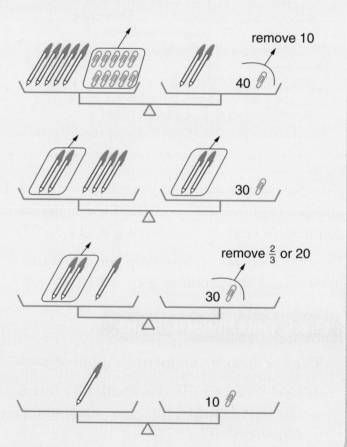

Pan-Balance Equations

You can think of equations as models for pan balance problems. The example on page 228 can be modeled by the equation $5P + 10C = 2P + 40C$. (C stands for the weight of one paper clip; P stands for the weight of one pen.)

| 5P | + | 10C | | 2P | + | 40C |

Step 1: Removing 10 paper clips from each pan will keep the pans balanced.

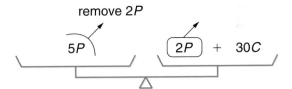

Step 2: Removing 2 pens from each pan will keep the pans balanced.

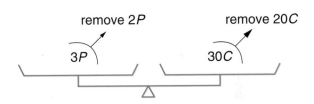

Step 3: Removing $\frac{2}{3}$ of the objects from each pan will keep the pans balanced.

Thus, 1 pen weighs the same as 10 paper clips.

So, $P = 10C$.

| P | | 10C |

Check Your Understanding

1. One cube weighs the same as _____ marbles.

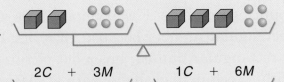

2. One cube weighs the same as _____ marbles.

$2C + 3M$ $1C + 6M$

Check your answers on page 441.

Number Patterns

You can use dot pictures to explore number patterns. The dot pictures below are for *counting numbers* (1, 2, 3, and so on).

Even Numbers

Even numbers are counting numbers that have a remainder of 0 when they are divided by 2. An even number has a dot picture with 2 equal rows.

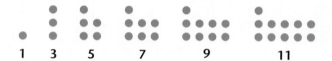

Odd Numbers

Odd numbers are counting numbers that have a remainder of 1 when they are divided by 2. An odd number has a dot picture with 2 equal rows plus 1 extra dot.

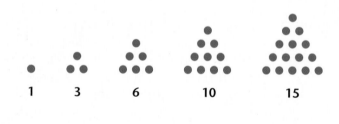

Triangular Numbers

Each of these dot pictures has a triangular shape, with the same number of dots on each side. Each row has 1 more dot than the row above it. Any number that has a dot picture like one of these is called a **triangular number.**

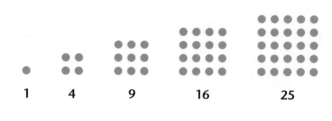

Square Numbers

A **square number** is the product of a counting number multiplied by itself. For example, 16 is a square number because 16 equals 4 * 4 or 4^2. A square number has a dot picture with a square shape, with the same number of dots in each row and column.

Rectangular Numbers

A **rectangular number** is a counting number that is the product of 2 smaller counting numbers. For example, 12 is a rectangular number because 12 = 3 * 4. A rectangular number has a dot picture with a rectangular shape, with at least 2 rows and at least 2 columns.

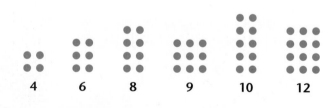

Function Machines and "What's My Rule?" Problems

A **function machine** in Everyday Mathematics is an imaginary machine that takes numbers in, uses a rule to change those numbers, and gives numbers out.

Here is a function machine with the rule "* 10 + 1." This machine will multiply any number put into it by 10, add 1, then give out the result.

If you put 3 into this "* 10 + 1" machine, it will multiply 3 * 10 and then add 1. The number 31 will come out. If you put 60 into this machine, it will multiply 60 * 10 and then add 1. The number 601 will come out.

If you put n, an unknown number, into the machine, it will multiply $n * 10$ and then add 1. The number $n * 10 + 1$ will come out.

To keep track of what goes in and what comes out, you can organize the "in" and "out" numbers in a table.

In previous grades, you solved many problems with function machines. You had to find the "out" numbers, or the "in" numbers, or a rule that fit the given "in" and "out" numbers. In *Everyday Mathematics,* these are called "What's My Rule?" problems.

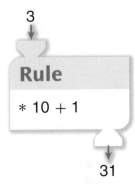

3

Rule

* 10 + 1

31

in	out
3	31
60	601
...	...
n	$n * 10 + 1$

Example Find the "out" numbers.

Rule

Add 7

The rule is: add 7 to each "in" number. Each "out" number must be 7 more than the "in" number.

If the variable n stands for an unknown number that is put into the machine, then the "out" number will be $n + 7$. (See the table.)

2 went in, so 2 + 7, or 9, came out.
8 went in, so 8 + 7, or 15, came out.
22 went in, so 22 + 7, or 29, came out.
50 went in, so 50 + 7, or 57, came out.

in	out
n	$n + 7$
2	
8	
22	
50	

Example   Find the "in" numbers.

Rule	The rule is to subtract 10 from the "in" numbers. So the "in" numbers must be 10 more than the "out" numbers.
Subtract 10	

in	out
r	r − 10
	2
	0
	−1

2 came out, so 10 + 2, or 12, went in.

0 came out, so 10 + 0, or 10, went in.

−1 came out, so 10 + (−1), or 9, went in.

Example   Use the table to find the rule.

Rule
?

When you have a table of "in" and "out" numbers, there are usually several different rules that will give those same "in" and "out" numbers. But it is sometimes hard to find any rule that works.

One rule that works for the table shown here is "Double and subtract 1."

in	out
1	1
2	3
3	5

Check Your Understanding

Copy and complete.

1.

Rule
Double and add 1

in	out
v	2 * v + 1
0	
1	
2	

2.

Rule
Multiply by 5

in	out
x	5x
	25
	45
	100

3.

Rule
?

in	out
10	5
20	10
2	1
100	50

Check your answers on page 441.

Rules, Tables, and Graphs

Relationships between variables can be shown by rules, by tables, or by graphs.

Did You Know?

The circumference of your head increases as you get older (up to age 18). The relationship between head size and age is hard to describe with a rule. The relationship is best shown by a table or graph.

Example   Lauren earns $4 per hour. Use a rule, a table, and a graph to show the relationship between how many hours Lauren works and how much she earns.

Rule: Lauren's earnings equal $4 times the number of hours she works.

Table: Make an in/out table for a function machine with the rule "*4."

Time (hours)	Earnings ($)
h	$h * 4$
0	0
1	4
2	8
3	12
4	16
...	...

Age (years)	Average Head Size (cm)
0	34
0.5	43
1	46
2	48.5
4	50
6	51
8	52
10	53
15	55
18	55.5

Graph: To draw a graph, plot the number pairs from the table: (0,0), (1,4), (2,8), (3,12), and (4,16). Plot each number pair as a point on the coordinate grid.

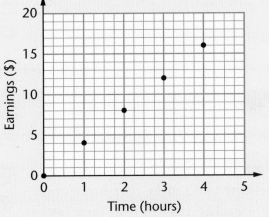

The points on the grid can be connected by a straight line. Draw this line to complete the graph.

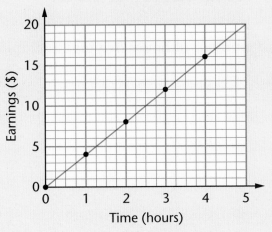

Average Head Size

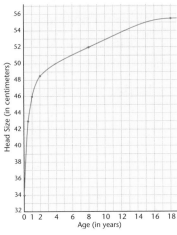

Example Use the table, the graph, and the rule to find how much Lauren earns if she works $3\frac{1}{2}$ hours.

One way to use the *table* to find Lauren's earnings is to think of $3\frac{1}{2}$ hours as 3 hours + $\frac{1}{2}$ hour. For 3 hours, Lauren earns $12. For $\frac{1}{2}$ hour, Lauren earns half of $4, or $2. In all, Lauren earns $12 + $2 = $14.

Another way to use the table is to note that $3\frac{1}{2}$ hours is halfway between 3 hours and 4 hours. So Lauren's earnings are halfway between $12 and $16, which is $14.

To use the *graph*, first find $3\frac{1}{2}$ hours on the horizontal axis. Then go straight up to the graph line. Turn left and go across to the vertical axis. You will end up at the same answer, $14, as you did when you used the table.

You can also use the *rule* to find Lauren's earnings: earnings = $4 * number of hours worked = $4 * $3\frac{1}{2}$ = $14.

Lauren earns $14 in $3\frac{1}{2}$ hours. The rule, table, and graph all give the same answer.

Time (hours)	Earnings ($)
h	*h* * 4
0	0
1	4
2	8
3	12
4	16
...	...

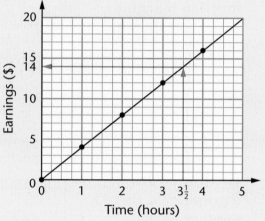

Check Your Understanding

1. Bakery cookies cost $4.00 a pound. Use the graph at the right to find the cost of $2\frac{1}{2}$ pounds.

2. Jim's average driving speed was 60 miles per hour on his trip to the mountains. Use the rule to find how far Jim drove in 6 hours.

Rule: Distance = 60 miles * number of hours driving

Check your answers on page 441.

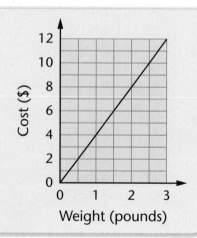

Polynesian Navigation

Because of their knowledge of the sea and their impressive ocean voyages, Polynesians have been compared to the Vikings. However, the Polynesians' journeys covered a much greater area of ocean over a much longer period of time. Historians think that as far back as 2500 B.C., they started migrating from Southeast Asia. By the eighth century A.D., they had repeatedly crossed thousands of miles of uncharted Pacific Ocean.

Ancient Navigators

◄ The navigators of these huge Polynesian dugout canoes were taught from childhood to find their way by using the positions of celestial bodies, ocean currents, waves, winds, and the habits of birds and other animals.

Using these navigation methods, the Polynesians were able to settle every habitable island in a huge triangle, with New Zealand on the southwest vertex, Easter Island on the southeast, and Hawaii on the north. ►

Recreating the Voyages

People have often wondered how the Polynesians were able to find tiny islands in a vast ocean. In 1976, the members of the Polynesian Voyaging Society, based in Hawaii, set out to find the answer. They built a replica of a Polynesian dugout canoe and have used traditional navigation methods to sail it several times from Hawaii to Tahiti, a distance of 2,800 miles.

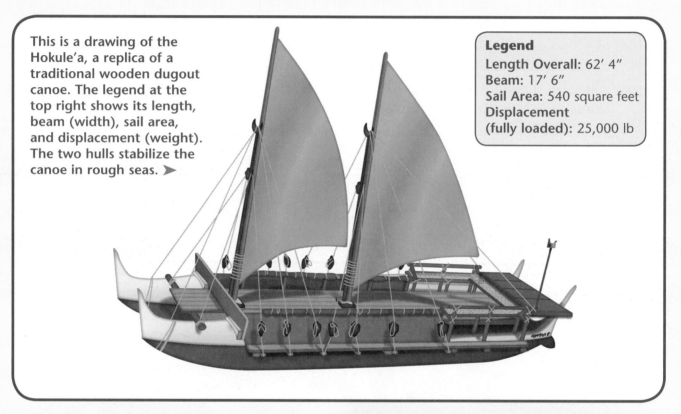

This is a drawing of the Hokule'a, a replica of a traditional wooden dugout canoe. The legend at the top right shows its length, beam (width), sail area, and displacement (weight). The two hulls stabilize the canoe in rough seas. ➤

Legend
Length Overall: 62' 4"
Beam: 17' 6"
Sail Area: 540 square feet
Displacement
(fully loaded): 25,000 lb

Here is the Hokule'a under sail. ▼

▲ Like the ancient Polynesians, navigators of the Hokule'a rely on many clues to help them find their way. For example, the position of the sun provides a directional point to steer by. The exact position of the sun changes each day, so navigators memorize and keep track of its changing location.

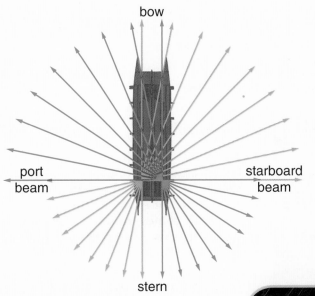

bow

port
beam

starboard
beam

stern

◀ To hold the course steady, navigators align the rising or setting sun to marks on the railings of the canoe. There are 8 marks on each side of the canoe, each paired with a single point at the stern (back).

At night, they use the stars to steer by. They know about 220 stars by name and have memorized their paths. The navigators keep a steady course by lining up the canoe to the rising and setting points of these stars. ▶

◀ Navigators know that the moon orbits the earth every 29.5 days. It also rises each night about 48 minutes later at a different point. By tracking the location of the rising and setting moon, they can use it to steer by.

Winds, Currents, and Swells

During the middle of the day and on cloudy days and nights when celestial bodies cannot be seen, navigators often rely on wind, current, and swells to estimate the canoe's speed and direction.

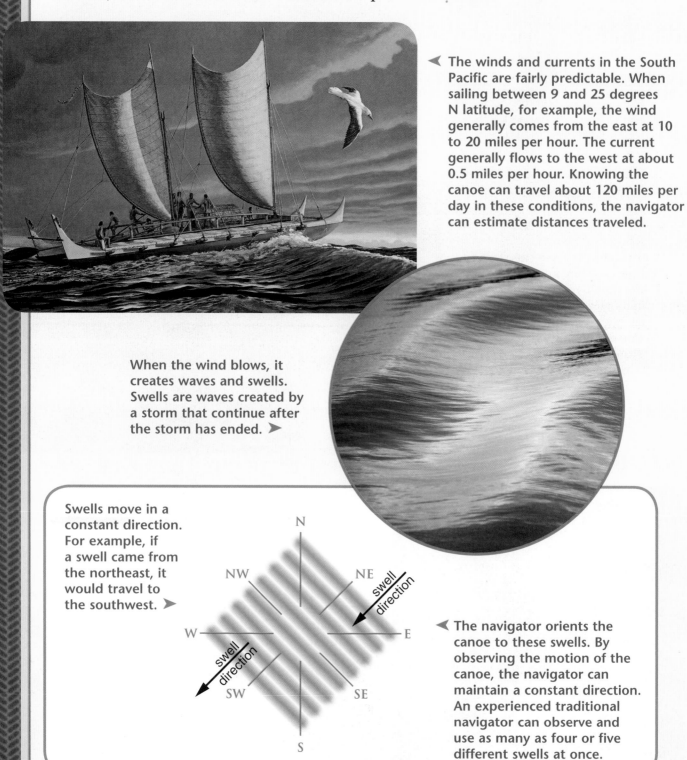

◄ The winds and currents in the South Pacific are fairly predictable. When sailing between 9 and 25 degrees N latitude, for example, the wind generally comes from the east at 10 to 20 miles per hour. The current generally flows to the west at about 0.5 miles per hour. Knowing the canoe can travel about 120 miles per day in these conditions, the navigator can estimate distances traveled.

When the wind blows, it creates waves and swells. Swells are waves created by a storm that continue after the storm has ended. ➤

Swells move in a constant direction. For example, if a swell came from the northeast, it would travel to the southwest. ➤

◄ The navigator orients the canoe to these swells. By observing the motion of the canoe, the navigator can maintain a constant direction. An experienced traditional navigator can observe and use as many as four or five different swells at once.

Seamarks and Signs of Land

When traveling on land, we often use landmarks to find our way. On the sea, navigators use mid-ocean clues, or **seamarks,** to check their canoe's daily position. Seamarks include schools of fish, flocks of birds, groups of driftwood, and the conditions of waves and the sky. Hundreds of seamarks have been discovered and passed on from generation to generation of navigators.

◄ Navigators of the Hokule'a have used a school of porpoises on their voyages from Hawaii to Tahiti. Finding this seamark indicates they have reached a point around 9° N latitude.

As the navigators get closer to their destination according to their estimation of course and distance, they start looking for land. ➤

◄ Islands in the Pacific are often in 300 to 400-mile-wide clusters. When navigators reach an island in a cluster, they identify it. Then they adjust their course to their destination.

Seabirds

Some seabirds go out to sea in the morning to feed on fish and return to land at night to rest. In the morning, navigators can sail in the direction the birds are coming from to find land. In the later afternoon, navigators can follow the birds as they return to land.

The white, or fairy, tern is a reliable indicator of nearby land. These birds venture up to 120 miles from their island nests. ➤

◀ A noddy tern can also indicate that land is nearby. These birds will fly about 40 miles from their nests in search of food. In general, a sighting of a large group of terns is a more reliable sign of land than one or two birds.

Navigating in the Pacific using traditional Polynesian methods is never exact. However, the crews of the Hokule'a have shown that it can be done successfully.

Why might it be worthwhile for people today to learn and use ancient navigation methods?

Problem Solving

Mathematical Modeling

A **mathematical model** is something mathematical that describes something in the real world. A sphere, for example, is a model of a basketball. The number sentence $5.00 - (3 * 0.89) = 2.33$ is a model for buying three notebooks that cost 89¢ each, paying with a $5 bill, and getting $2.33 change. The graph at the right is a model for the number of school days in several regions of the United States in 1900.

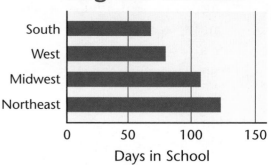

Days in School, 1900: Regional Medians

You have used mathematical models to solve problems for many years. In kindergarten and first grade, you used counters and drew pictures to solve simple problems. In later grades, you learned to use other models such as situation diagrams, graphs, and number models to solve problems. As you continue studying mathematics, you will learn to make and use more powerful mathematical models.

Everyday Mathematics has many different kinds of problems. Some problems ask you to find something. Other problems ask you to make something. When you get older, you will be asked to prove things, which means giving good reasons why they are true or correct.

Note

In the Geometry section of this book, you learn how to construct different geometric figures. When you are older, you will be asked to *prove* that the constructions are correct.

Problems that ask you to find something	Problems that ask you to make something
• The temperature at midnight was 5°F. The windchill temperature was −14°F. How much warmer was the actual temperature than the windchill temperature?	• This is $\frac{1}{4}$ of a shape: Draw a picture of the whole shape.
• What are the missing numbers? 1, 3, 7, 15, ____, 63, ____, 255	• Use a compass and straightedge to make a triangle with each side equal in length to segment *AB* below: A •————————————• B

Problems you already know how to solve offer good practice. But the problems that will help you learn the most are the ones you can't solve right away. Learning to be a good problem solver means learning what to do when you don't know what to do.

A Guide for Solving Number Stories

Learning to solve problems is the main reason for studying mathematics. One way you learn to solve problems is by solving number stories. A **number story** is a story with a problem that can be solved with arithmetic.

1. Understand the problem.
♦ Read the problem. Can you retell it in your own words?
♦ What do you want to find out?
♦ What do you know?
♦ Do you have all the information needed to solve the problem?

2. Plan what to do.
♦ Is the problem like one you solved before?
♦ Is there a pattern you can use?
♦ Can you draw a picture or a diagram?
♦ Can you write a number model or make a table?
♦ Can you use counters, base-10 blocks, or some other tool?
♦ Can you estimate the answer and check if you are right?

3. Carry out the plan.
♦ After you decide what to do, do it. Be careful.
♦ Make a written record of what you do.
♦ Answer the question.

4. Look back.
♦ Does your answer make sense?
♦ Does your answer agree with your estimate?
♦ Can you write a number model for the problem?
♦ Can you solve the problem in another way?

A Guide for Solving Number Stories

1. Understand the problem.
2. Plan what to do.
3. Carry out the plan.
4. Look back.

Note

Understanding the problem is an important step. Good problem solvers make sure they really understand the problem.

Note

Sometimes it's easy to know what to do. Other times you need to be creative.

Check Your Understanding

Use the Guide for Solving Number Stories to help you solve the following problems. Explain your thinking at each step. Explain your answer(s).

1. A store sells a certain brand of cereal in two sizes:
   • a 10-ounce box that costs $2.50
   • a 15-ounce box that costs $3.60
   Which box is the better buy? Why?

2. If you drive at an average speed of 50 miles per hour, how far will you travel for each length of time?
   a. 3 hours **b.** $\frac{1}{2}$ hour
   c. $2\frac{1}{2}$ hours **d.** 12 hours

Check your answers on page 441.

A Problem-Solving Diagram

Problems from everyday life, science, and business are often more complicated than the number stories you solve in school. Sometimes the steps in the Guide for Solving Number Stories may not be helpful.

The diagram below shows another way to think about problem solving. This diagram is more complicated than a list, but it shows more clearly what people do when they solve problems in science and business. The arrows connecting the boxes are meant to show that you don't always do things in the same order.

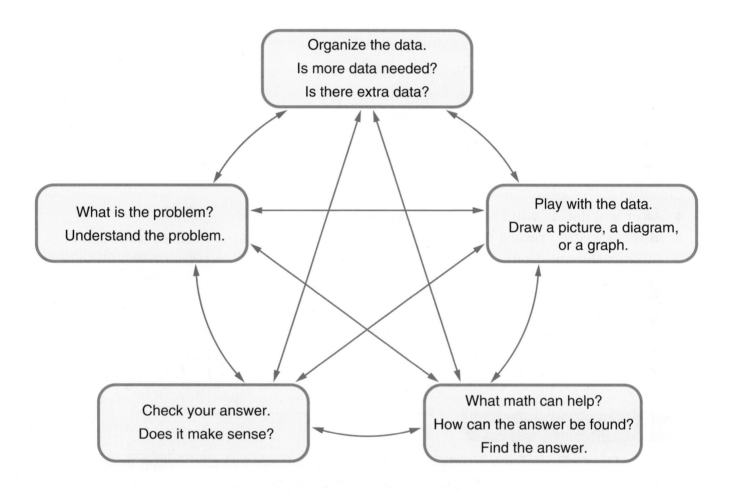

Using the diagram on the previous page as you solve problems may help you to be a better problem solver. Here are some things to try for each of the boxes in the diagram. Remember, these are not rules. They are only suggestions that might help.

♦ *What is the problem?* Try to understand the problem. Can you retell it in your own words? What do you want to find out? What do you know? Try to imagine what an answer might look like.

♦ *Organize the data.* Study the data you have and organize it in a list or in some other way. Look for more data if you need it. Get rid of any data that you don't need.

♦ *Play with the data.* Try drawing a picture, a diagram, or a graph. Can you write a number model? Can you model the problem with counters or blocks?

♦ *What math can help?* Can you use arithmetic, geometry, or other mathematics to find the answer? Do the math. Label the answer with units.

♦ *Check your answer.* Does it make sense? Compare your answer to a friend's answer. Try the answer in the problem. Can you solve the problem another way?

Did You Know?

Brainstorming is a method of shared problem solving by a group. Group members sit together, combine their skills, and let fly with ideas and possible solutions to the problem.

Check Your Understanding

Here is an up-and-down staircase that is 5 steps tall.

1. How many squares are needed for an up-and-down staircase that is 10 steps tall?

2. How many squares are needed for an up-and-down staircase that is 50 steps tall?

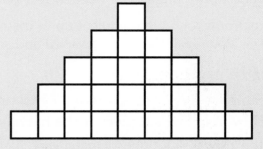

Check your answers on page 441.

Interpreting a Remainder in Division

Some number stories are solved by dividing whole numbers. You may need to decide what to do when there is a non-zero remainder.

Here are three possible choices:

♦ Ignore the remainder. Use the quotient as the answer.

♦ Round the quotient up to the next whole number.

♦ Rewrite the remainder as a fraction or decimal. Use this fraction or decimal as part of the answer.

To rewrite a remainder as a fraction:

1. Make the remainder the *numerator* of the fraction.

2. Make the divisor the *denominator* of the fraction.

Problem	Answer	Remainder Rewritten as a Fraction	Answer Written as a Mixed Number	Answer Written as a Decimal
367 / 4	91 R3	$\frac{3}{4}$	$91\frac{3}{4}$	91.75

Examples

- Suppose 3 people share 17 counters equally. How many counters will each person get? $17/3 \rightarrow 5$ R2

  Ignore the remainder. Use the quotient as the answer.

  Each person will have 5 counters and there will be 2 counters left over.

- Suppose 17 photos are placed in a photo album. How many pages are needed if 3 photos can fit on a page? $17/3 \rightarrow 5$ R2

  You must round the quotient up to the next whole number.
  The album will have 5 pages filled and another page only partially filled.

  So, 6 pages are needed.

- Suppose 3 friends share a 17-inch string of licorice. How long is each piece if the friends receive equal shares? $17/3 \rightarrow 5$ R2

  The answer, 5 R2, shows that if each friend receives 5 inches of licorice, 2 inches remain to be divided. Divide this 2-inch remainder into pieces that are $\frac{1}{3}$ inch long. Each friend receives two $\frac{1}{3}$-inch pieces, or $\frac{2}{3}$ inch.

  The remainder (2) has been rewritten as a fraction ($\frac{2}{3}$). Use this fraction as part of the answer.

  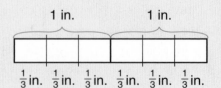

  Each friend will get a $5\frac{2}{3}$-inch piece of licorice.

Estimation

An **estimate** is an answer that should be close to an exact answer. You make estimates every day.

♦ You estimate how long it will take to walk to school.

♦ You estimate how much money you will need to buy some things at the store.

Sometimes you must estimate because it is impossible to know the exact answer. When you predict the future, for example, you have to estimate since it is impossible to know exactly what will happen. A weather forecaster's prediction is an estimate of what will happen in the future.

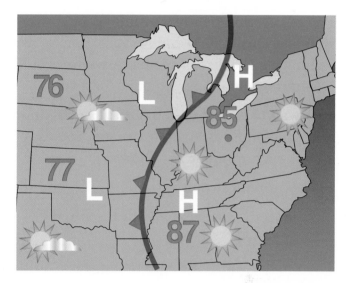

"Columbus, Ohio may *expect* sunny weather tomorrow. A high temperature of *about 85 degrees* is *predicted*."

Sometimes you estimate because finding an exact answer is not practical. For example, you could estimate the number of books in your school library instead of actually counting them.

Sometimes you estimate because finding an exact answer is not worth the trouble. For example, you might estimate the cost of several items at the store to be sure you have enough money. There is no need to find an exact answer until you pay for the items.

Estimation in Problem Solving

Estimation can be useful even when you need to find an exact answer. Making an estimate when you first start working on a problem may help you understand the problem better. Estimating before you solve a problem is like making a rough draft of a writing assignment.

Estimation can also be useful after you have found an answer for a problem. You can use the estimate to check whether your answer makes sense. If your estimate is not close to the exact answer you found, then you need to check your work.

Leading-Digit Estimation

The best estimators are usually people who are experts. Someone who lays carpet for a living, for example, would be very good at estimating the size of rooms. A waiter would be very good at estimating the proper amount for a tip.

One way to estimate is to adjust each number in a problem *before* you estimate. You may adjust each number as follows:

1. Keep the first non-zero digit of the number.

2. Replace the other digits of the number with zeros.

Exact number from a problem	Adjusted number used to make an estimate
6	6
68	60
429	400
8,578	8,000
125,718	100,000

Then make your estimate using these adjusted numbers.

This way of estimating is called **leading-digit estimation.**

Example What is the cost of 5 pounds of oranges at 74¢ per pound?

Use leading-digit estimation. The oranges cost about 70¢ per pound. The adjusted numbers are 5 and 70.

So, 5 pounds will cost about 5 * 70¢, or $3.50.

Leading-digit estimates are usually not very accurate. But they can be useful for checking calculations. If the estimate and the exact answer are not close, you should look for a mistake in your work.

Example Elisa added 694 + 415 + 382 and got 1,575. Was she correct?

The adjusted numbers are 600, 400, and 300. The leading-digit estimate using these adjusted numbers is 1,300. Since 1,300 is not close to 1,575, Elisa may not be correct. She should check her work.

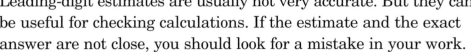

Exact numbers		Adjusted numbers and leading-digit estimate
694	→	600
415	→	400
+ 382	→	+ 300
		1,300

Check Your Understanding

Use leading-digit estimation to decide if the answers are correct.

1. Emily added 837 + 273 + 704 and got 1,054.   **2.** Luis said that 973 / 36 is 102.

Check your answers on page 441.

Rounding

Rounding is another way to adjust numbers and make them simpler and easier to work with. Estimation with rounded numbers is usually more accurate than leading-digit estimation.

Numbers are often rounded to the nearest multiple of 1,000, 100, 10, and so on. The next examples show the steps to follow in rounding a number.

Examples

1. Round 4,538 to the nearest hundred.
2. Round 26,781 to the nearest thousand.
3. Round 5,295 to the nearest ten.
4. Round 3.573 to the nearest tenth.

		Step 1: Find the digit in the place you are rounding to.	**Step 2:** Rewrite the number, replacing all digits to the right of this digit with zeros. This is the **lower number.**	**Step 3:** Add 1 to the digit in the place you are rounding to. If the sum is 10, write 0 and add 1 to the digit to its left. This is the **higher number.**	**Step 4:** Is the number you are rounding closer to the lower number or to the higher number?	**Step 5:** Round to the closer of the two numbers. If it is halfway between the lower and the higher number, round to the higher number.
1.	4,5̲38	4,5̲00	4,6̲00	lower number	4,500
2.	2̲6,781	2̲6,000	2̲7,000	higher number	27,000
3.	5,29̲5	5,29̲0	5,30̲0	halfway between	5,300
4.	3.5̲73	3.5̲00	3.6̲00	higher number	3.600 = 3.6

Thinking about a number line can also be helpful when you round numbers.

Example   Round 7,385 to the nearest thousand.

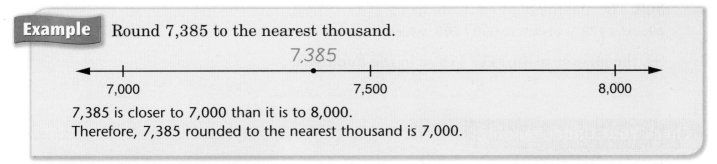

7,385 is closer to 7,000 than it is to 8,000.
Therefore, 7,385 rounded to the nearest thousand is 7,000.

Check Your Understanding

Round 25,795 to the nearest:

1. hundred
2. ten thousand
3. ten

Check your answers on page 441.

Making Other Estimates

Interval Estimates

An **interval estimate** is made up of a range of possible values. The exact value should be between the lowest and the highest value in the range.

Here is one way to give an interval estimate:

♦ Name a number you are sure is *less than the exact value.*

♦ Name a number you are sure is *greater than the exact value.*

The smaller the difference between the upper and lower numbers, the more useful an interval estimate is likely to be.

An interval estimate can be stated in various ways.

Did You Know ?

Each book in a library is coded according to its topic with a different decimal number. The codes used for mathematics books are between 510.000 and 519.000.

Examples   *Between* 225 and 300 people live in my apartment building.

The number of books in the school library is *greater than* 7,000, but *less than* 7,500.

There are *at least* 30 [12s] in 427, but *not more than* 40 [12s].

Magnitude Estimates

One kind of rough estimate is called a **magnitude estimate.** When making a magnitude estimate, ask yourself: "Is the answer in the tens? In the hundreds? In the thousands?" and so on. You can use magnitude estimates to check answers or to judge whether information you read or hear makes sense.

Example   Make a magnitude estimate. 49,741 / 178

49,741 / 178 is about 50,000 / 200, which is the same as 500 / 2, or 250.

So, the answer to 49,741 / 178 is in the hundreds.

Check Your Understanding

Make a magnitude estimate. Is the answer in the tens, hundreds, thousands, or ten thousands?

1. 520 * 32.1   **2.** 65,200 / 95   **3.** 3,456 / 1.21

Check your answers on page 441.

Calculators

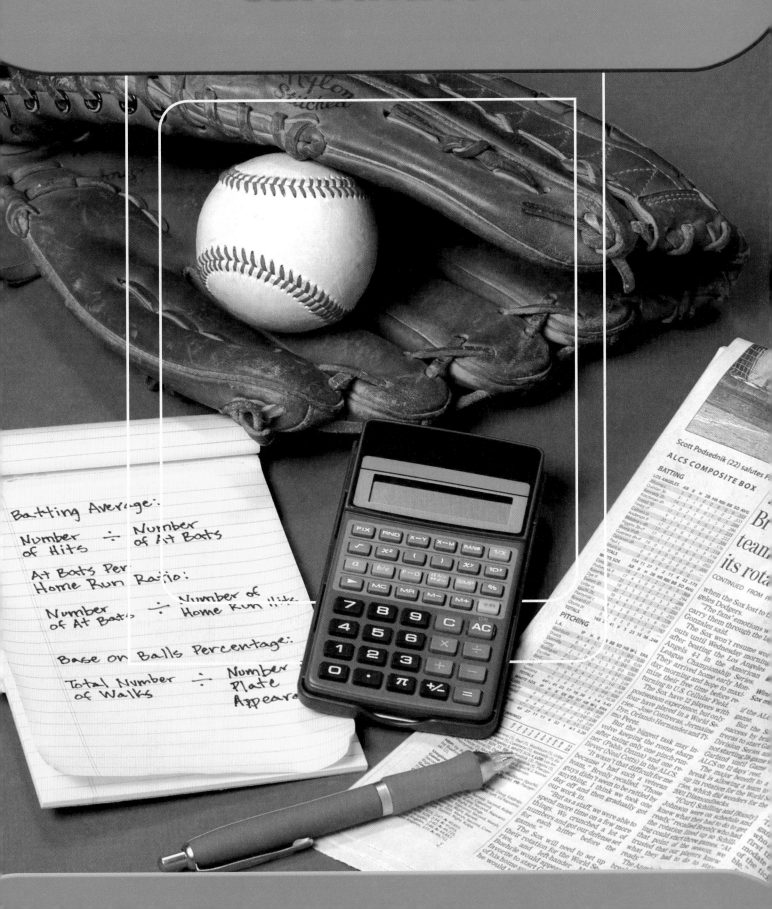

Batting Average:

Number of Hits ÷ Number of At Bats

At Bats Per Home Run Ratio:

Number of At Bats ÷ Number of Home Run Hits

Base on Balls Percentage:

Total Number of Walks ÷ Number Plate Appearances

About Calculators

You have used calculators to help you learn to count. Now you can use them for working with whole numbers, fractions, decimals, and percents.

As with any mathematical tool or strategy, you need to think about when and how to use a calculator. It can help you compute quickly and accurately when you have many problems to do in a short time. Calculators can help you compute with very large and very small numbers that may be hard to do in your head or with pencil and paper. Whenever you use a calculator, estimation should be part of your work. Always ask yourself if the number in the display makes sense.

There are many different kinds of calculators. **Four-function calculators** do little more than add, subtract, multiply, and divide whole numbers and decimals. More advanced **scientific calculators** let you find powers and reciprocals, and perform some operations with fractions. After elementary school, you may use **graphic calculators** that draw graphs, find data landmarks, and do even more complicated mathematics.

There are many calculators that work well with *Everyday Mathematics.* If the instructions in this book don't work for your calculator, or the keys on your calculator are not explained, you should refer to the directions that came with your calculator, or you can ask your teacher for help.

Calculator A

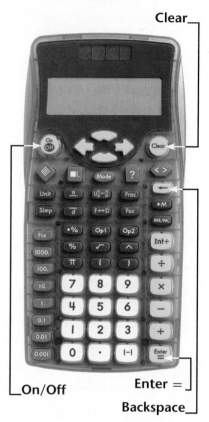

Calculator B

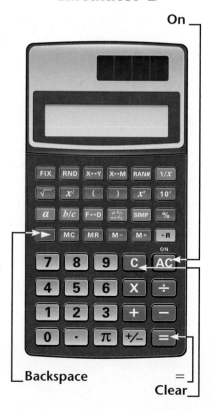

Basic Operations

You must take care of your calculator. Dropping it, leaving it in the sun, or other kinds of carelessness may break it or make it less reliable.

Many four-function and scientific calculators use light cells for power. If you press the ON key and see nothing on the display, hold the front of the calculator toward a light or a sunny window for a moment and then press ON again.

Entering and Clearing

Pressing a key on a calculator is called **keying in** or **entering.** In this book, calculator keys, except numbers and decimal points, are shown in rectangular boxes: $\boxed{+}$, $\boxed{=}$, $\boxed{\times}$, and so on. A set of instructions for performing a calculation is called a **key sequence.**

The simplest key sequences turn the calculator on and enter or clear numbers or other characters. These keys are labeled on the photos and summarized below.

> **Note**
>
> Calculators have two kinds of memory. **Short-term memory** is for the last number entered. The keys with an "M" are for **long-term memory** and are explained on pages 278–280.

Calculator A	Key Sequence	Purpose
	(On/Off)	Turn the display on.
	(Clear) and (On/Off) together	Clear the display and the short-term memory.
	(Clear)	Clear only the display.
	⬅	Clear the last digit.

Calculator B	Key Sequence	Purpose
	ON [AC]	Turn the display on.
	[AC]	Clear the display and the short-term memory.
	[C]	Clear a number if you make a mistake while entering a calculation.
	[▶]	Clear the last digit.

Always clear both the display and the memory each time you turn the calculator on.

Many calculators have a backspace key that will clear the last digit or digits you entered without re-entering the whole number.

Example Enter 123.444. Change it to 123.456.

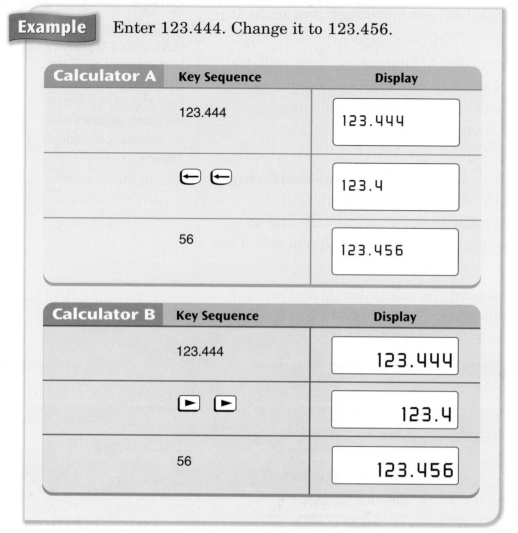

Try using the backspace key on your calculator.

Order of Operations and Parentheses

When you use a calculator for basic arithmetic operations, you enter the numbers and operations and press ⌑ or ⌑ to see the answer.

Try your calculator to see if it follows the rules for the order of operations. Key in 5 ⊞ 6 ⊠ 2 ⌑.

◆ If your calculator follows the order of operations, it will display 17.

◆ If it does not follow the order of operations, it will probably do the operations in the order they were entered: adding and then multiplying, displaying 22.

If you want the calculator to do operations in an order different from the order of operations, use the parentheses keys ⌑ and ⌑.

Note

Examples for arithmetic calculations are given in earlier sections of this book.

Example   Evaluate. $7 - (2 + 1)$

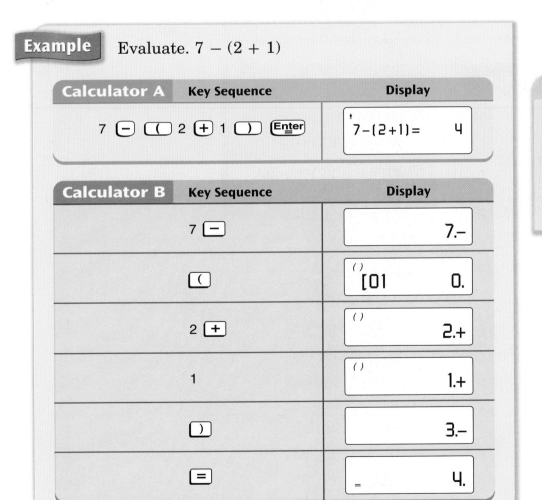

$7 - (2 + 1) = 4$

Note

If you see a tiny up or down arrow on the calculator display, you can use the up or down arrows to scroll the screen.

Sometimes expressions are shown without all of the multiplication signs. *Remember to press the multiplication key even when it is not shown.*

Example   Evaluate. $9 - 2(1 + 2)$

Calculator A   **Key Sequence**   **Display**

9 $\boxed{-}$ 2 $\boxed{\times}$ $\boxed{(}$ 1 $\boxed{+}$ 2 $\boxed{)}$ $\boxed{\text{Enter}}$

$$9-2\times(1+2)=\quad 3$$

Calculator B   **Key Sequence**   **Display**

9 $\boxed{-}$ 2 $\boxed{\times}$ $\boxed{(}$ 1 $\boxed{+}$ 2 $\boxed{)}$ $\boxed{=}$

$$= \qquad 3.$$

$9 - 2(1 + 2) = 3$

Check Your Understanding

Use your calculator to evaluate each expression.

1. $98 - (7 + 9)$ **2.** $64 - 6(2 + 8)$

3. $9(7 + 4.5) - 43$ **4.** $(34 - 22)/3 + 24$

Check your answers on page 441.

Negative Numbers

How you enter a negative number depends on your calculator. You will use the change sign key, either [+/−] or [(−)] depending on your calculator. Both keys change the sign of the number.

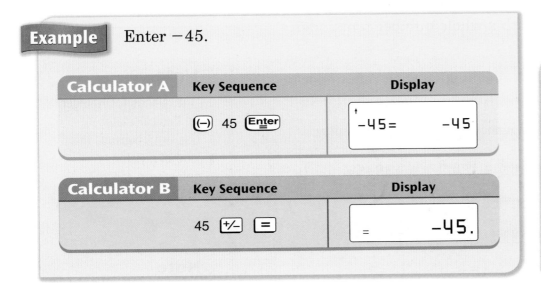

Example   Enter −45.

Calculator A	Key Sequence	Display
	[(−)] 45 [Enter]	−45 = −45

Calculator B	Key Sequence	Display
	45 [+/−] [=]	= −45.

Note

If the number on the display is positive, it becomes negative after you press [+/−]. If the number on the display is negative, it becomes positive after pressing [+/−]. Keys like this are called **toggles.**

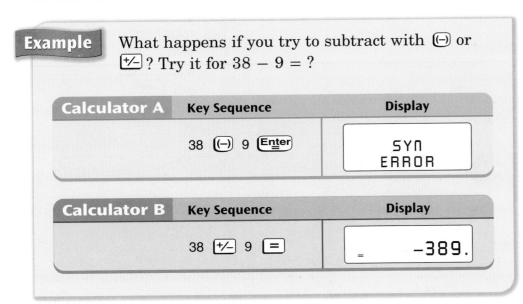

Example   What happens if you try to subtract with [(−)] or [+/−]? Try it for 38 − 9 = ?

Calculator A	Key Sequence	Display
	38 [(−)] 9 [Enter]	SYN ERROR

Calculator B	Key Sequence	Display
	38 [+/−] 9 [=]	= −389.

Note

"SYN" is short for "syntax," which means the ordering and meaning of keys in a sequence.

Note

If you try to subtract using [+/−] on this calculator, it just changes the sign of the first number and adds the digits of the second number to it.

Division with Remainders

The answer to a division problem with whole numbers does not always result in whole number answers. When this happens, most calculators display the answer as a decimal. Some calculators also have a second division key that displays the whole number quotient with a whole number remainder.

Example 39 ÷ 5 =? Use the division with remainder key.

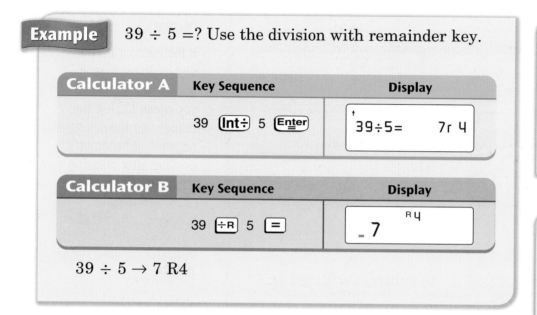

39 ÷ 5 → 7 R4

Note

"Int" stands for "integer" on this calculator. Use Int÷ because this kind of division is sometimes called "integer division."

Note

÷R means "divide with remainder." You can also divide positive fractions and decimals with ÷R.

Try the division with remainder in the previous example to see how your calculator works.

Check Your Understanding

Divide with remainder.

1. 102 ÷ 7   **2.** 221 ÷ 13   **3.** 33,333 ÷ 44

Check your answers on page 441.

Fractions and Percent

Some calculators let you enter, rewrite, and do operations with fractions. Once you know how to enter a fraction, you can add, subtract, multiply, or divide them just like whole numbers and decimals.

Entering Fractions and Mixed Numbers

Most calculators that let you enter fractions use similar key sequences. For proper fractions, always start by entering the numerator. Then press a key to tell the calculator to begin writing a fraction.

Example Enter $\frac{5}{8}$ as a fraction in your calculator.

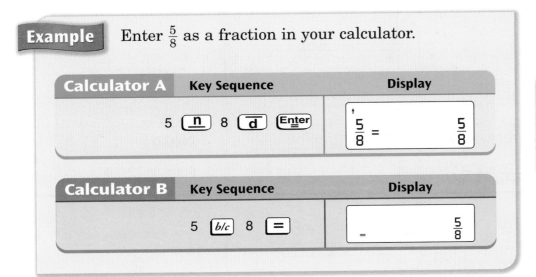

Note

Pressing $\boxed{d}$ after you enter the denominator is optional.

To enter a mixed number, enter the whole number part and then press a key to tell the calculator what you did.

Example Enter $73\frac{2}{5}$ as a fraction in your calculator.

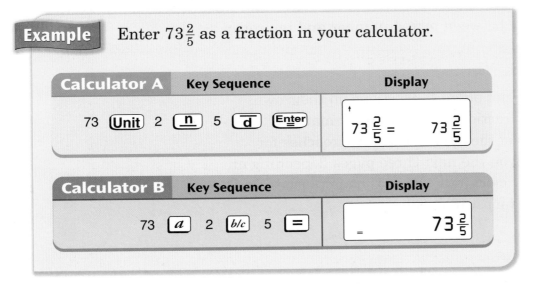

Try entering a mixed number on your calculator.

The keys to convert between mixed numbers and improper fractions are similar on all fraction calculators.

Example Convert $\frac{45}{7}$ to a mixed number with your calculator. Then change it back.

Calculator A	Key Sequence	Display
	45 [n] 7 [d] (Enter)	$\frac{45}{7} = \quad 6\frac{3}{7}$
	(U$\frac{n}{d}$↔$\frac{n}{d}$)	$\frac{45}{7}$
	(U$\frac{n}{d}$↔$\frac{n}{d}$)	$6\frac{3}{7}$

Note

Pressing (Enter) is *not* optional in this key sequence.

Calculator B	Key Sequence	Display
	45 [b/c] 7 [=]	$\frac{45}{7}$
	[a b/c ↔ d/c]	$6\frac{3}{7}$
	[a b/c ↔ d/c]	$\frac{45}{7}$

Note

Pressing [=] is optional in this key sequence.

Both (U$\frac{n}{d}$↔$\frac{n}{d}$) and [a b/c ↔ d/c] toggle between mixed number and improper fraction notation.

Simplifying Fractions

Ordinarily, calculators do not simplify fractions on their own. The steps for simplifying fractions are similar for many calculators, but the order of the steps varies. Approaches for two calculators are shown on the next three pages depending on the keys you have on your calculator. Read the approaches for the calculator having keys most like yours.

Simplifying Fractions on Calculator A

This calculator lets you simplify a fraction in two ways. Each way divides the numerator and denominator by a common factor. The first approach uses **Simp** to automatically divide by the smallest common factor, and **Fac** to display the factor.

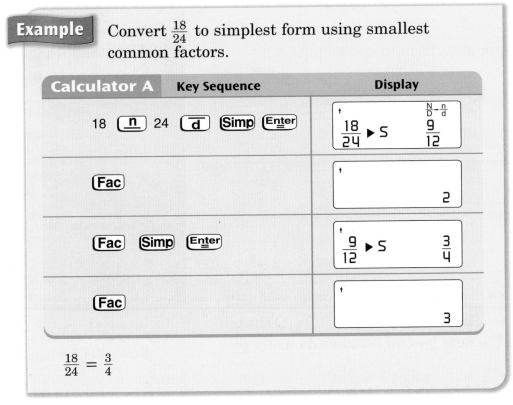

Example Convert $\frac{18}{24}$ to simplest form using smallest common factors.

Calculator A	Key Sequence	Display
18 **n** 24 **d** **Simp** **Enter**		$\frac{18}{24} \blacktriangleright S$ $\frac{9}{12}$
Fac		2
Fac **Simp** **Enter**		$\frac{9}{12} \blacktriangleright S$ $\frac{3}{4}$
Fac		3

$$\frac{18}{24} = \frac{3}{4}$$

In the second approach, you can simplify the fraction in one step by telling the calculator to divide by the greatest common factor of the numerator and the denominator.

Example Convert $\frac{18}{24}$ to simplest form in one step by dividing the numerator and the denominator by their greatest common factor, 6.

Calculator A	Key Sequence	Display
18 **n** 24 **d** **Simp** 6 **Enter**		$\frac{18}{24} \blacktriangleright S6$ $\frac{3}{4}$
Fac		6

$$\frac{18}{24} = \frac{3}{4}$$

Calculator A

Simp simplifies a fraction by a common factor.

Fac displays the common factor used to simplify a fraction.

$\frac{N}{D} \rightarrow \frac{n}{d}$ means that the fraction shown is not in simplest form.

Note

Pressing **Fac** toggles between the display of the factor and the display of the fraction.

Calculators

Simplifying Fractions on Calculator B

Calculators like the one shown here let you simplify fractions in three different ways. Each way divides the numerator and denominator by a common factor. The first approach uses $\boxed{=}$ to give the simplest form in one step. The word Simp in the display means that the fraction shown is not in simplest form.

Calculator B

$\boxed{\text{SIMP}}$ simplifies a fraction by a common factor.

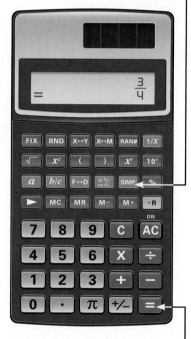

Press $\boxed{=}$ $\boxed{=}$ to display the fraction.

Example Convert $\frac{18}{24}$ to simplest form in one step.

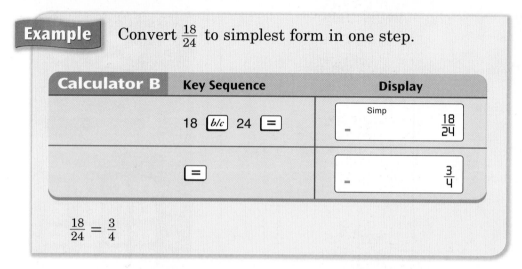

$\frac{18}{24} = \frac{3}{4}$

If you enter a fraction that is already in simplest form, you will not see Simp on the display. The one-step approach does not tell you the common factor as the next two approaches do using $\boxed{\text{SIMP}}$.

Example Convert $\frac{18}{24}$ to simplest form using smallest common factors.

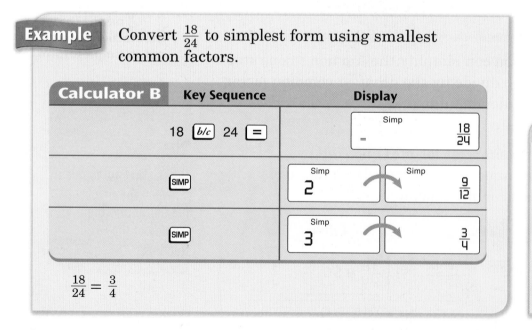

$\frac{18}{24} = \frac{3}{4}$

Note

Each time you press $\boxed{\text{SIMP}}$ in the smallest-common-factor approach you briefly see the common factor, then the simplified fraction. This can be done without pressing $\boxed{=}$ first.

In the last approach to simplifying fractions with this type of calculator you tell it what common factor to divide by. If you use the greatest common factor on the numerator and the denominator, you can simplify the fraction in one step.

Calculator A

% divides a number by 100.

Example Convert $\frac{18}{24}$ to simplest form by dividing the numerator and the denominator by their greatest common factor, 6.

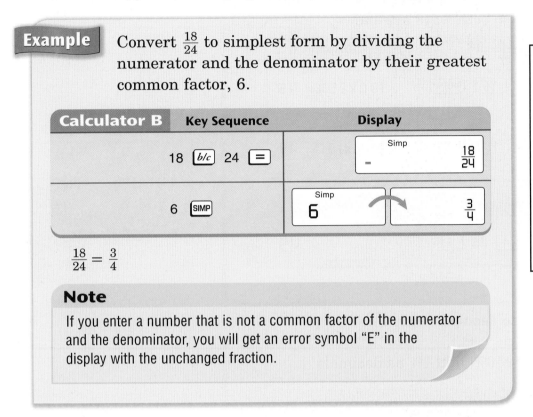

Calculator B	Key Sequence	Display
	18 [b/c] 24 [=]	Simp = $\frac{18}{24}$
	6 [SIMP]	Simp 6 → $\frac{3}{4}$

$$\frac{18}{24} = \frac{3}{4}$$

Note

If you enter a number that is not a common factor of the numerator and the denominator, you will get an error symbol "E" in the display with the unchanged fraction.

Try simplifying the fractions in the previous examples to see how your calculator works.

Percent

The calculators shown here, and many others, have a **%** key, but it is likely that they work differently. The best way to learn what your calculator does with percents is to read its manual.

Calculator B

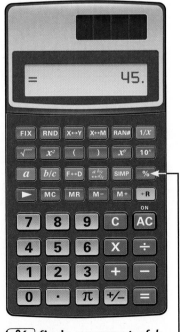

% finds *a* percent of *b*.

Calculators

Most calculators include $\boxed{\%}$ to solve "percent of" problems.

Example Calculate 25% of 180.

Using the first calculator, you can multiply 180 and 25% in either order, so both ways are shown.

Calculator A	Key Sequence	Display
	180 $\boxed{\times}$ 25 $\boxed{\%}$ (Enter)	180×25%= 45
	25 $\boxed{\%}$ $\boxed{\times}$ 180 (Enter)	25%×180= 45

Calculator B	Key Sequence	Display
	180 $\boxed{\times}$ 25 $\boxed{\%}$	45.

25% of 180 is 45.

You can change percents to decimals with $\boxed{\%}$.

Examples Display 85%, 250%, and 1% as decimals.

Calculator A	Key Sequence	Display
	85 $\boxed{\%}$ (Enter)	85%= 0.85
	250 $\boxed{\%}$ (Enter)	250%= 2.5
	1 $\boxed{\%}$ (Enter)	1%= 0.01

Calculator B	Key Sequence	Display
	1 $\boxed{\times}$ 85 $\boxed{\%}$	0.85
	1 $\boxed{\times}$ 250 $\boxed{\%}$	2.5
	1 $\boxed{\times}$ 1 $\boxed{\%}$	0.01

Note

To convert percents to decimals on this calculator, you calculate a percent of 1, like in the previous example.

85% = 0.85; 250% = 2.5; 1% = 0.01

You can also use %️ to convert percents to fractions.

On many calculators, first change the percent to a decimal as in the previous examples, then use F↔D to change to a fraction.

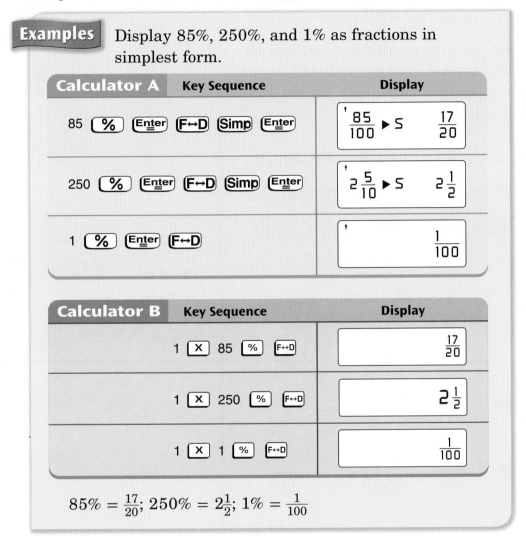

Examples Display 85%, 250%, and 1% as fractions in simplest form.

Calculator A	Key Sequence	Display
85 %️ Enter F↔D Simp Enter		↑ $\frac{85}{100}$ ▶ S $\frac{17}{20}$
250 %️ Enter F↔D Simp Enter		↑ $2\frac{5}{10}$ ▶ S $2\frac{1}{2}$
1 %️ Enter F↔D		↑ $\frac{1}{100}$

Note

You may need to use SIMP to simplify.

Calculator B	Key Sequence	Display
1 ✕ 85 %️ F↔D		$\frac{17}{20}$
1 ✕ 250 %️ F↔D		$2\frac{1}{2}$
1 ✕ 1 %️ F↔D		$\frac{1}{100}$

Note

This calculator simplifies automatically.

$85\% = \frac{17}{20}$; $250\% = 2\frac{1}{2}$; $1\% = \frac{1}{100}$

Try displaying some percents as fractions and decimals on your calculator.

Fraction/Decimal/Percent Conversions

Conversions of fractions to decimals and percents can be done on any calculator. For example, to rename $\frac{3}{5}$ as a decimal, simply enter 3 ⌈÷⌉ 5 ⌈=⌉. The display will show 0.6. To rename a decimal as a percent, just multiply by 100.

Conversions of decimals and percents to fractions can only be done on calculators that have special keys for fractions. Such calculators also have keys to change a fraction to its decimal equivalent or a decimal to an equivalent fraction.

Example   Convert $\frac{3}{8}$ to a decimal and back to a fraction.

Calculator A	Key Sequence	Display
	3 ⌈n⌉ 8 ⌈d⌉ ⌈Enter⌉	$\frac{3}{8} = \qquad \frac{3}{8}$
	⌈F↔D⌉	0.375
	⌈F↔D⌉	$\frac{375}{1000}$
	⌈Simp⌉ ⌈Enter⌉	$\frac{375}{1000} \blacktriangleright S \quad \frac{75}{200}$
	⌈Simp⌉ ⌈Enter⌉	$\frac{75}{200} \blacktriangleright S \quad \frac{15}{40}$
	⌈Simp⌉ ⌈Enter⌉	$\frac{15}{40} \blacktriangleright S \quad \frac{3}{8}$

Note

⌈F↔D⌉ toggles between fraction and decimal notation.

Calculator B	Key Sequence	Display
	3 ⌈b/c⌉ 8	$\frac{3}{8}$
	⌈F↔D⌉	0.375
	⌈F↔D⌉	$\frac{3}{8}$

$\frac{3}{8} = 0.375$

See how your calculator changes fractions to decimals.

The tables below show examples of various conversions. Although only one key sequence is shown for each conversion, there are often other key sequences that work as well.

Conversion	Starting Number	Calculator A Key Sequence	Display
Fraction to decimal	$\frac{3}{5}$	3 [n] 5 [d] [Enter] [F↔D]	0.6
Decimal to fraction	0.125	.125 [Enter] [F↔D]	$\frac{N}{D} \rightarrow \frac{n}{d}$ $\frac{125}{1000}$
Decimal to percent	0.75	.75 [▶%] [Enter]	0.75▶% 75%
Percent to decimal	125%	125 [%] [Enter]	125%= 1.25
Fraction to percent	$\frac{5}{8}$	5 [n] 8 [d] [▶%] [Enter]	$\frac{5}{8}$▶% 62.5%
Percent to fraction	35%	35 [%] [Enter] [F↔D]	$\frac{N}{D} \rightarrow \frac{n}{d}$ $\frac{35}{100}$

Conversion	Starting Number	Calculator B Key Sequence	Display
Fraction to decimal	$\frac{3}{5}$	3 [b/c] 5 [F↔D]	0.6
Decimal to fraction	0.125	.125 [F↔D]	$\frac{1}{8}$
Decimal to percent	0.75	.75 [X] 100 [=]	= 75.
Percent to decimal	125%	1 [X] 125 [%]	1.25
Fraction to percent	$\frac{5}{8}$	5 [b/c] 8 [F↔D] [X] 100 [=]	= 62.5
Percent to fraction	35%	1 [X] 35 [%] [F↔D]	$\frac{7}{20}$

Check Your Understanding

Use your calculator to convert between fractions, decimals, and percents.

1. $\frac{5}{16}$ to a decimal   **2.** 0.385 to a fraction   **3.** 0.009 to a percent

4. 458% to a decimal   **5.** $\frac{7}{32}$ to a percent   **6.** 58% to a fraction

Check your answers on page 441.

Other Operations

Your calculator can do more than simple arithmetic with whole numbers, fractions, and decimals. Every calculator does some things that other calculators cannot, or does them in different ways. See your owner's manual or ask your teacher to help you explore these things. The following pages explain some other things that many calculators can do.

Rounding

All calculators can round decimals. Decimals must be rounded to fit on the display. For example, if you key in 2 ⨰ 3 🟰

♦ Calculator A shows 11 digits and rounds to the nearest value: 0.6666666667.

♦ Calculator B shows 8 digits and rounds down to 0.6666666.

Try 2 ⨰ 3 🟰 on your calculator to see how big the display is and how it rounds.

All scientific calculators have a ⟨FIX⟩ key to set, or **fix,** the place value of decimals on the display. Fixing always rounds to the nearest value.

Note

To turn off fixed rounding on a calculator, press ⟨FIX⟩ ⟨•⟩.

Note

On this calculator you can fix the decimal places to the left of the decimal point, but are limited to the right of the decimal point to thousandths (0.001).

Examples Clear your calculator and fix it to round to tenths. Round each number 1.34; 812.79; and 0.06 to the nearest tenth.

Calculator A	Key Sequence	Display
	(Clear) (Fix) (0.1)	↑ Fix ◀
	1.34 (Enter)	↑ Fix 1.34 = 1.3
	812.79 (Enter)	↑ Fix 812.79 = 812.8
	.06 (Enter)	↑ Fix .06 = 0.1

1.34 rounds to 1.3; 812.79 rounds to 812.8; 0.06 rounds to 0.1.

Examples Clear your calculator and fix it to round to tenths. Round each number 1.34; 812.79; and 0.06 to the nearest tenth.

Note

This calculator only lets you fix places to the right of the decimal point.

Calculator B	Key Sequence	Display
	AC FIX	0-7
	1	FIX 0.0
	1.34 =	FIX = 1.3
	812.79 =	FIX = 812.8
	.06 =	FIX = 0.1

1.34 rounds to 1.3; 812.79 rounds to 812.8; 0.06 rounds to 0.1.

Note

You can fix either calculator to round without clearing the display first. It will round the number on the display.

Check Your Understanding

Use your calculator to round to the indicated place.

1. 0.78 to tenths

2. 234.66 to ones

3. 1258.3777 to thousandths

4. 0.9999 to hundredths

Check your answers on page 441.

Fixing the display to round to hundredths is helpful for solving problems about dollars and cents.

Example One CD costs $11.23 and another costs $14.67. Set your calculator to round to the nearest cent and calculate the total cost of the CDs.

Calculator A	Key Sequence	Display
	Fix 0.01	↑ Fix ◄
	11.23 + 14.67 Enter	↑ Fix 11.23+14.67 = 25.90

Calculator B	Key Sequence	Display
	FIX	0–7
	2	FIX 0.00
	11.23 + 14.67 =	FIX = 25.90

Together, the CDs cost $25.90.

On most calculators, if you find the total in the Example above with the "fix" turned off, the display reads 25.9. To show the answer in dollars and cents, fix the display to round to hundredths and you will see 25.90.

Powers, Reciprocals, and Square Roots

Powers of numbers can be calculated on all scientific calculators. Look at your calculator to see which key it has for finding powers of numbers.

◆ The key may look like x^y and is read "x to the y."

◆ The key may look like $\boxed{\wedge}$, and is called a **caret.**

To compute a number to a negative power, be sure to use the change-sign key $\boxed{(-)}$ or $\boxed{+/-}$, not the subtraction key $\boxed{-}$.

Calculator A

$\boxed{\wedge}$ finds powers and reciprocals.

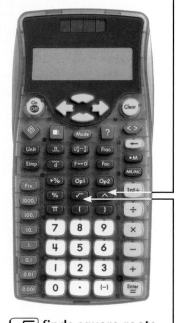

$\boxed{\sqrt{}}$ finds square roots.

Examples Find the values of 3^4 and 5^{-2}.

Calculator A	Key Sequence	Display
	3 $\boxed{\wedge}$ 4 $\boxed{\text{Enter}}$	3^4 = 81
	5 $\boxed{\wedge}$ $\boxed{(-)}$ 2 $\boxed{\text{Enter}}$	5^-2 = 0.04

Note

If you press $\boxed{(-)}$ after the 2, you will get an error message.

Calculator B

$\boxed{x^y}$ finds powers.

$\boxed{1/x}$ finds reciprocals.

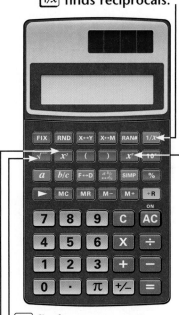

Calculator B	Key Sequence	Display
	3 $\boxed{x^y}$ 4 $\boxed{=}$	= 81.
	5 $\boxed{x^y}$ 2 $\boxed{+/-}$ $\boxed{=}$	= 0.04

Note

If you press $\boxed{+/-}$ before the 2, it will change the sign of the 5 and display the result of $(-5)^2 = 25$.

$3^4 = 81$; $5^{-2} = 0.04$

$\boxed{\sqrt{}}$ finds square roots.

$\boxed{x^2}$ is a shortcut to square numbers.

Most scientific calculators have a reciprocal key ▭1/x▭. On all scientific calculators you can find a reciprocal of a number by raising it to the −1 power.

Examples Find the reciprocals of 25 and $\frac{2}{3}$.

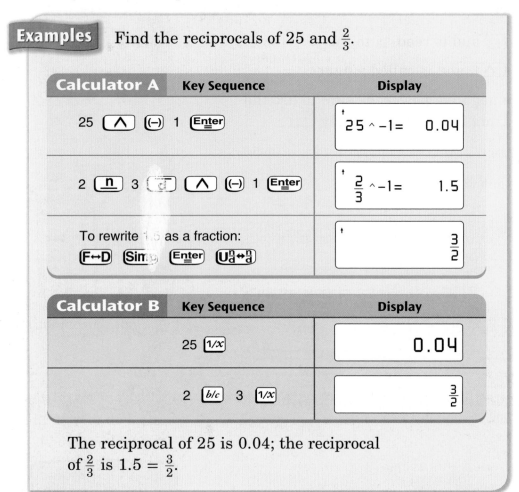

Calculator A	Key Sequence	Display
	25 ▭∧▭ ▭(−)▭ 1 ▭Enter▭	25^−1= 0.04
	2 ▭n▭ 3 ▭d▭ ▭∧▭ ▭(−)▭ 1 ▭Enter▭	$\frac{2}{3}$^−1= 1.5
	To rewrite 1.5 as a fraction: ▭F↔D▭ ▭Simp▭ ▭Enter▭ ▭Ud↔nd▭	$\frac{3}{2}$

Calculator B	Key Sequence	Display
	25 ▭1/x▭	0.04
	2 ▭b/c▭ 3 ▭1/x▭	$\frac{3}{2}$

The reciprocal of 25 is 0.04; the reciprocal of $\frac{2}{3}$ is 1.5 = $\frac{3}{2}$.

Note

You don't need to press the last key in the final step if your calculator is set to keep fractions in improper form. See the owner's manual for details.

Almost all calculators have a ▭√▭ key for finding square roots. It depends on the calculator whether you press ▭√▭ before or after entering a number.

Examples Find the square roots of 25 and 10,000.

Calculator A	Key Sequence	Display
	▭√▭ 25 ▭)▭ ▭Enter▭	√(25)= 5
	▭√▭ 10000 ▭)▭ ▭Enter▭	√(10000)= 100

Note

On this calculator you have to "close" the square root by pressing ▭)▭ after the number.

$\sqrt{25} = 5$; $\sqrt{10,000} = 100$

Examples Find the square roots of 25 and 10,000.

Calculator B	Key Sequence	Display
	25 $\sqrt{}$	**5.**
	10000 $\sqrt{}$	**100.**

$\sqrt{25} = 5$; $\sqrt{10,000} = 100$

Try finding the square roots of a few numbers.

Scientific Notation

Scientific notation is a way of writing very large or very small numbers. A number in scientific notation is shown as a product of a number between 1 and 10 and a power of 10. In scientific notation, the 9,000,000,000 bytes of memory on a 9-gigabyte hard drive is written $9 * 10^9$. On scientific calculators, numbers with too many digits to fit on the display are automatically shown in scientific notation like the bottom calculator in the margin.

Different calculators use different symbols for scientific notation. Your calculator may display raised exponents of 10, although most do not. Since the base of the power is always 10, most calculators leave out the 10 and simply put a space between the number and the exponent.

This calculator uses the caret $\wedge$ to display scientific notation.

This calculator shows $9 * 10^9$.

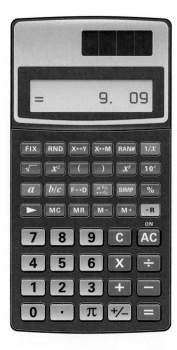

Examples Convert $7 * 10^4$, $4.35 * 10^5$ and $8 * 10^{-3}$ to decimal notation.

Calculator A	Key Sequence	Display
	7 ⊗ 10 ⌃ 4 Enter	7x10^4= 70000
	4.35 ⊗ 10 ⌃ 5 Enter	4.35x10^5= 435000
	8 ⊗ 10 ⌃ (−) 3 Enter	8x10^-3= 0.008

Calculator B	Key Sequence	Display
	7 ⊠ 4 10ˣ ▭	= 70'000.
	4.35 ⊠ 5 10ˣ ▭	= 435'000.
	8 ⊠ 3 +/− 10ˣ ▭	= 0.008

Note

Neither calculator displays large numbers in decimal notation with a comma like you do with pencil and paper. One uses an apostrophe; the other uses no symbol at all.

$7 * 10^4 = 70{,}000$; $4.35 * 10^5 = 435{,}000$; and $8 * 10^{-3} = 0.008$

How does your calculator display large numbers?

Check Your Understanding

Use your calculator to convert the following to standard notation:

1. $5.8 * 10^{-4}$ **2.** $7.6 * 10^7$ **3.** $4.389 * 10^{-6}$ **4.** $-1.1 * 10^{-5}$

Check your answers on page 441.

Calculators have different limits to the numbers they can display without scientific notation.

Note

A calculation resulting in a number larger than the limit is automatically displayed in scientific notation.

Example Write 123,456 * 654,321 in scientific notation. Then write the product in decimal notation.

Calculator A	Key Sequence	Display
	123456 ⊗ 654321 (Enter)	8.078x10^10

The product is $8.078 * 10^{10} = 80,780,000,000$.

Calculator B	Key Sequence	Display
	123456 ⊗ 654321 ⊟	8.0779 10

The product is $8.0779 * 10^{10} = 80,779,000,000$.

Note

Using (FIX) to round answers does not affect scientific notation on either calculator.

Example Write 1 * 2 * 3 * 4 * 5 * 6 * 7 * 8 * 9 * 10 * 11 * 12 * 13 * 14 * 15 in scientific notation.

Calculator A	Key Sequence	Display
	1⊗2⊗3⊗4⊗5⊗6⊗7⊗8 ⊗9⊗10⊗11⊗12⊗13⊗14⊗15(Enter)	1.308x10^12

The product is $1.308 * 10^{12}$, or 1,308,000,000,000.

Calculator B	Key Sequence	Display
	1⊗2⊗3⊗4⊗5⊗6⊗7⊗8⊗9⊗10⊗11⊗12⊗13⊗14⊗15⊟	1.3076 12

The product is $1.3076 * 10^{12}$, or 1,307,600,000,000.

Check Your Understanding

Write in scientific notation.

1. 995 * 7 * 54 * 65 * 659 * 807 * 468
2. 956 * 859 * 760 * 862
3. 527 * 32 * 987 * 424 * 77 * 145 * 195
4. 15^9 * 13 * 996 * 558
5. The number of different 5-card hands that can be drawn from a standard deck of 52 cards is: 52 * 51 * 50 * 49 * 48. How many hands is this in scientific notation? In decimal notation?

Check your answers on page 441.

Pi (π)

The formulas for the circumference and area of circles involve **pi (π)**. Pi is a number that is a little more than 3. The first nine digits of pi are 3.14159265. All scientific calculators have a pi key 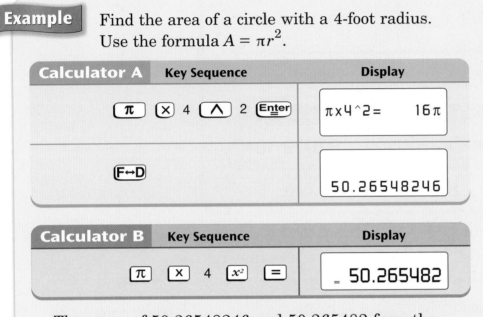 that gives an approximate value in decimal form. A few calculators display an exact value using the π symbol.

Example   Find the area of a circle with a 4-foot radius. Use the formula $A = \pi r^2$.

Calculator A	Key Sequence	Display
	π × 4 ∧ 2 Enter	π × 4 ^ 2 = 16 π
	F↔D	50.26548246

Calculator B	Key Sequence	Display
	π × 4 x² =	₌ 50.265482

The areas of 50.26548246 and 50.265482 from the two calculator displays look very precise. Because the decimal value of π is approximate, the decimal areas are also approximate, but still look accurate. In everyday life, the measure of the radius of a circle is probably approximate, and giving an area to 6 or 8 decimal places does not make sense.

So a good approximation of the area of the 4-foot radius circle is about 50 square feet.

Note

You can set the number of decimal places on your calculator's display to show 50 by pressing either Fix 1. or Fix 0 depending on the calculator.

Example Find the circumference of a circle with a 15-centimeter diameter to the nearest tenth of a centimeter. Use the formula $C = \pi d$.

Calculator A	Key Sequence	Display
	(Fix) (0.1)	↑ Fix
	(π) (×) 15 (Enter)	↑ Fix π×15= 15π
	(F↔D)	↑ Fix 47.1

Calculator B	Key Sequence	Display
	(FIX) 1	FIX 0.0
	(π) (×) 15 (=)	FIX = 47.1

The circumference is about 47.1 centimeters.

Note

When you are finished, remember to turn off the fixed rounding by pressing (FIX) (·).

Check Your Understanding

1. Find the area of a circle with a 90-foot radius. Display your answer to the nearest square foot.

2. To the nearest tenth of a centimeter, find the circumference of a circle with a 26.7-centimeter diameter.

Check your answers on page 441.

Using Calculator Memory

Many calculators let you save a number in **long-term memory** using keys with "M" on them. Later on, when you need the number, you can recall it from memory. Most calculators display an "M" or similar symbol when there is a number other than 0 in the memory.

Memory Basics

There are two main ways to enter numbers into long-term memory. Some calculators, including most 4-function calculators, have the keys in the table on page 280. If your calculator does not have at least the ⌐M+ and ⌐M- keys, see the examples on this page.

Memory on Calculator A

One way that calculators can put numbers in memory is using a key to **store** a value. On the first calculator the store key is ►M and only works on numbers that have been entered into the display with Enter.

Calculator A

►M stores the displayed number in memory.

MR/MC recalls and displays the number in memory. Press it twice to clear memory.

Calculator A	Key Sequence	Purpose
	MR/MC MR/MC	Clear the long-term memory. This should always be the first step to any key sequence using the memory. Afterward, there will be no "M" in the display. This tells you there is no number in memory.
	►M Enter	Store the number entered in the display in memory.
	MR/MC	Recall the number stored in memory and show it in the display.

Note

If you press MR/MC more than twice, you will recall and display the 0 that is now in memory. Press Clear to clear the display.

The following example first shows what happens if you don't
enter a number before trying to store it.

Example   Store 25 in memory and recall it to show that it
was saved.

Calculator A Key Sequence	Display
(MR/MC) (MR/MC) (Clear)	
25 (▶M) (Enter)	MEM ERROR

Oops. Start again. First, press (Clear) twice.

Calculator A Key Sequence	Display
(MR/MC) (MR/MC) (Clear)	
25 (Enter) (▶M) (Enter)	M 25 = 25
(Clear)	M
(MR/MC)	M 25

If your calculator is like this one, try the Check Your
Understanding problem on page 280.

Memory on Calculator B

Calculator B

Calculators put a 0 into memory when MC is pressed. To store a single number in a cleared memory, simply enter the number and press M+.

MC clears the memory.

MR recalls the number in memory and displays it.

Calculator B	Key Sequence	Purpose
	MC	Clear the long-term memory. This should always be the first step to any key sequence using the memory. Afterward, there will be no "M" in the display. This tells you there is no number in memory.
	MR	Recall the number stored in memory and show it in the display.
	M+	Add the number on the display to the number in memory.
	M−	Subtract the number on the display from the number in memory.

M− subtracts the number on the display from the number in memory.

M+ adds the number on the display to the number in memory.

Example   Store 25 in memory and recall it to show that it was saved.

Calculator B	Key Sequence	Display
	AC MC	0.
	25 M+	M 25.
	AC	M 0.
	MR	M 25.

Note

When this calculator turns off, the display clears, but a value in memory is *not* erased.

Check Your Understanding

Store π in the long-term memory. Clear the display. Then compute the area A of a circle whose radius r is 14 feet, without pressing the π key. ($A = \pi r^2$)

Check your answer on page 442.

Using Memory in Problem Solving

A common use of memory in calculators is to solve problems that have two or more steps in the solution.

Example   Compute a 15% tip on a $25 bill. Store the tip in the memory, then find the total bill.

Calculator A	Key Sequence	Display
	(MR/MC) (MR/MC) (Clear)	
	15 (%) (×) 25 (Enter)	15%×25= 3.75
	(▶M) (Enter)	M 15%×25= 3.75
	25 (+) (MR/MC) (Enter)	M 25+3.75= 28.75

Note

Always be sure to clear the memory after solving one problem and before beginning another.

Calculator B	Key Sequence	Display
	(AC) (MC)	0.
	25 (×) 15 (%) (=)	3.75
	(M+)	M 3.75
	25 (+) (MR) (=)	M = 28.75.

Calculator B	Key Sequence	Display
	(AC) (MC)	0.
	25 (×) 15 (%) (+)	28.75

The total bill is $28.75.

Note

The second solution shows how this calculator solves the problem by using memory automatically.

Check Your Understanding

Compute a 15% tip on a $75 bill. Then find the total bill.

Check your answer on page 442.

Calculators

Example Marguerite ordered the following food at the food court: 2 hamburgers at $1.49 each and 3 hot dogs at $0.89 each. How much change will she receive from a $10 bill?

Calculator A	Key Sequence	Display
(MR/MC) (MR/MC) (Clear)		
2 (×) 1.49 (Enter) (▶M) (Enter)		M 2x1.49= 2.98
3 (×) .89 (Enter) (▶M) (+)		M 3x.89= 2.67
10 (−) (MR/MC) (Enter)		M 10−5.65= 4.35

Note

The key sequence (▶M) (+) is a shortcut to add the displayed number to memory. Similarly, (▶M) (−) subtracts a number from memory.

Calculator B	Key Sequence	Display
(AC) (MC)		0.
2 (×) 1.49 (=) (M+)		M 2.98
3 (×) .89 (=) (M+)		M 2.67
10 (−) (MR) (=)		M = 4.35

Marguerite will receive $4.35 in change.

Example Mr. Beckman bought 2 adult tickets at $8.25 each and 3 child tickets at $4.75 each. He redeemed a $5 gift certificate. How much did he pay for the tickets?

Note

If you fix the rounding to hundredths, all the values will be displayed as dollars and cents.

Calculator A	Key Sequence	Display
(MR/MC) (MR/MC) (Clear)		
2 (×) 8.25 (Enter) (▶M) (Enter)		M 2×8.25= 16.5
3 (×) 4.75 (Enter) (▶M) (+)		M 3×4.75= 14.25
(MR/MC) (−) 5 (Enter)		M 30.75−5= 25.75

Calculator B	Key Sequence	Display
(AC) (MC)		0.
2 (×) 8.25 (=) (M+)		M 16.5
3 (×) 4.75 (=) (M+)		M 14.25
(MR) (−) 5 (=)		M = 25.75

Mr. Beckman paid $25.75 for the tickets.

Example | Juan bought the following tickets to a baseball game for himself and 6 friends: 2 bleacher seats at $15.25 each and 5 mezzanine seats at $27.50 each. If everyone intends to split the costs evenly seven ways how much does each person owe Juan?

Note

If you fix the rounding to hundredths, all the values will be displayed as dollars and cents.

Calculator A	Key Sequence	Display
(MR/MC) (MR/MC) (Clear)		
2 (×) 15.25 (Enter) (▶M) (Enter)		M 2×15.25= 30.5
5 (×) 27.50 (Enter) (▶M) (+)		M 5×27.50= 137.5
(MR/MC) (÷) 7 (Enter)		M 168÷7= 24

Calculator B	Key Sequence	Display
(AC) (MC)		0.
2 (×) 15.25 (=) (M+)		M 30.5
5 (×) 27.50 (=) (M+)		M 137.5
(MR) (÷) 7 (=)		M = 24.

Each friend owes Juan $24.00 for the tickets.

Check Your Understanding

1. How much would 2 shirts and 2 hats cost if shirts cost $18.50 each and hats cost $13.25 each?

2. How much would it cost to take a family of 2 adults and 4 children to a matinee if tickets cost $6.25 for adults and $4.25 for children?

Check your answers on page 442.

Skip Counting on a Calculator

In earlier grades, you may have been using a 4-function calculator to skip-count.

Recall that the program needs to tell the calculator:

1. What number to count by;
2. Whether to count up or down;
3. What number to start at;
4. When to count.

Here's how to program each calculator.

Example   Starting at 3, count by 7s on this calculator.

Calculator A

Purpose	Key Sequence	Display
Tell the calculator to count up by 7. **[Op1]** is programmed to do any operation with any number that you enter between presses of **[Op1]**.	**[Op1]** **[+]** 7 **[Op1]**	↑ Op1 +7
Tell the calculator to start at 3 and do the first count.	3 **[Op1]**	↑ Op1 3+7 1 10
Tell the calculator to count again.	**[Op1]**	↑ Op1 10+7 2 17
Keep counting by pressing **[Op1]**.	**[Op1]**	↑ Op1 17+7 3 24

To count back by 7, begin with **[Op1]** **[−]** 7 **[Op1]** .

Note

You can use **[Op2]** to define a second constant operation. **[Op2]** works in exactly the same way as **[Op2]** .

Note

The number in the lower left corner of the display shows how many counts you have made.

Example   Starting at 3, count by 7s on this calculator.

Calculator B		
Purpose	**Key Sequence**	**Display**
Tell the calculator to count up by 7. The "K" on the display means you have successfully programmed the "constant," as the count-by number is sometimes called.	7 [+] [+]	_K_ **7.+**
Tell the calculator to start at 3 and do the first count.	3 [=]	_K_ **10.+**
Tell the calculator to count again.	[=]	_K_ **17.+**
Keep counting by pressing [=].	[=]	_K_ **24.+**

To count back by 7, begin with 7 [−] [−].

Check Your Understanding

Use your calculator to do the following counts. Write five counts each.

1. Starting at 11, count on by 7s.
2. Starting at 120, count back by 13s.

Check your answers on page 442.

American Currency

American currency has a long and complicated history. Bills, notes, coins, wampum, dollars, and Pieces of Eight are just a few of the things Americans have used as money.

Early Currency

A common form of currency used around the time of the American Revolution came from the Spanish. After the Spanish came to the new world, they discovered huge silver deposits in Mexico and Peru. In 1772, they started minting a new silver coin, the Spanish dollar.

◄ The Spanish dollar became plentiful in the colonies as trade with Spain increased. The coins became known as "Pieces of Eight" because they were often cut into eight pie-shaped "bits" to make change. Thus, whole coins became known as "Pieces of Eight."

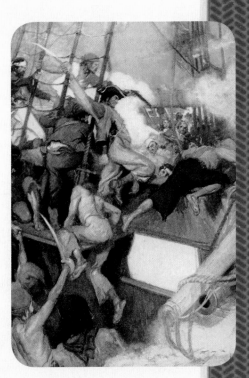

▲ "Pieces of Eight" have long been a part of pirate lore. Here pirates seek riches as they board a Spanish galleon.

The "Continental," ► issued by the Continental Congress, was used to pay for the American Revolution. Those who traded coins, gold, or silver for the note were promised re-payment after victory by the Patriots.

Coins

In 1792, the first U.S. Mint was built in Philadelphia, and soon after began producing coins for the new country. Today, the U.S. Mint can produce 1.8 million coins an hour, 32 million coins per day, 13.5 billion coins each year.

◀ It is said that Martha and George Washington gave some of their silverware to the Mint to use for the first U.S. dollars. Because silver was rare in the country, few of these coins were made.

Here a person inspects pennies at the U.S. Mint. ▶

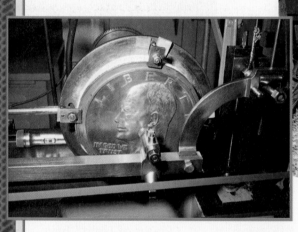

⬆ "Striking" is the process of adding designs and letters to a coin blank.

⬆ Many of our coins show portraits of famous historical figures. Who are the people portrayed on these coins?

Notes

At one time in the United States, 1,600 state banks printed their own notes, and there were 7,000 different kinds of notes in circulation. Today, the Bureau of Engraving and Printing has only two locations and produces only 1, 2, 5, 10, 20, 50, and 100 dollar notes.

Notes are produced in large sheets and then cut apart. These paper notes are made of 75% cotton and 25% linen. ▼

▲ About 37 million notes with a value of about $696 million are produced each day in America. Approximately 95% of the notes printed are used to replace notes already in use.

◄ An inspector checks over a sheet of 100-dollar notes.

A $1 note lasts about 22 months. This pile of worn out notes will be shredded. ➤

Other Forms of Currency

Throughout American history, many valuable things have been used as trade for goods.

Gold has always been scarce in this country, making it a valuable form of currency. ➤

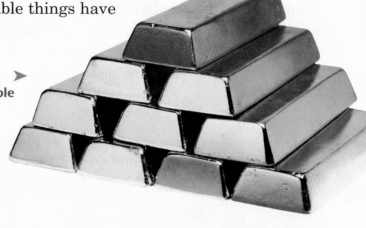

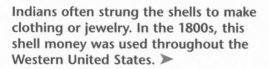

◀ Dentalia shells were used as currency by North American Indians for 2,500 years. They held their value because they came from a small region and were difficult to harvest.

Indians often strung the shells to make clothing or jewelry. In the 1800s, this shell money was used throughout the Western United States. ➤

◀ Wheat, tobacco, and other crops have been used for trade throughout American history. For example, in the 1800s, farmers would trade wheat flour for goods and services.

Funny Money

Counterfeiting of money was a huge problem during the 1800s because banks issued their own currency and there were too many unreliable notes in circulation. By the end of the Civil War, between $\frac{1}{3}$ and $\frac{1}{2}$ of all currency was counterfeit.

▲ The United States adopted a national currency in the 1860s to try to resolve the problem. But soon, the new national currency was being widely counterfeited. So, in 1865, the Secret Service was created to crack down on counterfeiting, which soon dropped sharply.

▲ This Secret Service officer is holding counterfeit gold coins.

▲ The Federal Bureau of Engraving uses a "watermark" to help prevent counterfeiting. If you hold up an authentic note to the light, you will see the watermark on the far right side. The watermark is a small portrait of the president whose face is on the note.

This hollow nickel was part of a famous spy case in 1953. When the nickel split apart in the hands of a young boy, he noticed a tiny photograph of a series of numbers. After four years of investigating, the Federal Bureau of Investigation (FBI) determined that the numbers were secret codes to be delivered to a Russian spy known as Rudolf Abel. ➤

Checks and Plastic

Checks, credit cards and debit cards are common today. These forms of payment can be used in place of money.

◄ People write checks to pay for things. The person to whom a check is written can either cash the check or deposit it into a bank account. As a result, the amount of the check is deducted from the check writer's account.

Inserting a debit card into an automatic teller machine (ATM) enables a person to withdraw money from his or her bank account. ATMs are conveniently located throughout the world. ►

◄ Credit cards enable people to borrow money to buy the things they want and need. Credit card debt must be paid off each month, or interest must be paid on the loan.

What types of currency do you and your family use?

Games

Games

Throughout the year, you will play games that help you practice important math skills. Playing mathematics games gives you a chance to practice math skills in a way that is different and enjoyable. We hope that you will play often and have fun!

In this section of your *Student Reference Book*, you will find the directions for many games. The numbers in most games are generated randomly. This means that the games can be played over and over without repeating the same problems.

Many students have created their own variations of these games to make them more interesting. We encourage you to do this too.

Materials

You need a deck of number cards for many of the games. You can use an Everything Math Deck, a deck of regular playing cards, or make your own deck out of index cards.

An Everything Math Deck includes 54 cards. There are 4 cards each for the numbers 0–10. And there is 1 card for each of the numbers 11–20.

You can also use a deck of regular playing cards after making a few changes. A deck of playing cards includes 54 cards (52 regular cards, plus 2 jokers). To create a deck of number cards, use a permanent marker to mark the cards in the following way:

◆ Mark each of the 4 aces with the number 1.

◆ Mark each of the 4 queens with the number 0.

◆ Mark the 4 jacks and 4 kings with the numbers 11 through 18.

◆ Mark the 2 jokers with the numbers 19 and 20.

For some games you will have to make a game board, a score sheet, or a set of cards that are not number cards. The instructions for doing this are included with the game directions. More complicated game boards and card decks are available from your teacher.

Angle Tangle

Materials ☐ 1 protractor

☐ 1 straightedge

☐ several blank sheets of paper

Players 2

Skill Estimating and measuring angle size

Object of the game To estimate angle sizes accurately and have the lower total score.

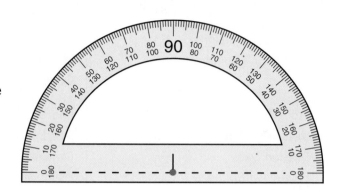

Directions

In each round:

1. Player 1 uses a straightedge to draw an angle on a sheet of paper.

2. Player 2 estimates the degree measure of the angle.

3. Player 1 measures the angle with a protractor. Players agree on the measure.

4. Player 2's score is the difference between the estimate and the actual measure of the angle. (The difference will be 0 or a positive number.)

5. Players trade roles and repeat Steps 1–4.

Players add their scores at the end of five rounds.
The player with the lower total score wins the game.

Example

	Player 1			Player 2		
	Estimate	**Actual**	**Score**	**Estimate**	**Actual**	**Score**
Round 1	120°	108°	12	50°	37°	13
Round 2	75°	86°	11	85°	87°	2
Round 3	40°	44°	4	15°	19°	4
Round 4	60°	69°	9	40°	56°	16
Round 5	135°	123°	12	150°	141°	9
Total score			**48**			**44**

Player 2 has the lower total score. Player 2 wins the game.

Baseball Multiplication (1 to 6 Facts)

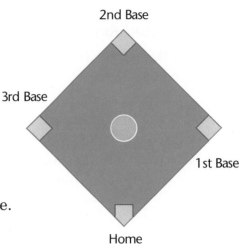

2nd Base

3rd Base

1st Base

Home

Materials ☐ 1 *Baseball Multiplication* Game Mat
(*Math Masters*, p. 445)

☐ 2 six-sided dice

☐ 4 counters

☐ 1 calculator or a multiplication/division table

Players 2 teams of one or more players each

Skill Multiplication facts 1 to 6

Object of the game To score more runs in a 3-inning game.

Directions

1. Draw a diamond and label *Home, 1st base, 2nd base,* and *3rd base*. Make a Scoreboard sheet like the one shown at the right.

2. Teams take turns being the *pitcher* and the *batter*. The rules are similar to the rules of baseball, but this game lasts only 3 innings.

3. The batter puts a counter on home plate. The pitcher rolls the dice. The batter multiplies the numbers rolled and gives the answer. The pitcher checks the answer and may use a calculator to do so.

4. If the answer is correct, the batter looks up the product in the Hitting Table at the right. If it is a hit, the batter moves all counters on the field the number of bases shown in the table. The pitcher tallies each out on the scoreboard.

5. An incorrect answer is a strike and another pitch (dice roll) is thrown. Three strikes make an out.

6. A run is scored each time a counter crosses home plate. The batter tallies each run scored on the Scoreboard.

7. After each hit or out, the batter puts a counter on home plate. The batting and pitching teams switch roles after the batting team has made 3 outs. The inning is over when both teams have made 3 outs.

The team with more runs at the end of 3 innings wins the game. If the game is tied at the end of 3 innings, play continues into extra innings until one team wins.

Scoreboard					
Inning		1	2	3	Total
Team 1	outs				
	runs				
Team 2	outs				
	runs				

Hitting Table 1 to 6 Facts	
1 to 9	Out
10 to 19	Single (1 base)
20 to 29	Double (2 bases)
30 to 35	Triple (3 bases)
36	Home Run (4 bases)

two hundred ninety-seven

Baseball Multiplication (Advanced Versions)

Skill   Multiplication facts through 12s, extended facts

Object of the game To score more runs in a 3-inning game.

1 to 10 Facts

Materials   ☐ number cards 1–10 (4 of each)

Follow the basic rules. The pitcher draws 2 cards from the deck. The batter finds their product and uses the Hitting Table at the right to find out how to move the counters.

2 to 12 Facts

Materials   ☐ 4 six-sided dice

Follow the basic rules. The pitcher rolls 4 dice. The batter separates them into 2 pairs, adds the numbers in each pair, and multiplies the sums. Use the Hitting Table at the right.

How you pair the numbers can determine the kind of hit you get or whether you get an out. For example, suppose you roll a 1, 2, 3, and 5. You could add pairs in different ways and multiply as follows:

one way	a second way	a third way
$1 + 2 = 3$	$1 + 3 = 4$	$1 + 5 = 6$
$3 + 5 = 8$	$2 + 5 = 7$	$2 + 3 = 5$
$3 * 8 = 24$	$4 * 7 = 28$	$6 * 5 = 30$
Out	Single	Single

Three-Factors Game

Materials   ☐ 3 six-sided dice

The pitcher rolls 3 dice. The batter multiplies the 3 numbers (factors) and uses the Hitting Table at the right.

10s * 10s Game

Materials   ☐ 4 six-sided dice

The rules for this game are the same as for the **2 to 12 Facts** game with two exceptions:

1. A sum of 2 through 9 represents 20 through 90. A sum of 10 through 12 represents itself. For example:
   Roll 1, 2, 3, and 5. Get sums 6 and 5. Multiply $60 * 50$.
   Roll 3, 4, 6, and 6. Get sums 12 and 7. Multiply $12 * 70$.

2. Use the Hitting Table at the right.

Hitting Table 1 to 10 Facts	
1 to 21	Out
24 to 45	Single (1 base)
48 to 70	Double (2 bases)
72 to 81	Triple (3 bases)
90 to 100	Home Run (4 bases)

Hitting Table 2 to 12 Facts	
4 to 24	Out
25 to 49	Single (1 base)
50 to 64	Double (2 bases)
66 to 77	Triple (3 bases)
80 to 144	Home Run (4 bases)

Hitting Table Three-Factors Game	
1 to 54	Out
60 to 90	Single (1 base)
96 to 120	Double (2 bases)
125 to 150	Triple (3 bases)
180 to 216	Home Run (4 bases)

Hitting Table 10s * 10s Game	
100 to 2,000	Out
2,100 to 4,000	Single (1 base)
4,200 to 5,400	Double (2 bases)
5,600 to 6,400	Triple (3 bases)
7,200 to 8,100	Home Run (4 bases)

Beat the Calculator
Multiplication Facts

Materials ☐ number cards 1–10 (4 of each)

☐ 1 calculator

Players 3

Skill Mental multiplication skills

Object of the game To multiply numbers without a calculator faster than a player using one.

Directions

1. One player is the "Caller," one is the "Calculator," and one is the "Brain."

2. Shuffle the deck and place it number-side down on the table.

3. The Caller draws 2 cards from the number deck and asks for their product.

4. The Calculator solves the problem using a calculator. The Brain solves it without a calculator. The Caller decides who got the answer first.

5. The Caller continues to draw 2 cards at a time from the number deck and ask for their product.

6. Players trade roles every 10 turns or so.

Example The Caller draws a 10 and a 5 and calls out "10 times 5." The Brain and the Calculator each solve the problem. The Caller decides who got the answer first.

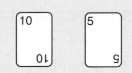

Extended Multiplication Facts

In this version of the game, the Caller:

♦ Draws 2 cards from the number deck.

♦ Attaches a 0 to either one of the factors, or to both factors, before asking for the product.

Example If the Caller turns over a 4 and an 8, he or she may make up any one of the following problems:

4 * 80   40 * 8   40 * 80

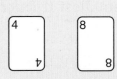

The Brain and the Calculator solve the problem. The Caller decides who got the answer first.

Build-It

Materials ☐ 1 *Build-It* Card Deck (*Math Masters*, p. 446)

☐ 1 *Build-It* Gameboard for each player (*Math Masters*, p. 447)

Players 2

Skill Comparing and ordering fractions

Object of the game To be the first player to arrange 5 fraction cards in order from smallest to largest.

Directions

1. Shuffle the fraction cards. Deal 1 card number-side down on each of the 5 spaces on the 2 *Build-It* gameboards.

2. Put the remaining cards number-side down for a **draw pile.** Turn the top card over and place it number-side up in a **discard pile.**

3. Players turn over the 5 cards on their gameboards. Do not change the order of the cards at any time during the game.

4. Players take turns. When it is your turn:

   ◆ Take either the top card from the draw pile or the top card from the discard pile.

   ◆ Decide whether to keep this card or put it on the discard pile.

   ◆ If you keep the card, it must replace 1 of the 5 cards on your *Build-It* gameboard. Put the replaced card on the discard pile.

5. If all the cards in the draw pile are used, shuffle the discard pile. Place them number-side down in a draw pile. Turn over the top card to start a new discard pile.

6. The winner is the first player to have all 5 cards on his or her gameboard in order from the smallest fraction to the largest.

Build-It Card Deck

$\frac{5}{9}$	$\frac{1}{3}$	$\frac{11}{12}$	$\frac{1}{12}$
$\frac{7}{12}$	$\frac{3}{8}$	$\frac{1}{4}$	$\frac{1}{5}$
$\frac{2}{3}$	$\frac{3}{7}$	$\frac{4}{7}$	$\frac{3}{4}$
$\frac{3}{5}$	$\frac{4}{5}$	$\frac{7}{9}$	$\frac{5}{6}$

Build-It Gameboard

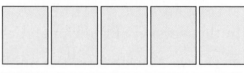

Closest to 0   Closest to 1

Credits/Debits Game (Advanced Version)

Materials
☐ 1 complete deck of number cards

☐ 1 penny

☐ 1 *Credits/Debits Game* Recording Sheet for each player (*Math Masters*, p. 450)

Players 2

Skill Addition and subtraction of positive and negative numbers

Object of the game To have more money after adding and subtracting credits and debits.

		Recording Sheet		
	Start	**Change**		**End, and next start**
		Addition or Subtraction	**Credit or Debit**	
1	+$10			
2				
3				
4				
5				
6				
7				
8				
9				
10				

Directions

You are an accountant for a business. Your job is to keep track of the company's **current balance.** The current balance is also called the **bottom line.**

1. Shuffle the deck and lay it number-side down between the players. The black-numbered cards are the "credits," and the blue- or red-numbered cards are the "debits."

2. The heads side of the coin tells you to **add** a credit or debit to the bottom line. The tails side of the coin tells you to **subtract** a credit or debit from the bottom line.

3. Each player begins with a bottom line of +$10.

4. Players take turns. On your turn, do the following:

   ◆ Flip the coin. This tells you whether to add or subtract.

   ◆ Draw a card. The card tells you what amount in dollars (positive or negative) to add or subtract from your bottom line. Red or blue numbers are negative numbers. Black numbers are positive numbers.

   ◆ Record the results in your table.

Scoring

After 10 turns each, the player with more money is the winner of the round. If both players have negative dollar amounts, the player whose amount is closer to 0 wins.

Examples
Max has a "Start" balance of $5. His coin lands heads up and he records + in the "Addition or Subtraction" column.

He draws a red 9 and records −$9 in the "Credit or Debit" column. Max adds: $5 + (−$9) = −$4. He records −$4 in the "End" balance column and also in the "Start" column on the next line.

Beth has a "Start" balance of −$20. Her coin lands tails up, which means subtract. She draws a black 1 (+$1). She subtracts: −$20 − (+$1) = −$21. Her "End" balance is −$21.

Divisibility Dash

Materials □ number cards 0–9 (4 of each)

□ number cards: 2, 3, 5, 6, 9, and 10 (2 of each)

Players 2 or 3

Skill Recognizing multiples, using divisibility tests

Object of the game To discard all cards.

Directions

1. Shuffle the divisor cards and place them number-side down on the table. Shuffle the draw cards and deal 8 to each player. Place the remaining draw cards number-side down next to the divisor cards.

2. For each round, turn the top divisor card number-side up. Players take turns. When it is your turn:

   ♦ Use the cards in your hand to make 2-digit numbers that are multiples of the divisor card. Make as many 2-digit numbers that are multiples as you can. A card used to make one 2-digit number may not be used again to make another number.

   ♦ Place all the cards you used to make 2-digit numbers in a discard pile.

   ♦ If you cannot make a 2-digit number that is a multiple of the divisor card, you must take a card from the draw pile. Your turn is over.

3. If a player disagrees that a 2-digit number is a multiple of the divisor card, that player may challenge. Players use the divisibility test for the divisor card value to check the number in question. Any numbers that are not multiples of the divisor card must be returned to the player's hand.

4. If the draw pile or divisor cards have all been used, they can be reshuffled and put back into play.

5. The first player to discard all of his or her cards is the winner.

Note

The number cards 0–9 (4 of each) are the **draw cards**. This set of draw cards is also called the **draw pile**.

The number cards 2, 3, 5, 6, 9, and 10 (2 of each) are the **divisor cards**.

| **Example** | Andrew's cards: 1 2 5 5 7 8 | Divisor card: 3 |

Andrew uses his cards to make 2 numbers that are multiples of 3: 1 5 5 7

He discards these 4 cards and holds the 2 and 8 for the next round of play.

Division Dash

Materials ☐ number cards 1–9 (4 of each)

☐ 1 score sheet

Players 1 or 2

Skill Division of 2-digit by 1-digit numbers

Object of the game To reach 100 in the fewest divisions possible.

Directions

1. Prepare a score sheet like the one shown at the right.

2. Shuffle the cards and place the deck number-side down on the table.

3. Each player follows the instructions below:

   ♦ Turn over 3 cards and lay them down in a row, from left to right. Use the 3 cards to generate a division problem. The 2 cards on the left form a 2-digit number. This is the *dividend*. The number on the card at the right is the *divisor*.

   ♦ Divide the 2-digit number by the 1-digit number and record the result. This result is your quotient. Remainders are ignored. Calculate mentally or on paper.

   ♦ Add your quotient to your previous score and record your new score. (If this is your first turn, your previous score was 0.)

4. Players repeat Step 3 until one player's score is 100 or more. The first player to reach at least 100 wins. If there is only one player, the Object of the game is to reach 100 in as few turns as possible.

Player 1		Player 2	
Quotient	Score	Quotient	Score

Example

Turn 1: Bob draws 6, 4, and 5. He divides 64 by 5. Quotient = 12. Remainder is ignored. The score is 12 + 0 = 12.

Turn 2: Bob then draws 8, 2, and 1. He divides 82 by 1. Quotient = 82. The score is 82 + 12 = 94.

Turn 3: Bob then draws 5, 7, and 8. He divides 57 by 8. Quotient = 7. Remainder is ignored. The score is 7 + 94 = 101.

Bob has reached 100 in 3 turns and the game ends.

64 is the dividend. 5 is the divisor.

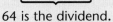

Quotient	Score
12	12
82	94
7	101

Estimation Squeeze

Materials ☐ 1 calculator

Players 2

Skill Estimating square roots

Object of the game To estimate the square root of a number without using the $\sqrt{}$ key on the calculator.

Directions

1. Pick a number that is less than 600 and is NOT a perfect square. (See the table at the right.) This is the **target number.** Record the target number.

2. Players take turns. When it is your turn:

   ◆ Estimate the square root of the target number, and enter the estimate on the calculator.

   ◆ Find the square of the estimate with the calculator, and record it.

3. The first player who makes an estimate whose square is within 0.1 of the target number wins the game. For example, if the target number is 139, the square of the estimate must be greater than 138.9 and less than 139.1.

Perfect Squares		
1	81	289
4	100	324
9	121	361
16	144	400
25	169	441
36	196	484
49	225	529
64	256	576

A perfect square is the square of a whole number.
$1 = 1 * 1$, $64 = 8 * 8$, $400 = 20^2$

Example Use your calculator to square the number 13.5.

Press 13 $\boxed{\cdot}$ 5 $\boxed{x^2}$.

Or press 13 $\boxed{\cdot}$ 5 $\boxed{\times}$ 13 $\boxed{\cdot}$ 5 $\boxed{=}$.

Answer: 182.25.

4. Do not use the $\sqrt{}$ key. This key provides the best estimate of a square root that the calculator can calculate.

Example Target Number: 139

	Estimate	Square of Estimate	
Nick	12	144	too large
Erin	11	121	too small
Nick	11.5	132.25	too small
Erin	11.8	139.24	too large
Nick	11.75	138.0625	too small
Erin	11.79	139.0041	between 138.9 and 139.1

Erin wins.

Exponent Ball

Materials ☐ 1 *Exponent Ball* Gameboard (*Math Masters,* p. 451)

☐ 1 six-sided die

☐ 1 counter

☐ 1 calculator

Players 2

Skill Converting exponential notation to standard notation, comparing probabilities

Object of the game To score more points in 4 turns.

Directions

1. The game is similar to U.S. football. Player 1 first puts the ball (counter) on one of the 20-yard lines. The objective is to reach the opposite goal line, 80 yards away. A turn consists of 4 chances to advance the counter to the goal line and score.

2. The first 3 chances must be runs on the ground. To run, Player 1 rolls the die twice. The first roll names the **base,** the second roll names the **exponent.** For example, rolls of 5 and 4 name the number $5^4 = 625$.

3. Player 1 calculates the value of the roll and uses Table 1 on the gameboard page to find how far to move the ball forward (+) or backward (−).

4. If Player 1 does not score in the first 3 chances, he or she may choose to run or kick on the fourth chance. To kick, the player rolls the die once and multiplies the number shown by 10. The result is the distance the ball travels (Table 2 on the gameboard).

5. If the ball reaches the goal line on a run, the player scores 7 points. If the ball reaches the goal line on a kick, the player scores 3 points.

6. If the ball does not reach the goal line in 4 chances, Player 1's turn ends. Player 2 starts where Player 1 stopped and moves toward the opposite goal line.

7. If Player 1 scores, Player 2 puts the ball on one of the 20-yard lines and follows the directions above.

8. Players take turns. A round consists of 4 turns for each player. The player with more points wins.

Exponent Ball Gameboard

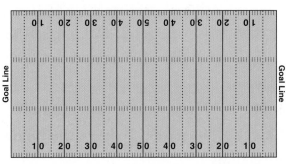

Table 1: Runs		
Value of Roll	**Move Ball**	**Chances of Gaining on the Ground**
1	−15 yd	−15 yards: 1 out of 6 or about 17%
2 to 6	+10 yd	10 yards or more: 5 out of 6 or about 83%
8 to 81	+20 yd	20 yards or more: 4 out of 6 or about 67%
in the 100s	+30 yd	30 yards or more: 13 out of 36 or about 36%
in the 1,000s	+40 yd	40 yards or more: 7 out of 36 or about 19%
in the 10,000s	+50 yd	50 yards: 1 out of 18 or about 6%

Table 2: Kicks		
Value of Roll	**Move Ball**	**Chances of Kicking**
1	+10 yd	10 yards or more: 6 out of 6 or 100%
2	+20 yd	20 yards or more: 5 out of 6 or about 83%
3	+30 yd	30 yards or more: 4 out of 6 or about 67%
4	+40 yd	40 yards or more: 3 out of 6 or about 50%
5	+50 yd	50 yards or more: 2 out of 6 or about 33%
6	+60 yd	60 yards: 1 out of 6 or about 17%

Note

If a backward move should carry the ball behind the goal line, the ball (counter) is put on the goal line.

Factor Captor

Materials ☐ 1 calculator for each player

☐ paper and pencil for each player

☐ 1 *Factor Captor* Grid—either Grid 1 or Grid 2
(*Math Masters,* pp. 453 and 454)

☐ coin-size counters (48 for Grid 1; 70 for Grid 2)

Players 2

Skill Finding factors of a number

Object of the game To have the higher total score.

Directions

1. To start the first round, Player 1 chooses a 2-digit number on the number grid, covers it with a counter, and records the number on scratch paper. This is Player 1's score for the round.

2. Player 2 covers all of the factors of Player 1's number. Player 2 finds the sum of the factors and records it on scratch paper. This is Player 2's score for the round.

A factor may only be covered once during a round.

3. If Player 2 missed any factors, Player 1 can cover them with counters and add them to his or her score.

4. In the next round, players switch roles. Player 2 chooses a number that is not covered by a counter. Player 1 covers all factors of that number.

5. Any number that is covered by a counter is no longer available and may not be used again.

6. The first player in a round may not cover a number that is less than 10, unless no other numbers are available.

7. Play continues with players trading roles after each round, until all numbers on the grid have been covered. Players then use their calculators to find their total scores. The player with the higher score wins the game.

Grid 1 (Beginning Level)

1	2	2	2	2	2
2	3	3	3	3	3
3	4	4	4	4	5
5	5	5	6	6	7
7	8	8	9	9	10
10	11	12	13	14	15
16	18	20	21	22	24
25	26	27	28	30	32

Grid 2 (Advanced Level)

1	2	2	2	2	2	3
3	3	3	3	4	4	4
4	5	5	5	5	6	6
6	7	7	8	8	9	9
10	10	11	12	13	14	15
16	17	18	19	20	21	22
23	24	25	26	27	28	30
32	33	34	35	36	38	39
40	42	44	45	46	48	49
50	51	52	54	55	56	60

Example

Round 1: James covers 27 and scores 27 points. Emma covers 1, 3, and 9, and scores 1 + 3 + 9 = 13 points.

Round 2: Emma covers 18 and scores 18 points. James covers 2, 3, and 6, and scores 2 + 3 + 6 = 11 points. Emma covers 9 with a counter, because 9 is also a factor of 18. Emma adds 9 points to her score.

Factor Top-It

Materials ☐ number cards 0–9 (4 of each)

☐ paper and pencil for each player

Players   2 to 4

Skill   Finding factors of a number

Object of the game To score the most points in 5 rounds.

Directions

1. Shuffle the deck and place it number-side down on the table.

2. In each round, players take turns. When it is your turn:
   ♦ Draw 2 cards from the top of the deck.
   ♦ Use the cards to make a 2-digit number.
   ♦ Record the number and all of its factors on a piece of paper.
   ♦ Find the sum of all the factors. This is your score for the round.

3. Play 5 rounds.

4. The winner is the player with the most points at the end of 5 rounds.

Example   Find each player's score for the round.

Player 1
Cards 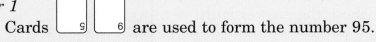 are used to form the number 95.

Factors: 1, 5, 19, 95
Score: $1 + 5 + 19 + 95 = 120$

Player 2
Cards are used to form the number 88.

Factors: 1, 2, 4, 8, 11, 22, 44, 88
Score: $1 + 2 + 4 + 8 + 11 + 22 + 44 + 88 = 180$

Player 3
Cards are used to form the number 52.

Factors: 1, 2, 4, 13, 26, 52
Score: $1 + 2 + 4 + 13 + 26 + 52 = 98$

Player 2 scored the most points for this round.

First to 100

Materials ☐ one set of *First to 100* Problem Cards
(*Math Masters,* pp. 456 and 457)

☐ 2 six-sided dice

☐ 1 calculator

Players 2 to 4

Skill Variable substitution, solving equations

Object of the game To collect 100 points by solving problems.

Directions

1. Shuffle the Problem Cards and place them word-side down on the table.

2. Players take turns. When it is your turn:

   ♦ Roll 2 dice and find the product of the numbers.

   ♦ Turn over the top Problem Card and substitute the product for the variable x in the problem on the card.

   ♦ Solve the problem mentally or use paper and pencil. Give the answer. (You have 3 chances to use a calculator to solve difficult problems during a game.) Other players check the answer with a calculator.

   ♦ If the answer is correct, you win the number of points equal to the product that was substituted for the variable x. Some Problem Cards require 2 or more answers. In order to win any points, you must answer all parts of the problem correctly.

   ♦ Put the used Problem Card at the bottom of the deck.

3. The first player to get at least 100 points wins the game.

First to 100 Problem Cards

How many inches are there in x feet? How many centimeters are there in x meters? **1**	How many quarts are there in x gallons? **2**	What is the smallest number of x's you can add to get a sum greater than 100? **3**	Is $50 * x$ greater than 1,000? Is $\frac{x}{10}$ less than 1? **4**
$\frac{1}{2}$ of x = ? $\frac{1}{10}$ of x = ? **5**	$1 - x$ = ? $x + 998$ = ? **6**	If x people share 1,000 stamps equally, how many stamps will each person get? **7**	What time will it be x minutes from now? What time was it x minutes ago? **8**
It is 102 miles to your destination. You have gone x miles. How many miles are left? **9**	What whole or mixed number equals x divided by 2? **10**	Is x a prime or a composite number? Is x divisible by 2? **11**	The time is 11:05 A.M. The train left x minutes ago. What time did the train leave? **12**
Bill was born in 1939. Freddy was born the same day, but x years later. In what year was Freddy born? **13**	Which is larger: $2 * x$ or $x + 50$? **14**	There are x rows of seats. There are 9 seats in each row. How many seats are there in all? **15**	Sargon spent x cents on apples. If she paid with a $5 bill, how much change should she get? **16**

The temperature was 25°F. It dropped x degrees. What was the new temperature? **17**	Each story in a building is 10 ft high. If the building has x stories, how tall is it? **18**	Which is larger: $2 * x$ or $\frac{100}{x}$? **19**	$20 * x$ = ? **20**
Name all the whole-number factors of x. **21**	Is x an even or an odd number? Is x divisible by 9? **22**	Shalanda was born on a Tuesday. Linda was born x days later. On what day of the week was Linda born? **23**	Will had a quarter plus x cents. How much money did he have in all? **24**
Find the perimeter and area of this square. x cm x cm **25**	What is the median of these weights? 5 pounds 21 pounds x pounds What is the range? **26**	 **27**	x^2 = ? 50% of x^2 = ? **28**
$(3x + 4) - 8$ = ? **29**	x out of 100 students voted for Ruby. Is this more than 25%, less than 25%, or exactly 25% of the students? **30**	There are 200 students at Wilson School. x% speak Spanish. How many students speak Spanish? **31**	People answered a survey question either Yes or No. x% answered Yes. What percent answered No? **32**

Example Alice rolls a 3 and a 4. The product is 12.

She turns over a Problem Card: $20 * x$ = ?
She substitutes 12 for x and answers 240.
The answer is correct. Alice wins 12 points.

Frac-Tac-Toe

2-4-5-10 Frac-Tac-Toe

Materials ☐ number cards 0–10 (4 of each)

☐ 1 *Frac-Tac-Toe* Number-Card Board (*Math Masters,* p. 472)

☐ 1 *2-4-5-10 Frac-Tac-Toe* (Decimals) gameboard (*Math Masters,* p. 474)

☐ counters (2 colors), or pennies (one player using heads; the other, tails)

☐ 1 calculator

Players 2

Skill Renaming fractions as decimals and percents

Object of the game To cover 3 squares in a row, in any direction (horizontal, vertical, diagonal).

Advance Preparation Separate the cards into 2 piles on the Number-Card Board—a numerator pile and a denominator pile. For a *2-4-5-10* game, place 2 each of the 2, 4, 5, and 10 cards in the denominator pile. All other cards are placed on the numerator pile.

Shuffle the cards in each pile. Place the piles number-side down in the left-hand spaces. When the numerator pile is completely used, reshuffle that pile, and place it number-side down in the left-hand space. When the denominator pile is completely used, turn it over and place it number-side down in the left-hand space without reshuffling it.

Directions

1. Players take turns. When it is your turn:

   ◆ Turn over the top card from each pile to form a fraction (numerator card above denominator card).

   ◆ Try to match the fraction shown with one of the grid squares on the gameboard. If a match is found, cover that grid square with your counter and your turn is over. If no match is found, your turn is over.

2. To change the fraction shown by the cards to a decimal, players may use either a calculator or the *Table of Decimal Equivalents for Fractions* on page 400.

Frac-Tac-Toe
Number-Card Board

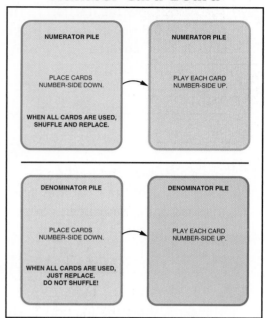

Gameboard for *2-4-5-10* *Frac-Tac-Toe* (Decimals)

>1.0	0 or 1	>2.0	0 or 1	>1.0
0.1	0.2	0.25	0.3	0.4
>1.5	0.5	>1.5	0.5	>1.5
0.6	0.7	0.75	0.8	0.9
>1.0	0 or 1	>2.0	0 or 1	>1.0

Examples

The cards show the fraction $\frac{4}{5}$. The player may cover the 0.8 square, unless that square has already been covered.

The cards show the fraction $\frac{0}{5}$. The player may cover any 1 of the 4 squares labeled "0 or 1" that has not already been covered.

The cards show the fraction $\frac{4}{2}$. The player may cover any square labeled ">1.0" or "> 1.5" that has not already been covered. The player may not cover a square labeled "> 2.0," because $\frac{4}{2}$ is equal to, but not greater than, 2.0.

3. The first player to cover 3 squares in a row in any direction (horizontal, vertical, diagonal) is the winner of the game.

Variation

Play a version of the *2-4-5-10* game using the percent gameboard shown at the right. Use *Math Masters*, page 476.

Gameboard for *2-4-5-10*
***Frac-Tac-Toe* (Percents)**

>100%	0% or 100%	>200%	0% or 100%	>100%
10%	20%	25%	30%	40%
>100%	50%	>200%	50%	>100%
60%	70%	75%	80%	90%
>100%	0% or 100%	>200%	0% or 100%	>100%

2-4-8 and 3-6-9 Frac-Tac-Toe

Play the *2-4-8* or the *3-6-9* version of the game. Gameboards for the different versions are shown below.

♦ For a *2-4-8* game, place 2 each of the 2, 4, and 8 cards in the denominator pile. Use *Math Masters*, page 478 or 480.

♦ For a *3-6-9* game, place 2 each of the 3, 6, and 9 cards in the denominator pile. Use *Math Masters*, page 482 or 484.

2-4-8 Frac-Tac-Toe (Decimals)

>2.0	0 or 1	>1.5	0 or 1	>2.0
1.5	0.125	0.25	0.375	1.5
>1.0	0.5	0.25 or 0.75	0.5	>1.0
2.0	0.625	0.75	0.875	2.0
>2.0	0 or 1	1.125	0 or 1	>2.0

2-4-8 Frac-Tac-Toe (Percents)

>200%	0% or 100%	>150%	0% or 100%	>200%
150%	$12\frac{1}{2}$%	25%	$37\frac{1}{2}$%	150%
>100%	50%	25% or 75%	50%	>100%
200%	$62\frac{1}{2}$%	75%	$87\frac{1}{2}$%	200%
>200%	0% or 100%	$112\frac{1}{2}$%	0% or 100%	>200%

3-6-9 Frac-Tac-Toe (Decimals)

>1.0	0 or 1	$0.\overline{1}$	0 or 1	>1.0
$0.1\overline{6}$	$0.\overline{2}$	$0.\overline{3}$	$0.\overline{3}$	$0.\overline{4}$
>2.0	$0.\overline{5}$	>1.0	$0.\overline{6}$	>2.0
$0.\overline{6}$	$0.\overline{7}$	$0.8\overline{3}$	$0.\overline{8}$	$1.\overline{3}$
>1.0	0 or 1	$1.\overline{6}$	0 or 1	>1.0

3-6-9 Frac-Tac-Toe (Percents)

>100%	0% or 100%	11.1%	0% or 100%	>100%
$16\frac{2}{3}$%	22.2%	$33\frac{1}{3}$%	33.3%	44.4%
>200%	55.5%	>100%	66.6%	>200%
$66\frac{2}{3}$%	77.7%	$83\frac{1}{3}$%	88.8%	$133\frac{1}{3}$%
>100%	0% or 100%	$166\frac{2}{3}$%	0% or 100%	>100%

Fraction Action, Fraction Friction

Materials ☐ 1 *Fraction Action, Fraction Friction* Card
Deck (*Math Masters*, p. 459)

☐ 1 or more calculators

Players 2 or 3

Skill Estimating sums of fractions

Object of the game To collect a set of fraction cards with
a sum as close as possible to 2, without going over 2.

Directions

1. Shuffle the deck and place it number-side down on
   the table between the players.

2. Players take turns.

   ◆ On each player's first turn, he or she takes a card
     from the top of the pile and places it number-side
     up on the table.

   ◆ On each of the player's following turns, he or she
     announces one of the following:

"Action" This means that the player wants an additional
card. The player believes that the sum of the fraction cards
he or she already has is *not* close enough to 2 to win the
hand. The player thinks that another card will bring the
sum of the fractions closer to 2, without going over 2.

"Friction" This means that the player does not want an
additional card. The player believes that the sum of the
fraction cards he or she already has *is* close enough to 2 to
win the hand. The player thinks there is a good chance that
taking another card will make the sum of the fractions
greater than 2.

**Once a player says "Friction," he or she cannot say
"Action" on any turn after that.**

3. Play continues until all players have announced "Friction"
   or have a set of cards whose sum is greater than 2. The
   player whose sum is closest to 2 without going over 2 is
   the winner of that round. Players may check each other's
   sums on their calculators.

4. Reshuffle the cards and begin again. The winner of the
   game is the first player to win 5 rounds.

**Fraction Action, Fraction
Friction Card Deck**

$\frac{1}{2}$	$\frac{1}{3}$	$\frac{2}{3}$	$\frac{1}{4}$
$\frac{3}{4}$	$\frac{1}{6}$	$\frac{1}{6}$	$\frac{5}{6}$
$\frac{1}{12}$	$\frac{1}{12}$	$\frac{5}{12}$	$\frac{5}{12}$
$\frac{7}{12}$	$\frac{7}{12}$	$\frac{11}{12}$	$\frac{11}{12}$

Fraction Of

Materials ☐ 1 deck of *Fraction Of* Fraction Cards
(*Math Masters*, pp. 464 and 465)

☐ 1 deck of *Fraction Of* Set Cards
(*Math Masters*, p. 469)

☐ 1 *Fraction Of* Gameboard and Record Sheet
for each player (*Math Masters*, p. 466)

Players 2

Skill Multiplication of fractions and whole numbers

Object of the game To score more points by solving "fraction of" problems.

Directions

1. Shuffle each deck separately. Place both decks number-side down on the table.

2. Players take turns. On your turn, draw 1 card from each deck. Use the cards to create a "fraction of" problem on your gameboard.

♦ The Fraction Card indicates what fraction of the set you must find.

♦ The Set Card offers 3 possible choices. Choose a set that will result in a "fraction of" problem with a whole-number solution.

♦ Solve the "fraction of" problem and set the 2 cards aside. The solution is your point score for the turn.

Example

Player 1 draws $\frac{1}{10}$ and [28 counters / 35 counters / 30 counters].

$\frac{1}{10}$ of 28 will *not* result in a whole-number solution.
$\frac{1}{10}$ of 28 counters is 2.8 counters.

$\frac{1}{10}$ of 35 will *not* result in a whole-number solution.
$\frac{1}{10}$ of 35 counters is 3.5 counters.

$\frac{1}{10}$ of 30 *will* result in a whole-number solution.
$\frac{1}{10}$ of 30 counters is 3 counters.

Player 1 chooses 30 counters as the set for the "fraction of" problem.

Example

$\frac{1}{2}$ ⃝

| 12 counters |
| 30 counters |
| 25 counters |

Player 2 draws [ ] and [ ] .

Player 2 could choose 12 or 30 counters as the set.

Player 2 chooses 30 counters (since they will give more points than 12 counters will), finds $\frac{1}{2}$ of 30, and earns 15 points.

3. Play continues until all of the cards in the fraction pile or the set pile have been used. The player with more points wins.

Variation

For a basic version of the game, use only the fraction cards marked with a hexagon in the corner.

	WHOLE (Choose 1 of these sets.)
Fraction Card of	Set Card

Round	Fraction-of Problem	Points
Sample	$\frac{1}{5}$ of 25	5
1		
2		
3		
4		
5		
6		
7		
8		
Total score		

Fraction/Percent Concentration

Materials ☐ 1 set of *Fraction/Percent Concentration* Tiles
(*Math Masters,* pp. 467 and 468)

☐ 1 calculator

Players 2 or 3

Skill Recognizing fractions and percents that are
equivalent

Object of the game To collect the most tiles by matching
equivalent fraction and percent tiles.

Advance Preparation Make a 2-sided single-sheet copy of
Math Masters, pages 467 and 468.

Directions

1. Spread the tiles out number-side down on the table.
   Create 2 separate piles—a fraction pile and a percent
   pile. Mix up the tiles in each pile. The 12 fraction tiles
   should have the fraction $\frac{a}{b}$ showing. The 12 percent
   tiles should have the percent symbol % showing.

2. Players take turns. At each turn, a player turns over
   both a fraction tile and a percent tile. If the fraction
   and the percent are equivalent, the player keeps the
   tiles. If the fraction and the percent are not equivalent,
   the player turns the tiles number-side down.

3. Players may use a calculator to check each other's
   matches.

4. The game ends when all tiles have been taken. The
   player with the most tiles wins.

**Fraction/Percent
Concentration Tiles
(number-side up)**

10%	20%	25%	30%
40%	50%	60%	70%
75%	80%	90%	100%
$\frac{1}{2}$	$\frac{1}{4}$	$\frac{3}{4}$	$\frac{1}{5}$
$\frac{2}{5}$	$\frac{3}{5}$	$\frac{4}{5}$	$\frac{1}{10}$
$\frac{3}{10}$	$\frac{7}{10}$	$\frac{9}{10}$	$\frac{2}{2}$

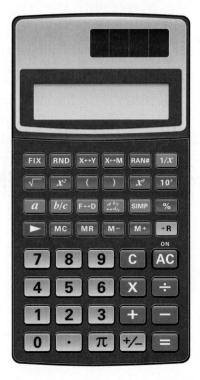

Fraction Top-It

Materials ☐ 1 set of Fraction Cards 1 and 2
(*Math Masters*, pp. 462 and 463)

Players   2 to 4

Skill   Comparing fractions

Object of the game To collect the most cards.

Advance Preparation Before beginning the game, write
the fraction for the shaded part on the back of each card.

Directions

1. Deal the same number of cards, fraction-side up, to
    each player:

    ◆ If there are 2 players, 16 cards each.

    ◆ If there are 3 players, 10 cards each.

    ◆ If there are 4 players, 8 cards each.

2. Players spread their cards out, fraction-side up, so
    that all of the cards may be seen.

3. Starting with the dealer and going in a clockwise
    direction, each player plays one card. Place the
    cards fraction-side up on the table.

4. The player with the largest fraction wins the round
    and takes the cards. Players may check who has
    the largest fraction by turning over the cards and
    comparing the amounts shaded.

5. If there is a tie for the largest fraction, each player
    plays another card. The player with the largest
    fraction takes all the cards from both plays.

6. The player who takes the cards starts the next
    round. The game is over when all cards have
    been played.

7. The player who takes the most cards wins.

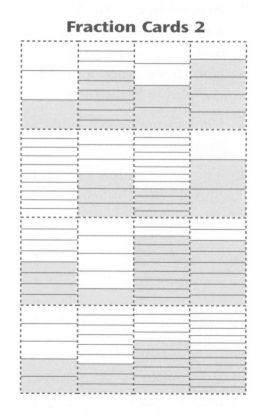

Fraction Cards 1

Fraction Cards 2

Fraction/Whole Number Top-It

Materials □ number cards 1–10 (4 of each)

□ 1 calculator (optional)

Players 2 to 4

Skill Multiplication of whole numbers and fractions

Object of the game To collect the most cards.

Directions

1. Shuffle the cards and place them number-side down on the table.

2. Each player turns over 3 cards. The card numbers are used to form 1 whole number and 1 fraction.

   ♦ The first card drawn is placed number-side up on the table. This card number is the whole number.

   ♦ The second and third cards drawn are used to form a fraction and are placed number-side up next to the first card. The fraction that these cards form must be less than or equal to 1.

3. Each player calculates the product of their whole number and fraction and calls it out as a mixed number. The player with the largest product takes all the cards. Players may use a calculator to compare their products.

4. In case of a tie for the largest product, each tied player repeats Steps 2 and 3. The player with the largest product takes all the cards from both plays.

5. The game ends when there are not enough cards left for each player to have another turn. The player with the most cards wins.

Example Amy turns over a 3, a 9, and a 5, in that order.

Roger turns over a 7, a 2, and an 8, in that order.

Amy's product is $3 * \frac{5}{9} = \frac{15}{9}$.

Roger's product is $7 * \frac{2}{8} = \frac{14}{8}$.

$\frac{15}{9} = \frac{5}{3} = 1\frac{2}{3}$. $\frac{14}{8} = \frac{7}{4} = 1\frac{3}{4}$.

Roger's product is larger, so he takes all of the cards.

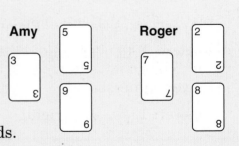

Advanced Version

Each player turns over 4 cards and forms 1 fraction from their first 2 cards and a second fraction from their last 2 cards. (All fractions must be less than or equal to 1.) Each player calculates the product of their fractions, and the player with the largest product takes all the cards.

Getting to One

Materials ☐ 1 calculator

Players 2

Skill Estimation

Object of the game To correctly guess a mystery number in as few tries as possible.

Directions

1. Player 1 chooses a mystery number that is between 1 and 100.

2. Player 2 guesses the mystery number.

3. Player 1 uses a calculator to divide Player 2's guess by the mystery number. Player 1 then reads the answer in the calculator display. If the answer has more than 2 decimal places, only the first 2 decimal places are read.

4. Player 2 continues to guess until the calculator result is 1. Player 2 keeps track of the number of guesses.

5. When Player 2 has guessed the mystery number, players trade roles and follow Steps 1–4 again. The player who guesses their mystery number in the fewest number of guesses wins the round. The first player to win 3 rounds wins the game.

Note

For a decimal number, the places to the right of the decimal point with digits in them are called *decimal places*.

For example, 4.56 has 2 decimal places, 123.4 has 1 decimal place, and 0.789 has 3 decimal places.

Example   Player 1 chooses the mystery number 65.

Player 2 guesses: 45.   Player 1 keys in: 45 ÷ 65 Enter.
Answer: 0.69   Too small.

Player 2 guesses: 73.   Player 1 keys in: 73 ÷ 65 Enter.
Answer: 1.12   Too big.

Player 2 guesses: 65.   Player A keys in: 65 ÷ 65 Enter.
Answer: 1.   Just right!

Advanced Version

Allow mystery numbers up to 1,000.

Hidden Treasure

Materials □ 1 sheet of *Hidden Treasure* Gameboards for each player
(Math Masters, p. 485)

□ 2 pencils

□ 1 red pen or crayon

Players 2

Skill Plotting ordered pairs, developing a search strategy

Object of the game To find the other player's hidden point on a coordinate grid.

Grid 1

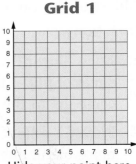

Hide your point here.

Directions

1. Each player uses 2 grids. Players sit so they cannot see what the other is writing.

2. Each player secretly marks a point on his or her Grid 1. Use the red pen or crayon. These are the "hidden" points.

3. Player 1 guesses the location of Player 2's hidden point by naming an ordered pair. To name the ordered pair (1,2), say "1 comma 2."

4. If Player 2's hidden point is at that location, Player 1 wins.

5. If the hidden point is not at that location, Player 2 marks the guess in pencil on his or her Grid 1. Player 2 counts the least number of "square sides" needed to travel from the hidden point to the guessed point and tells it to Player 1. Repeat Steps 3–5 with Player 2 guessing and Player 1 answering.

6. Play continues until one player finds the other's hidden point.

Grid 2

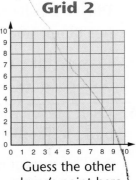

Guess the other player's point here.

Example Player 1 marks a hidden point at (2,5).

Player 2 marks a hidden point at (3,7).

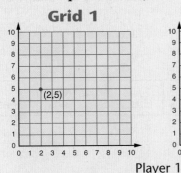

Player 1

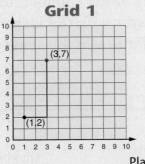

Player 2

- Player 1 guesses that Player 2's hidden point is at (1,2) and marks it on Grid 2 in pencil.

- Player 2 marks the point (1,2) in pencil on Grid 1 and tells Player 1 that (1,2) is 7 units (square sides) away from the hidden point.

- Player 1 writes "7" next to the point (1,2) on his or her Grid 2. Player 1's turn is over, and Player 2 makes a guess.

High-Number Toss

Materials ☐ 1 six-sided die

☐ 1 sheet of paper for each player

Players 2

Skill Place value, exponential notation

Object of the game To make the largest numbers possible.

Directions

1. Each player draws 4 blank lines on a sheet of paper to record the numbers that come up on the rolls of the die.

   Player 1: ____ ____ ____ | ____

   Player 2: ____ ____ ____ | ____

2. Player 1 rolls the die and writes the number on any of his or her 4 blank lines. It does not have to be the first blank—it can be any of them. *Keep in mind that the larger number wins!*

3. Player 2 rolls the die and writes the number on one of his or her blank lines.

4. Players take turns rolling the die and writing the number 3 more times each.

5. Each player then uses the 4 numbers on his or her blanks to build a number.

   ◆ The numbers on the first 3 blanks are the first 3 digits of the number the player builds.

   ◆ The number on the last blank tells the number of zeros that come after the first 3 digits.

6. Each player reads his or her number. (See the place-value chart below.) The player with the larger number wins the round. The first player to win 4 rounds wins the game.

Note

If you don't have a die, you can use a deck of number cards. Use all cards with the numbers 1 through 6. Instead of rolling the die, draw the top card from the facedown deck.

Hundred Millions	Ten Millions	Millions	,	Hundred Thousands	Ten Thousands	Thousands	,	Hundreds	Tens	Ones

Example

                    First three digits   Number of zeros

Player 1: _1_ _3_ _2_ | _6_ = 132,000,000 (132 million)

Player 2: _3_ _5_ _6_ | _4_ = 3,560,000 (3 million, 560 thousand)

Player 1 wins.

High-Number Toss: Decimal Version

Materials ☐ number cards 0–9 (4 of each)
☐ scorecard for each player

Players 2

Skill Decimal place value, subtraction, and addition

Object of the game To make the largest decimal numbers possible.

Directions

1. Each player makes a scorecard like the one at the right. Players fill out their own scorecards at each turn.

2. Shuffle the cards and place them number-side down on the table.

3. In each round:

   ♦ Player 1 draws the top card from the deck and writes the number on any of the 3 blanks on the scorecard. It need not be the first blank—it can be any of them.

   ♦ Player 2 draws the next card from the deck and writes the number on one of his or her blanks.

   ♦ Players take turns doing this 2 more times. The player with the larger number wins the round.

4. The winner's score for a round is the difference between the two players' numbers. (Subtract the smaller number from the larger number.) The loser scores 0 points for the round.

Scorecard

Game 1

Round 1	Score
0. __ __ __	_____

Round 2

| 0. __ __ __ | _____ |

Round 3

| 0. __ __ __ | _____ |

Round 4

| 0. __ __ __ | _____ |

Total:
_____

Example   Player 1: 0 . <u>6</u> <u>5</u> <u>4</u>
Player 2: 0 . <u>7</u> <u>5</u> <u>3</u>

Player 2 has the larger number and wins the round.

Since 0.753 − 0.654 = 0.099, Player 2 scores 0.099 point for the round. Player 1 scores 0 points.

5. Players take turns starting a round. At the end of 4 rounds, they find their total scores. The player with the larger total score wins the game.

Mixed-Number Spin

Materials ☐ 1 *Mixed-Number Spin* Record Sheet (*Math Masters,* p. 489)

☐ 1 *Mixed-Number* Spinner (*Math Masters,* p. 488)

☐ 1 large paper clip

Players 2

Skill Addition and subtraction of fractions and mixed numbers, solving inequalities

Object of the game To complete 10 number sentences that are true.

Directions

1. Each player writes his or her name in one of the boxes on the Record Sheet.

2. Players take turns. When it is your turn:
   ♦ Anchor the paper clip to the center of the spinner with the point of your pencil and use your other hand to spin the paper clip.
   ♦ Once the paper clip stops, write the fraction or mixed number it most closely points to on any one of the blanks below your name.

3. The first player to complete 10 true number sentences is the winner.

Mixed-Number Spin Record Sheet

name		name	
___ + ___ < 3		___ + ___ < 3	
___ + ___ > 3		___ + ___ > 3	
___ − ___ < 1		___ − ___ < 1	
___ − ___ < $\frac{1}{2}$		___ − ___ < $\frac{1}{2}$	
___ + ___ > 1		___ + ___ > 1	
___ + ___ < 1		___ + ___ < 1	
___ + ___ < 2		___ + ___ < 2	
___ − ___ = 3		___ − ___ = 3	
___ − ___ > 1		___ − ___ > 1	
___ + ___ > $\frac{1}{2}$		___ + ___ > $\frac{1}{2}$	
___ + ___ < 3		___ + ___ < 3	
___ + ___ > 2		___ + ___ > 2	

Example Ella has filled in 2 blanks in different sentences.

On her next turn, Ella spins $1\frac{3}{8}$. She has 2 choices for where she can write this mixed number. She can place it on a line where there are 2 blanks.

Or, she can use it to form the true number sentence $1\frac{3}{8} - 1\frac{1}{8} < \frac{1}{2}$.

She cannot use it on the first line because $2 + 1\frac{3}{8}$ is not < 3.

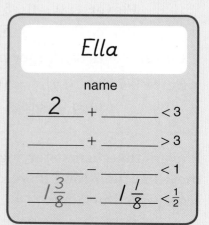

Ella		Ella		Ella	
name		name		name	
__2__ + ___ < 3		__2__ + ___ < 3		__2__ + ___ < 3	
___ + ___ > 3		___ + ___ > 3		___ + ___ > 3	
___ − ___ < 1		$1\frac{3}{8}$ − ___ < 1		___ − ___ < 1	
___ − $1\frac{1}{8}$ < $\frac{1}{2}$		___ − $1\frac{1}{8}$ < $\frac{1}{2}$		$1\frac{3}{8}$ − $1\frac{1}{8}$ < $\frac{1}{2}$	

Multiplication Bull's-Eye

Materials ☐ number cards 0–9 (4 of each)

☐ 1 six-sided die

☐ 1 calculator

Players 2

Skill Estimating products of 2- and 3-digit numbers

Object of the game To score more points.

Directions

1. Shuffle the deck and place it number-side down on the table.

2. Players take turns. When it is your turn:

   ◆ Roll the die. Look up the target range of the product in the table at the right.

   ◆ Take 4 cards from the top of the deck.

   ◆ Use the cards to try to form 2 numbers whose product falls within the target range. **Do not use a calculator.**

   ◆ Multiply the 2 numbers on your calculator to determine whether the product falls within the target range. If it does, you have hit the bull's-eye and score 1 point. If it doesn't, you score 0 points.

   ◆ Sometimes it is impossible to form 2 numbers whose product falls within the target range. If this happens, you score 0 points for that turn.

3. The game ends when each player has had 5 turns.

4. The player scoring more points wins the game.

Number on Die	Target Range of Product
1	500 or less
2	501 – 1,000
3	1,001 – 3,000
4	3,001 – 5,000
5	5,001 – 7,000
6	more than 7,000

Example   Tom rolls a 3, so the target range of the product is from 1,001 to 3,000. He turns over a 5, a 7, a 9, and a 2.

Tom uses estimation to try to form 2 numbers whose product falls within the target range—for example, 97 and 25.

He then finds the product on the calculator: 97 * 25 = 2,425.

Since the product is between 1,001 and 3,000, Tom has hit the bull's-eye and scores 1 point.

Some other possible winning products from the 5, 7, 2, and 9 cards are:
25 * 79, 27 * 59, 9 * 257, and 2 * 579.

Multiplication Wrestling

Materials ☐ number cards 0-9 (4 of each)

Players 2

Skill Partial-Products Algorithm

Object of the game To get the larger product of two 2-digit numbers.

Directions

1. Shuffle the deck and place it number-side down on the table.

2. Each player draws 4 cards and forms two 2-digit numbers. Players should form their 2 numbers so that their product is as large as possible.

3. Players create 2 "wrestling teams" by writing each of their numbers as a sum of 10s and 1s.

4. Each player's 2 teams wrestle. Each member of the first team (for example, 70 and 5) is multiplied by each member of the second team (for example, 80 and 4). Then the 4 products are added.

5. The player with the larger product wins the round and receives 1 point.

6. To begin a new round, each player draws 4 new cards to form 2 new numbers. A game consists of 3 rounds.

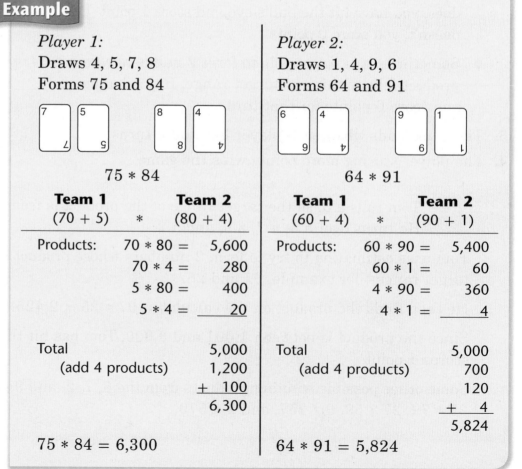

Example

Player 1:
Draws 4, 5, 7, 8
Forms 75 and 84

75 * 84

Team 1		**Team 2**
(70 + 5)	*	(80 + 4)

Products:	70 * 80 =	5,600
	70 * 4 =	280
	5 * 80 =	400
	5 * 4 =	20

Total	5,000
(add 4 products)	1,200
	+ 100
	6,300

75 * 84 = 6,300

Player 2:
Draws 1, 4, 9, 6
Forms 64 and 91

64 * 91

Team 1		**Team 2**
(60 + 4)	*	(90 + 1)

Products:	60 * 90 =	5,400
	60 * 1 =	60
	4 * 90 =	360
	4 * 1 =	4

Total	5,000
(add 4 products)	700
	120
	+ 4
	5,824

64 * 91 = 5,824

Name That Number

Materials   ☐ 1 complete deck of number cards

Players   2 or 3

Skill   Naming numbers with expressions

Object of the game   To collect the most cards.

Directions

1. Shuffle the deck and deal 5 cards to each player. Place the remaining cards number-side down on the table between the players. Turn over the top card and place it beside the deck. This is the **target number** for the round.

2. Players try to match the target number by adding, subtracting, multiplying, or dividing the numbers on as many of their cards as possible. A card may only be used once.

3. Players write their solutions on a sheet of paper. When players have written their best solutions:

    ◆ Each player sets aside the cards they used to match the target number.

    ◆ Each player replaces the cards they set aside by drawing new cards from the top of the deck.

    ◆ The old target number is placed on the bottom of the deck.

    ◆ A new target number is turned over, and another round is played.

4. Play continues until there are not enough cards left to replace all of the players' cards. The player who has set aside the most cards wins the game.

Example   Target number: 16

Player 1's cards: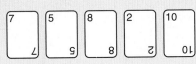

Some possible solutions:

$10 + 8 - 2 = 16$ (3 cards used)

$7 * 2 + 10 - 8 = 16$ (4 cards used)

$8 / 2 + 10 + 7 - 5 = 16$ (all 5 cards used)

The player sets aside the cards used to make a solution and draws the same number of cards from the top of the deck.

Number Top-It (7-Digit Numbers)

Materials ☐ number cards 0–9 (4 of each)

☐ one Place-Value Mat
(*Math Masters*, pp. 491 and 492)

Players   2 to 5

Skill   Place value for whole numbers

Object of the game   To make the largest 7-digit numbers.

Directions

1. Shuffle the cards and place the deck number-side down on the table.

2. Each player uses one row of boxes on the place-value mat.

3. In each round, players take turns turning over the top card from the deck and placing it on any one of their empty boxes. Each player takes a total of 7 turns, and places 7 cards on his or her row of the game mat.

4. At the end of each round, players read their numbers aloud and compare them to the other players' numbers. The player with the largest number for the round scores 1 point. The player with the next-largest number scores 2 points, and so on.

5. Players play 5 rounds for a game. Shuffle the deck between each round. The player with the *smallest* total number of points at the end of 5 rounds wins the game.

Example   Andy and Barb played 7-digit *Number Top-It*. Here is the result for one complete round of play.

		Place-Value Mat				
Millions	**Hundred Thousands**	**Ten Thousands**	**Thousands**	**Hundreds**	**Tens**	**Ones**
Andy 7	6	4	5	2	0	1
Barb 4	9	7	3	5	2	4

Andy's number is larger than Barb's number. So Andy scores 1 point for this round, and Barb scores 2 points.

Number Top-It (3-Place Decimals)

Materials ☐ number cards 0–9 (4 of each)

☐ 1 Place-Value Mat (Decimals)

Players 2 or more

Skill Place value for decimals

Object of the game To make the largest 3-digit decimal numbers.

Directions

1. This game is played the same as *Number Top-It* (7-Digit Numbers). The only difference is that players use a place-value mat for decimals. Make a decimal place-value mat like the one shown below.

2. Players take turns turning over the top card from the deck and placing it on any of their empty boxes. Each player takes 3 turns, and places 3 cards on his or her row of the game mat.

3. Players play 5 rounds for a game. Shuffle the deck between each round. The player with the smallest total number of points at the end of the 5 rounds wins the game.

> **Example** Phil and Claire played *Number Top-It* using the place-value mat for decimals. Here is the result.
>
> | **Place-Value Mat (Decimals)** | | | | |
> |---|---|---|---|---|
> | | Ones | · | Tenths | Hundredths | Thousandths |
> | Phil | 0 | · | 3 | 5 | 8 |
> | Claire | 0 | · | 6 | 4 | 2 |
>
> Claire's number is larger than Phil's number. So Claire scores 1 point for this round, and Phil scores 2 points.

Polygon Capture

Materials ☐ 1 set of *Polygon Capture* Pieces
(*Math Journal 1,* Activity Sheet 3)

☐ 1 set of *Polygon Capture* Property Cards
(*Math Journal 1,* Activity Sheet 4)

Players 2, or two teams of 2

Skill Properties of polygons

Object of the game To collect more polygons.

Directions

1. Spread the polygons out on the table. Shuffle the Property Cards and sort them writing-side down into ANGLE-card and SIDE-card piles. (The cards are labeled on the back.)

2. Players take turns. When it is your turn:

  ◆ Draw the top card from each pile of Property Cards.

  ◆ Take all of the polygons that have **both** of the properties shown on the Property Cards in your hand.

  ◆ If there are no polygons with both properties, draw one additional Property Card—either an ANGLE- or a SIDE-card. Look for polygons that have this new property and one of the properties already drawn. Take these polygons.

  ◆ At the end of a turn, if you have not captured a polygon that you could have taken, the other player may name and capture it.

3. When all the Property Cards in either pile have been drawn, shuffle *all* of the Property Cards. Sort them writing-side down into ANGLE-card and SIDE-card piles. Continue play.

4. The game ends when there are fewer than 3 polygons left.

5. The winner is the player who has captured more polygons.

Polygon Capture Pieces

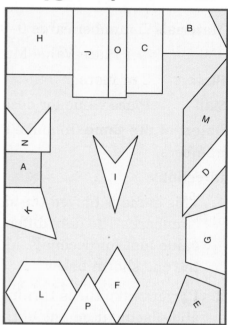

Polygon Capture Property Cards (writing-side up)

There is only one right angle.	There are one or more right angles.	All angles are right angles.	There are no right angles.
There is at least one acute angle.	At least one angle is more than 90°.	All angles are right angles.	There are no right angles.
All opposite sides are parallel.	Only one pair of sides is parallel.	There are no parallel sides.	All sides are the same length.
All opposite sides are parallel.	Some sides have the same length.	All opposite sides have the same length.	**Wild Card:** Pick your own side property.

Example Liz has these Property Cards: "All angles are right angles," and "All sides are the same length." She can take all the squares (polygons A and H). Liz has "captured" these polygons.

Scientific Notation Toss

Materials ☐ 2 six-sided dice

Players 2

Skill Converting from scientific notation to standard notation

Object of the game To create the largest number written in scientific notation.

Directions

When a player rolls 2 dice, either number may be used to name a power of 10, such as 10^2 or 10^4. The other number is used to multiply that power of 10.

Examples	A 5 and a 4 are rolled.	A 2 and a 3 are rolled.
	Either $4 * 10^5$ or $5 * 10^4$ can be written.	Either $2 * 10^3$ or $3 * 10^2$ can be written.

1. Each player rolls the dice 3 times and writes each result in scientific notation (as shown above).

2. Players convert their numbers from scientific notation to standard notation. Then they order the numbers from largest to smallest.

3. Players compare lists. The player who has the largest number wins. In case of a tie, they roll a fourth time.

Example

Ann	rolls:	2 and 4	5 and 3	1 and 6
	writes:	$2 * 10^4$	$3 * 10^5$	$1 * 10^6$
		$= 2 * 10,000$	$= 3 * 100,000$	$= 1 * 1,000,000$
		$= 20,000$	$= 300,000$	$= 1,000,000$
	orders:	1,000,000; 300,000; 20,000		

Keith	rolls:	5 and 5	2 and 1	4 and 3
	writes:	$5 * 10^5$	$1 * 10^2$	$3 * 10^4$
		$= 5 * 100,000$	$= 1 * 100$	$= 3 * 10,000$
		$= 500,000$	$= 100$	$= 30,000$
	orders:	500,000; 30,000; 100		

Ann's largest number is greater than Keith's largest number. Ann wins.

Spoon Scramble

Materials ☐ one set of *Spoon Scramble* Cards
(*Math Masters*, p. 512)

☐ 3 spoons

Players 4

Skill Fraction, decimal, and percent multiplication

Object of the game To avoid getting all the letters in the word *SPOONS*.

Directions

1. Place the spoons in the center of the table.
2. The dealer shuffles the deck and deals 4 cards number-side down to each player.
3. Players look at their cards. If a player has 4 cards of equal value, proceed to Step 5 below. Otherwise, each player chooses a card to discard and passes it, number-side down, to the player on the left.
4. Each player picks up the new card and repeats Step 3. The passing of the cards should proceed quickly.
5. As soon as a player has 4 cards of equal value, the player places the cards number-side up on the table and grabs a spoon.
6. The other players then try to grab one of the 2 remaining spoons. The player left without a spoon is assigned a letter from the word *SPOONS*, starting with the first letter. If a player incorrectly claims to have 4 cards of equal value, that player receives a letter instead of the player left without a spoon.
7. A new round begins. Players put the spoons back in the center of the table. The dealer shuffles and deals the cards (Step 2 above).
8. Play continues until 3 players each get all the letters in the word *SPOONS*. The player who does not have all the letters is the winner.

Spoon Scramble Cards

$\frac{1}{4}$ of 24	$\frac{3}{4} * 8$	50% of 12	0.10 * 60
$\frac{1}{3}$ of 21	$3\frac{1}{2} * 2$	25% of 28	0.10 * 70
$\frac{1}{5}$ of 40	$2 * \frac{16}{4}$	1% of 800	0.10 * 80
$\frac{3}{4}$ of 12	$4\frac{1}{2} * 2$	25% of 36	0.10 * 90

Variations

◆ For 3 players: Eliminate one set of 4 equivalent *Spoon Scramble* Cards. Use only 2 spoons.

◆ Players can make their own deck of *Spoon Scramble* Cards. Each player writes 4 computation problems that have equivalent answers on 4 index cards. Check to be sure the players have all chosen different values.

Subtraction Target Practice

Materials ☐ number cards 0–9 (4 of each)

☐ 1 calculator for each player

Players 1 or more

Skill 2- and 3-digit subtraction

Object of the game To get as close to 0 as possible, without going below 0.

Directions

1. Shuffle the deck and place it number-side down on the table. Each player starts with a score of 250.

2. Players take turns. Each player has 5 turns in the game. When it is your turn, do the following:

   ♦ *Turn 1:* Turn over the top 2 cards and make a 2-digit number. (You can place the cards in either order.) Subtract this number from 250 on scratch paper. Check the answer on a calculator.

   ♦ *Turns 2–5:* Take 2 cards and make a 2-digit number. Subtract this number from the result obtained in your previous subtraction. Check the answer on a calculator.

3. The player whose final result is closest to 0, without going below 0, is the winner. If the final results for all players are below 0, no one wins.

If there is only one player, the Object of the game is to get as close to 0 as possible without going below 0.

Example

Turn 1: Draw 4 and 5. Subtract 45 or 54.	$250 - 45 = 205$
Turn 2: Draw 0 and 6. Subtract 6 or 60.	$205 - 60 = 145$
Turn 3: Draw 4 and 1. Subtract 41 or 14.	$145 - 41 = 104$
Turn 4: Draw 3 and 2. Subtract 32 or 23.	$104 - 23 = 81$
Turn 5: Draw 6 and 9. Subtract 69 or 96.	$81 - 69 = 12$

Variation

Each player starts at 100 instead of 250.

3-D Shape Sort

Materials ☐ 1 set of *3-D Shape Sort* Shape Cards
(*Math Masters,* p. 507

☐ 1 set of *3-D Shape Sort* Property Cards
(*Math Masters,* pp. 505 and 506)

Players 2, or 2 teams of 2

Skill Properties of 3-D shapes

Object of the game To collect more Shape Cards.

Advance Preparation Make a 2-sided single-sheet copy of *Math Masters,* pp. 505 and 506.

Directions

1. Spread out the Shape Cards writing-side up on the table. Shuffle the Property Cards and sort them writing-side down into VERTEX/EDGE-card and SURFACE-card piles. (The cards are labeled on the back.)

2. Players take turns. When it is your turn:

   ♦ Draw the top card from each pile of Property Cards.

   ♦ Take all the Shape Cards that have **both** of the properties shown on the Property Cards.

   ♦ If there are no Shape Cards with **both** properties, draw 1 additional Property Card—either a VERTEX/EDGE Card or a SURFACE Card. Look for Shape Cards that have the new property and one of the properties drawn before. Take those Shape Cards.

   ♦ At the end of a turn, if you have not taken a Shape Card that you could have taken, the other player may name and take it.

3. When all of the Property Cards in either pile have been drawn, shuffle *all* of the Property Cards. Sort them writing-side down into VERTEX/EDGE-card and SURFACE-card piles. Continue play.

4. The game ends when there are fewer than 3 Shape Cards left.

5. The winner is the player with more Shape Cards.

Shape Cards

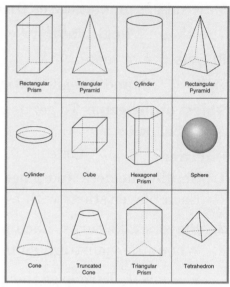

Rectangular Prism	Triangular Pyramid	Cylinder	Rectangular Pyramid
Cylinder	Cube	Hexagonal Prism	Sphere
Cone	Truncated Cone	Triangular Prism	Tetrahedron

Property Cards
(writing-side up)

I have an even number of vertices.	I have no vertices.	I have at least 2 edges that are parallel to each other.	I have an odd number of edges.
One of my vertices is formed by an even number of edges.	I have at least one curved edge.	I have fewer than 6 vertices.	I have at least 2 edges that are perpendicular to each other.
All of my surfaces are polygons.	I have at least one face (flat surface).	I have at least one curved surface.	All of my faces are triangles.
All of my faces are regular polygons.	At least one of my faces is a circle.	I have at least one pair of faces that are parallel to each other.	**Wild Card:** Pick your own surface property.

Top-It Games

The materials, number of players, and object of the game are the same for all *Top-It* games.

Materials ☐ number cards 1–10 (4 of each)

☐ 1 calculator (optional)

Players 2 to 4

Skill Addition, subtraction, multiplication, and division facts

Object of the game To collect the most cards.

Addition Top-It
Directions

1. Shuffle the deck and place it number-side down on the table.

2. Each player turns over 2 cards and calls out the sum of the numbers. The player with the largest sum takes all the cards. In case of a tie for the largest sum, each tied player turns over 2 more cards and calls out the sum of the numbers. The player with the largest sum takes all the cards from both plays.

3. Check answers using an Addition Table or a calculator.

4. The game ends when there are not enough cards left for each player to have another turn.

5. The player with the most cards wins.

Variation Each player turns over 3 cards and finds their sum.

Advanced Version

Use only the number cards 1–9. Each player turns over 4 cards, forms two 2-digit numbers, and finds the sum. Players should carefully consider how they form their numbers since different arrangements have different sums. For example, 74 + 52 has a greater sum than 47 + 25.

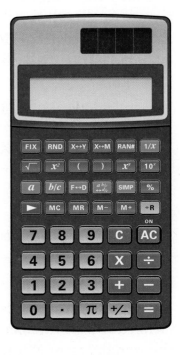

Subtraction Top-It
Directions

1. Each player turns over 3 cards, finds the sum of any 2 of the numbers, then finds the difference between the sum and the third number.

2. The player with the largest difference takes all the cards.

Example | A 4, an 8, and a 3 are turned over. There are three ways to form the numbers. Always subtract the smaller number from the larger one.

$$4 + 8 = 12 \quad \text{or} \quad 3 + 8 = 11 \quad \text{or} \quad 3 + 4 = 7$$
$$12 - 3 = 9 \qquad\qquad 11 - 4 = 7 \qquad\qquad 8 - 7 = 1$$

Advanced Version

Use only the number cards 1–9. Each player turns over 4 cards, forms two 2-digit numbers, and finds their difference. Players should carefully consider how they form their numbers. For example, $75 - 24$ has a greater difference than $57 - 42$ or $74 - 25$.

Multiplication Top-It

Directions

1. The rules are the same as for *Addition Top-It*, except that players find the product of the numbers instead of the sum.

2. The player with the largest product takes all the cards. Answers can be checked with a Multiplication Table or a calculator.

Variation

Use only the number cards 1–9. Each player turns over 3 cards, forms a 2-digit number, then multiplies the 2-digit number by the remaining number.

Division Top-It

Directions

1. Use only the number cards 1–9. Each player turns over 3 cards and uses them to generate a division problem as follows:

   ◆ Choose 2 cards to form the dividend.

   ◆ Use the remaining card as the divisor.

   ◆ Divide and drop any remainder.

2. The player with the largest quotient takes all the cards.

Advanced Version

Use only the number cards 1–9. Each player turns over 4 cards, chooses 3 of them to form a 3-digit number, then divides the 3-digit number by the remaining number. Players should carefully consider how they form their 3-digit numbers. For example, $462/5$ is greater than $256/4$.

Top-It Games with Positive and Negative Numbers

Materials □ 1 complete deck of number cards

□ 1 calculator (optional)

Players 2 to 4

Skill Addition and subtraction of positive and negative numbers

Object of the game To collect the most cards.

Addition Top-It with Positive and Negative Numbers
Directions

The color of the number on each card tells you if a card is a positive number or a negative number.

♦ Black cards (spades and clubs) are positive numbers.

♦ Red cards (hearts and diamonds) or blue cards (Everything Math Deck) are negative numbers.

1. Shuffle the deck and place it number-side down on the table.

2. Each player turns over 2 cards and calls out the sum of the numbers. The player with the largest sum takes all the cards.

3. In case of a tie, each tied player turns over 2 more cards and calls out the sum of the numbers. The player with the largest sum takes all the cards from both plays. If necessary, check answers with a calculator.

4. The game ends when there are not enough cards left for each player to have another turn. The player with the most cards wins.

Example Lindsey turns over a red 3 and a black 6.

$-3 + 6 = 3$

Fred turns over a red 2 and a red 5.

$-2 + (-5) = -7$

$3 > -7$ Lindsey takes all 4 cards because 3 is greater than -7.

Variation

Each player turns over 3 cards and finds the sum.

Subtraction Top-It with Positive and Negative Numbers
Directions

The color of the number on each card tells you if a card is a positive number or a negative number.

♦ Black cards (spades and clubs) are positive numbers.

♦ Red cards (hearts and diamonds) or blue cards (Everything Math Deck) are negative numbers.

1. Shuffle the deck and place it number-side down on the table.

2. Each player turns over 2 cards, one at a time, and subtracts the second number from the first number. The player who calls out the largest difference takes all the cards.

3. In case of a tie, each tied player turns over 2 more cards and calls out the difference of the numbers. The player with the largest answer takes all the cards from both plays. If necessary, check answers with a calculator.

4. The game ends when there are not enough cards left for each player to have another turn. The player with the most cards wins.

Example | Lindsey turns over a black 2 first, then a red 3.

$$+2 - (-3) = 5$$

Fred turns over a red 5 first, then a black 8.

$$-5 - (+8) = -13$$

$5 > -13$ Lindsey takes all 4 cards because 5 is greater than -13.

American Tour

Introduction

This section of the *Student Reference Book* is called the "American Tour." It uses mathematics to explore the history, people, and environment of the United States.

As you read the American Tour, you will learn how to use and interpret its maps, graphs, and tables. You will see that mathematics is a powerful tool for learning about and understanding our nation.

How to Use the American Tour

Throughout the year, you will examine the American Tour with the whole class or in small groups. You should also read and analyze this section of the *Student Reference Book* on your own. As you read the American Tour, do the following:

1. Examine the information.

Ask yourself—

What am I being told? How is the information being reported? Is it a count, a measurement, a ratio, or a rate?

How exact are the numbers? Are these old or recent data?

Is this a rough estimate, an actual count, or a measure?

Are the numbers medians, averages, or ranges? Or are the numbers based on just *one* count or measurement?

2. Use the information.

Ask yourself—

What patterns and trends do I see? If I organize or display the data another way, what else will I find?

How can I use mathematics to study the data and learn something else?

3. Question the information.

Ask yourself—

Can I be sure this information is correct?

How might I check this information?

Would another count or measure show similar results?

Old Faithful Geyser at Yellowstone National Park. Yellowstone is the oldest national park in the United States.

Willis Tower in Chicago, Illinois is the tallest building in the nation.

The Grand Canyon is the largest land gorge in the world.

The First Americans

How did the first Americans get here and when did they come? There are several possible theories to explain the first migration into the Americas. Scientists cannot agree that any one of these theories provides the best explanation.

Scientists do not agree on when the first migration began. Some believe that people first entered the Americas between 14,000 and 16,000 years ago. Others believe that the first migration was at least 20,000 years ago.

Scientists also do not agree on where the first Americans came from. Siberia (northeast Asia), Australia, and Europe have all been proposed. There is also no agreement on whether the entry point was in North America or South America.

Land Bridge Theory

This theory of migration supposes that the first Americans wandered from Siberia into Alaska while tracking big game animal herds. They were able to cross between Asia and Alaska by a land bridge called the Bering Strait. The land bridge had formed during the last Ice Age and was at least 1,000 miles wide.

Theories of Migration to the New World

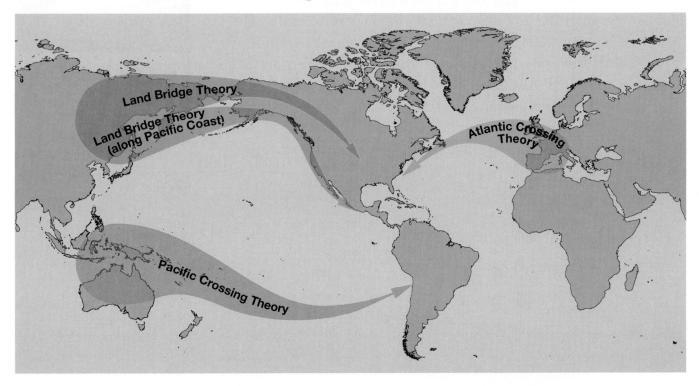

After crossing into Alaska, people moved south. It was once thought that they could enter the south through an ice-free pathway between the glaciers that covered Canada. Geologists now believe that such a pathway did not exist. It is more likely that people used boats or rafts and traveled south, following the Pacific coastline.

Ice Age stone tool

Pacific Crossing Theory

The land bridge theory supposes that the first Americans entered the New World through Alaska and then migrated south. One of the problems with this theory is that tools and pottery shards have been discovered in South America that are older than any found in North America. (It is very likely that some of the clay pieces found in Chile were made by humans more than 30,000 years ago.) If humans migrated south, we might expect that the oldest tools discovered should be from North America, not South America.

The Pacific crossing theory assumes that people entered South America before North America. This theory of migration supposes that the first Americans used boats or large rafts to cross the Pacific Ocean. The likely starting point for the migration was Australia or the South Sea Islands.

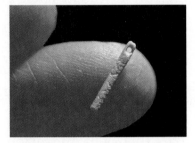

Ice Age sewing needle

Atlantic Crossing Theory

North American stone and bone tools (dated about 14,000 years ago) are very similar to tools of the same age from Spain. The Atlantic crossing theory was developed to explain the similarity of tools. This theory supposes that hunters and fishers from Europe reached the eastern shores of Canada and the United States. They could have migrated by boat along the edges of the ice sheets that covered the northern part of the Atlantic Ocean. And if they did, they might have been the first Americans.

Ice Age stone tool

European settlers began to arrive in North America in the early 1600s. We can be fairly certain that there were at least 1 million Native Americans living in North America at that time. The map below shows that some areas contained many more Native Americans than other areas did.

Native American Population Density, 1600

Population density is a measure of how many people live within a certain area. The key for this map gives density as the number of people per 100 square miles. (See below.)

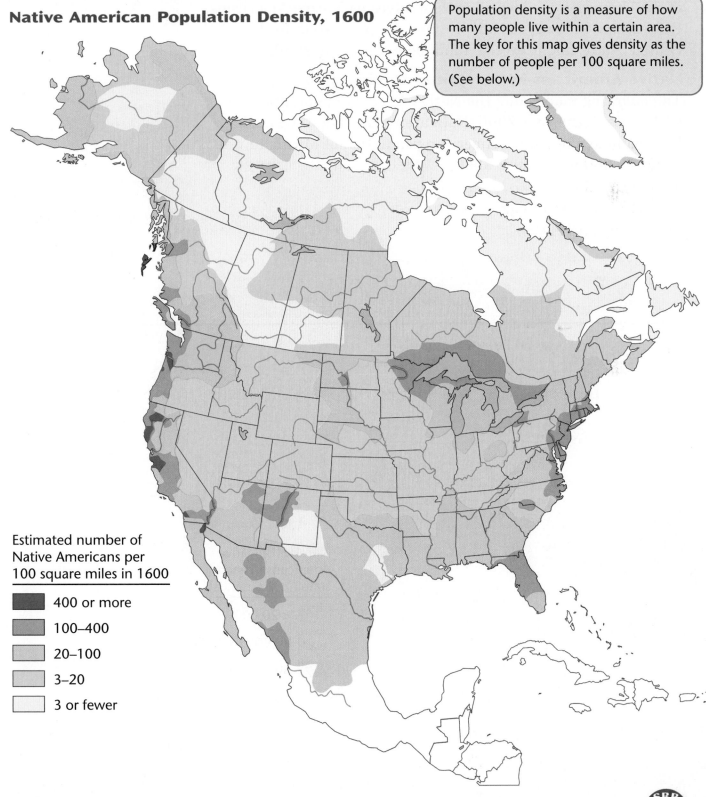

Estimated number of Native Americans per 100 square miles in 1600

- 400 or more
- 100–400
- 20–100
- 3–20
- 3 or fewer

From 1500 to 1900, disease and war greatly reduced the number of Native Americans. According to the 1900 census, only about 250,000 Native Americans lived inside the United States at that time. This trend was reversed during the twentieth century. By the year 2000, about 2,500,000 citizens of the United States identified themselves as Native Americans. It is estimated that the Native American population in the United States might exceed 4 million by the year 2050.

Native Americans in the United States, 2000

The following map shows the Native American population for each state in the year 2000. Data are reported in thousands. For example, the Native American population of Michigan was about 58,000 in the year 2000.

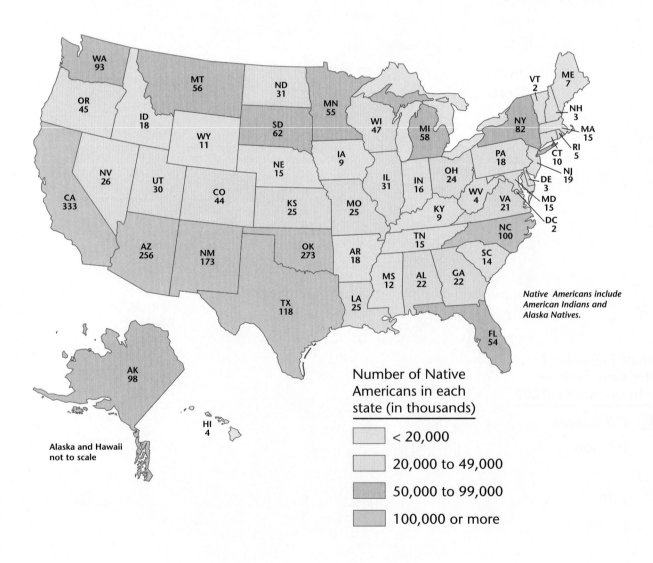

Native Americans include American Indians and Alaska Natives.

Alaska and Hawaii not to scale

Number of Native Americans in each state (in thousands)

	< 20,000
	20,000 to 49,000
	50,000 to 99,000
	100,000 or more

A Diverse Nation

The year 1788 is often considered to be when the United States of America became a nation. In that year, 11 of the 13 original states agreed to accept the new Constitution.

Almost 70 percent of the new nation's people had English or West African ancestors. The remainder had roots in Scotland, Ireland, Wales, Germany, the Netherlands, France, or Sweden. Native Americans are not counted in these estimates, nor shown on the graph at the right.

Ethnic Groups, 1790

- English 48.2%
- Welsh 3.5%
- Scottish 4.3%
- Scotch-Irish 8.5%
- Irish 4.7%
- French 1.7%
- German 7.2%
- Dutch 2.5%
- Swedish-Finnish 0.2%
- African 19.3%

At least nine out of every ten Africans in the United States in 1790 were slaves. They endured terrible cruelty and hardship. They also played a major role in building the new nation. They cleared land, made roads, raised crops, and built houses. Some were skilled craftspeople. During the Revolutionary War, more than 5,000 African Americans fought alongside the colonists against the British.

Most African Americans did not gain their freedom until after the Civil War. In 1865, the Thirteenth Amendment to the Constitution was adopted. It states that "neither slavery nor involuntary servitude. . .shall exist within the United States."

African American Population		
Year	Number	Percent of U.S. Total Population
1790	757,000	19%
1850	3,639,000	16%
1900	8,834,000	12%
1950	15,042,000	10%
2000	34,658,000	12%

Number of Immigrants

An **immigrant** is a person who moves permanently from one country to another country. Millions of immigrants have come to the United States in search of a better life.

The graph below shows the number of immigrants who entered the United States each year, beginning in 1820. The total number of immigrants entering between 1820 and 2000 was approximately 65 million.

Immigration Rate in Peak Years	
Year	Immigrants per 1,000 residents
1854	16.0
1882	15.2
1907	14.8
1921	7.4
1991	7.2

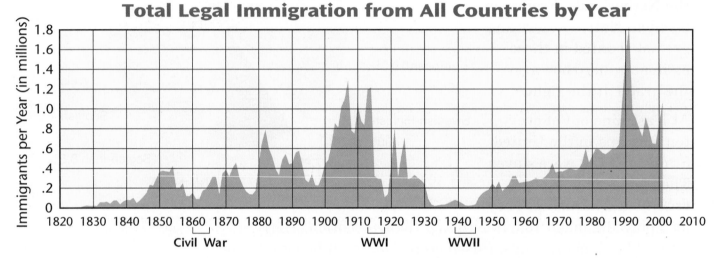

Total Legal Immigration from All Countries by Year

NOTE: The 1989–1991 figures include people already residing in the U.S. who were granted permanent resident status.

Ellis Island

Ellis Island is a small island in Upper New York Bay, about one mile southwest of Manhattan Island. The early Dutch settlers used it as a picnic ground. When the Immigration Bureau was created in 1891, Ellis Island was converted into a major U.S. immigration station. New immigrants were examined there, in a large area known as the Great Hall, and were either admitted or deported. Twelve million immigrants passed through Ellis Island station from 1892 to 1954. Ellis Island station was able to process one million people a year.

The Great Hall of Ellis Island

The Statue of Liberty was erected in Upper New York Bay in 1886, less than one-half mile from Ellis Island. The statue, together with the Main Building and Great Hall on Ellis Island, were renovated as part of the 1986 Statue of Liberty centennial celebration. On July 4, 1986, Chief Justice Warren Burger swore in 5,000 new citizens on Ellis Island, and 20,000 others across the country were sworn in at the same time through a satellite telecast.

Foreign-Born Population

About 10 percent (1 in 10) of the current U.S. population was not born in the United States. Mexico is the most common country of birth among those who were born in other countries.

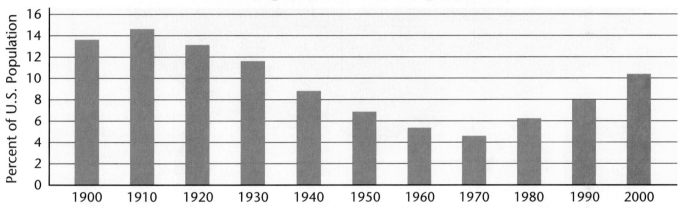

Foreign-Born U.S. Population

Leading Countries of Birth								
1920			**1960**			**2000**		
Country	Number (million)	%	Country	Number (million)	%	Country	Number (million)	%
Germany	1.7	12.1	Italy	1.3	12.9	Mexico	7.8	27.6
Italy	1.6	11.6	Germany	1.0	10.2	Philippines	1.2	4.3
Soviet Union	1.4	10.1	Canada	1.0	9.8	China/Hong Kong	1.1	3.8
Poland	1.1	8.2	Great Britain	0.8	7.9	India	1.0	3.5
Canada	1.1	8.2	Poland	0.7	7.7	Cuba	1.0	3.4
Great Britain	1.1	8.2	Soviet Union	0.7	7.1	Vietnam	0.9	3.0
Ireland	1.0	7.5	Mexico	0.6	5.9	El Salvador	0.8	2.7
Sweden	0.6	4.5	Ireland	0.3	3.5	Korea	0.7	2.5
Austria	0.6	4.1	Austria	0.3	3.1	Dominican Rep.	0.7	2.4
Mexico	0.5	3.5	Hungary	0.2	2.5	Canada	0.7	2.4

Example In 1920, 500,000 (0.5 million) persons living in the U.S. were born in Mexico. They accounted for 3.5% of all foreign-born persons living in the U.S.

By 2000, 7.8 million persons living in the U.S. were born in Mexico. They accounted for more than one-fourth (27.6%) of all foreign-born persons living in the U.S.

Non-English Speaking Population

About 18 percent of the U.S. population speaks a language other than English at home. More than half this number speak Spanish.

People at Least 5 Years Old Who Speak a Language Other Than English at Home (subdivisions within states are counties)

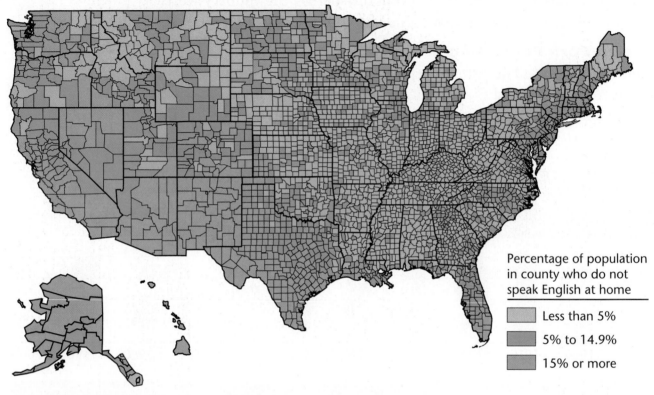

Percentage of population in county who do not speak English at home

	Less than 5%
	5% to 14.9%
	15% or more

Alaska and Hawaii not to scale

In the United States, each state is divided into smaller areas called *counties*. There are slightly more than 3,000 counties in the United States. On the map above, the heavy black lines are state boundaries and the thin black lines are county boundaries.

Each county is colored orange, green, or purple. The color indicates the percent of population in that county who speak a language other than English at home. For example, the map key shows that in the orange-colored counties, less than 5% of the people speak a language other than English at home.

Arizona

Example There are 15 counties in the state of Arizona, and 13 of them are colored purple. At least 15% of the people in each of these 13 counties speak a language other than English at home. The other two counties in Arizona are colored green. In both of those counties, between 5% and 14.9% of the people speak a language other than English at home.

Westward Expansion

In 1790, almost all of the people in the United States lived within 200 miles of the Atlantic Ocean. In the 1800s, the nation grew westward. Native Americans lived in the new areas that were added. Settlers from France, Spain, and Mexico also lived in these areas. By 1900, the land area of the United States had quadrupled (become four times as large). The population had begun a migration to the West that continues today.

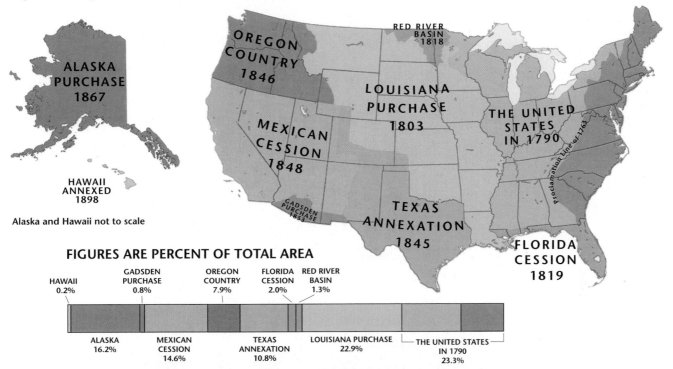

FIGURES ARE PERCENT OF TOTAL AREA

Alaska and Hawaii not to scale

| HAWAII 0.2% | GADSDEN PURCHASE 0.8% | OREGON COUNTRY 7.9% | FLORIDA CESSION 2.0% | RED RIVER BASIN 1.3% |
| ALASKA 16.2% | MEXICAN CESSION 14.6% | TEXAS ANNEXATION 10.8% | LOUISIANA PURCHASE 22.9% | THE UNITED STATES IN 1790 23.3% |

U.S. Territorial Expansion

	Date	Area[1]
The 48 states		
Territory in 1790	1790	842,432
Louisiana Purchase	1803	827,192
Red River Basin	1818	46,253
Florida Cession	1819	72,003
Texas Annexation	1845	390,143
Oregon Country	1846	285,580
Mexican Cession	1848	529,017
Gadsden Purchase	1853	29,640
Alaska and Hawaii		
Alaska	1867	586,412
Hawaii	1898	6,450
U.S. Commonwealths		
Puerto Rico	1899	3,435
Northern Mariana Islands	1976	179
United States Total		**3,618,736**

[1]Total land and water area in square miles

Note

The boundaries of the new areas that were added were often not clear. For example, some historians say that the Red River Basin was part of the Louisiana Purchase. Others say that it was acquired from Great Britain.

19th Century Settlement Patterns

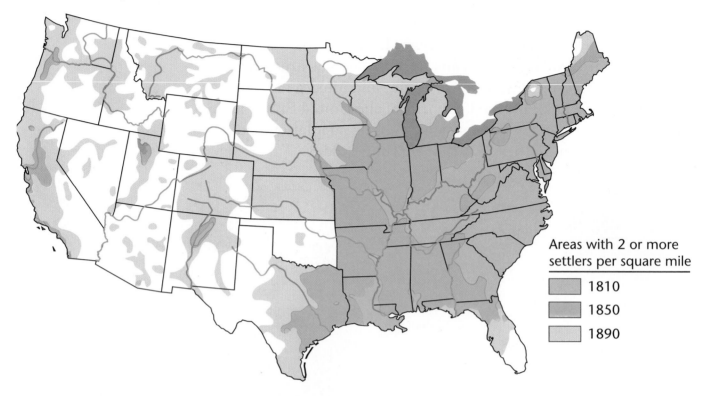

Areas with 2 or more
settlers per square mile

1810

1850

1890

The Center of Population Moving West

Imagine a map of the United States that is thin, flat, and rigid (can't bend). Suppose that a 1-ounce weight is set on the map for each person in the United States at the place where that person lives. The **center of population** is the point on the map where the map would balance.

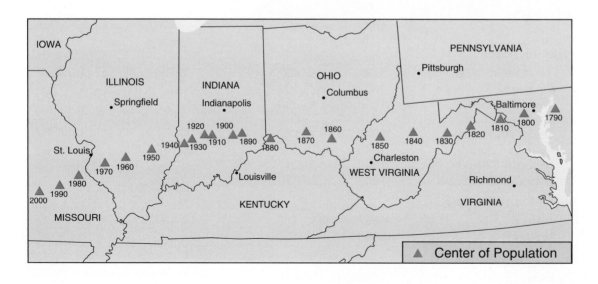

European Exploration, Settlement, and Statehood

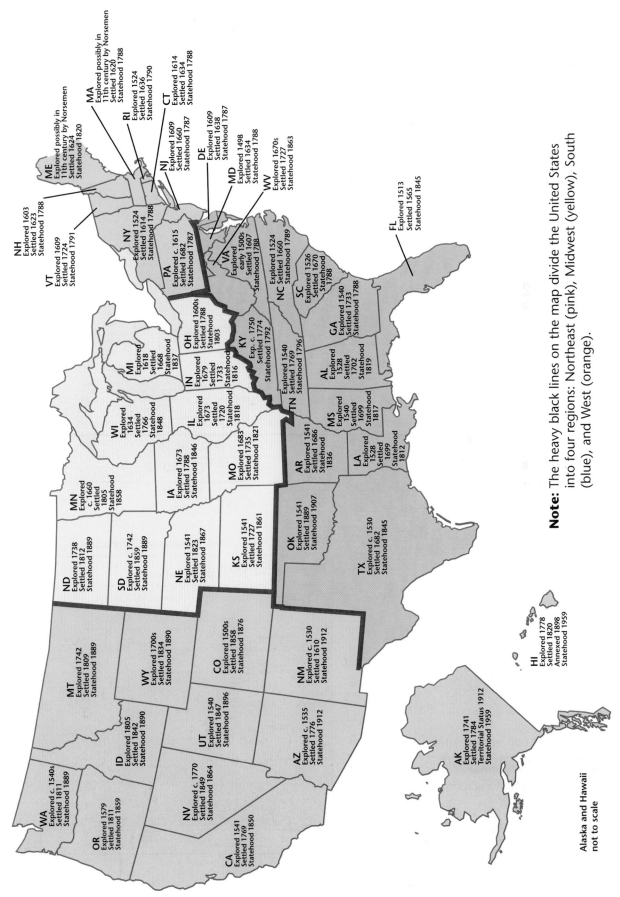

Note: The heavy black lines on the map divide the United States into four regions: Northeast (pink), Midwest (yellow), South (blue), and West (orange).

WA
Explored c. 1540s
Settled 1811
Statehood 1889

OR
Explored 1579
Settled 1811
Statehood 1859

CA
Explored 1541
Settled 1769
Statehood 1850

NV
Explored c. 1770
Settled 1849
Statehood 1864

ID
Explored 1805
Settled 1842
Statehood 1890

UT
Explored 1540
Settled 1847
Statehood 1896

AZ
Explored c. 1535
Settled 1776
Statehood 1912

MT
Explored 1742
Settled 1809
Statehood 1889

WY
Explored 1700s
Settled 1834
Statehood 1890

CO
Explored 1500s
Settled 1858
Statehood 1876

NM
Explored c. 1530
Settled 1610
Statehood 1912

ND
Explored 1738
Settled 1812
Statehood 1889

SD
Explored c. 1742
Settled 1859
Statehood 1889

NE
Explored 1541
Settled 1823
Statehood 1867

KS
Explored 1541
Settled 1727
Statehood 1861

OK
Explored 1541
Settled 1889
Statehood 1907

TX
Explored c. 1530
Settled 1682
Statehood 1845

MN
Explored c. 1660
Settled 1805
Statehood 1858

WI
Explored 1634
Settled 1766
Statehood 1848

IA
Explored 1673
Settled 1788
Statehood 1846

MO
Explored 1683
Settled 1735
Statehood 1821

AR
Explored 1541
Settled 1686
Statehood 1836

LA
Explored 1528
Settled 1699
Statehood 1812

MI
Explored 1618
Settled 1668
Statehood 1837

IL
Explored 1673
Settled 1720
Statehood 1818

IN
Explored 1679
Settled 1733
Statehood 1816

OH
Explored 1600s
Settled 1788
Statehood 1803

KY
Exp. c. 1750
Settled 1774
Statehood 1792

TN
Explored 1540
Settled 1769
Statehood 1796

MS
Explored 1540
Settled 1699
Statehood 1817

AL
Explored 1528
Settled 1702
Statehood 1819

GA
Explored 1540
Settled 1733
Statehood 1788

SC
Explored 1526
Settled 1670
Statehood 1788

NC
Explored 1524
Settled 1660
Statehood 1789

VA
Explored early 1500s
Settled 1607
Statehood 1788

WV
Explored 1670s
Settled 1727
Statehood 1863

MD
Explored 1498
Settled 1634
Statehood 1788

DE
Explored 1609
Settled 1638
Statehood 1787

FL
Explored 1513
Settled 1565
Statehood 1845

PA
Explored c. 1615
Settled 1682
Statehood 1787

NY
Explored 1524
Settled 1614
Statehood 1788

NJ
Explored 1609
Settled 1660
Statehood 1787

CT
Explored 1614
Settled 1634
Statehood 1788

RI
Explored 1524
Settled 1636
Statehood 1790

MA
Explored possibly in 11th century by Norsemen
Settled 1620
Statehood 1788

VT
Explored 1609
Settled 1724
Statehood 1791

NH
Explored 1603
Settled 1623
Statehood 1788

ME
Explored possibly in 11th century by Norsemen
Settled 1624
Statehood 1820

HI
Explored 1778
Settled 1820
Annexed 1898
Statehood 1959

AK
Explored 1741
Settled 1784
Territorial Status 1912
Statehood 1959

Alaska and Hawaii
not to scale

The United States in 1790

Area Map

Area: 842,000 square miles

Percent of 2000 Area: 23%

Population: 3,929,000

Percent of 2000 Population: 1.5%

The white region on the map above and below is included in the area total but not in the population total.

Population Distribution

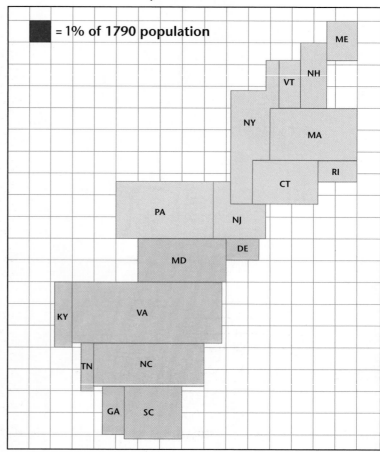

= 1% of 1790 population

The United States in 1850

Area Map

Area: 2,993,000 square miles

Percent of 2000 Area: 83%

Population: 23,192,000

Percent of 2000 Population: 8%

Population Distribution

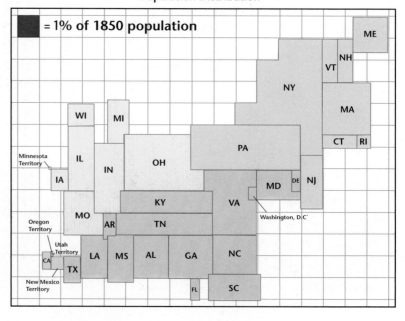

= 1% of 1850 population

The United States in 1900

Area Map

Area: 3,619,000 square miles

Percent of 2000 Area: 100%

Population: 76,212,000

Percent of 2000 Population: 27%

The white region on the map is included
in the area total but not in the population total.

Population Distribution

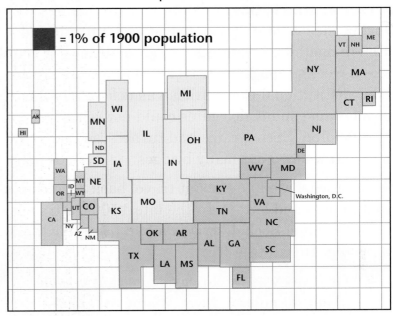

■ = 1% of 1900 population

The United States in 2000

Area Map

Area: 3,619,000 square miles

Percent of 2000 Area: 100%

Population: 281,422,000

Percent of 2000 Population: 100%

Population Distribution

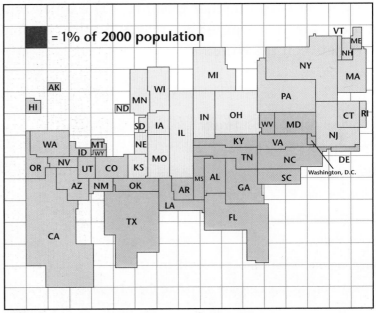

■ = 1% of 2000 population

Travel

Travel in the 1700s

During the colonial period, travel was difficult and often dangerous. Even under good conditions, it could take a week or longer to travel 200 miles. In 1787, the Constitutional Convention was delayed for 11 days because many delegates could not reach Philadelphia. Spring rains had turned the roads to mud and washed out many bridges.

Much of what we know about travel back then comes from diaries and letters. In the late 1700s, a traveler named Samuel Beck described a trip from New York to Boston. He wrote—

> One way was by clumsy stage that travels about 40 miles a day...Rising at 3 or 4 o'clock and prolonging the day's ride into the night, one made to reach Boston in six days.

Another source of information is the speed of mail delivery. The table below shows how fast mail could be delivered when travel conditions were good. Using a system created by Benjamin Franklin, a series of riders took turns carrying the mail. When one rider got tired, another took over. The times reported below are much faster than the time it would take one person to travel between the cities.

Travel Times for Postal Riders, 1775	
From New York City to...	
Boston, Massachusetts	2 to 4 days
Philadelphia, Pennsylvania	2 to 4 days
Baltimore, Maryland	4 to 8 days
Williamsburg, Virginia	8 to 12 days
Wilmington, North Carolina	12 to 16 days
Charleston, South Carolina	More than 16 days

Travel in the 1800s

During the nineteenth century, ordinary Americans traveled farther and more often than people in any other country. On foot and on horseback, by stagecoach, wagon, steamboat, and railroad, Americans were on the move.

In 1828, a Boston newspaper reported: "There is more traveling in the United States than in any part of the world. Here, the whole population is in motion, whereas, in old countries, there are millions [of people] who have never been beyond the sound of the parish [church] bell."

Between 1775 and the mid-1800s, the speed and comfort of travel increased. Roads and stagecoaches were improved. The steamboat and railroad were invented. Travel was typically faster in the more developed eastern United States. Travel was slower in the West, which had more trails than roads. In addition, the Rocky and Sierra Nevada Mountains made travel difficult.

Stagecoach travel

Top Travel Speeds	
East of the Mississippi River, 1800–1840	
Foot	25 to 35 miles per day
Horseback	60 to 70 miles per day
Stagecoach	8 to 9 miles per hour
Railroad	15 to 25 miles per hour[1]
Transcontinental Travel, 1840–1860	
Wagon Train	2,000 miles in 150 to 180 days
Stagecoach	3,000 miles in 130 to 150 days
Clipper Ship	New York to San Francisco via Cape Horn (about 17,000 miles) in 90 to 120 days
Train Travel, 1860–1900	
1860 New York to Chicago	Less than 2 days
1880 New York to San Francisco	8 days
1900 New York to to San Francisco	Less than 5 days

[1]Railroads first appeared in the 1830s. Until the 1850s, service was limited.

Elevation along the 39th Parallel

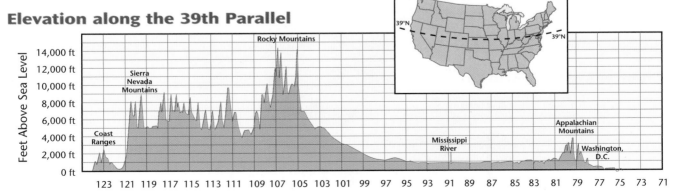

Longitude (degrees W)

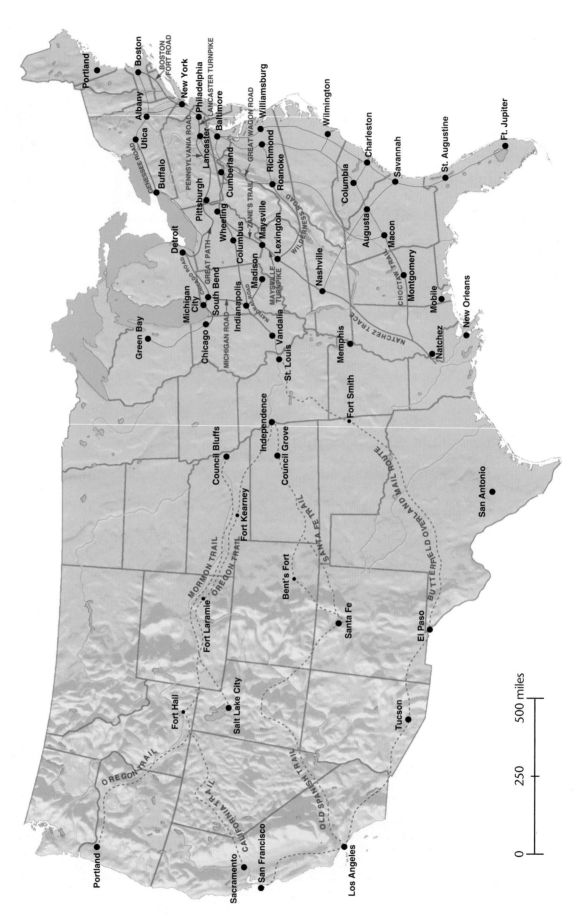

Main Roads and Trails, 1840–1860

0   250   500 miles

Travel from 1870 to the Present

The first railroad connecting the east and west coasts of the United States was completed on May 10, 1869. Twenty-six years later, in 1895, the first practical American automobile was manufactured. In 1903, the Wright brothers flew the first successful airplane. The flight lasted just 12 seconds and covered only 120 feet, but it was the start of modern aviation.

These three events—these three "firsts"—marked the beginning of the modern era of travel and transportation. Trains, automobiles, trucks, and airplanes have made it easy for people and goods to move from one corner of the nation to another. As a result, everyday life has changed dramatically.

Airline Schedule, Chicago to New York

Departure	Arrival
6:00 A.M.	8:59 A.M.
6:20 A.M.	1:12 P.M.[1]
7:00 A.M.	9:56 A.M.
7:00 A.M.	10:04 A.M.
8:00 A.M.	11:00 A.M.
8:45 A.M.	2:00 P.M.[1]
9:00 A.M.	12:00 P.M.
10:00 A.M.	12:58 P.M.
10:20 A.M.	3:19 P.M.[1]
11:00 A.M.	1:55 P.M.
12:00 P.M.	3:00 P.M.
1:00 P.M.	3:55 P.M.
1:20 P.M.	4:21 P.M.
1:20 P.M.	6:45 P.M.[1]
1:30 P.M.	4:39 P.M.
2:00 P.M.	5:09 P.M.
3:00 P.M.	6:04 P.M.
4:00 P.M.	7:00 P.M.
4:14 P.M.	9:19 P.M.[1]
4:40 P.M.	7:30 P.M.
5:00 P.M.	8:01 P.M.
6:00 P.M.	9:02 P.M.
7:00 P.M.	10:00 P.M.

[1]Flight makes other stops

Train Schedule between New York and Chicago

Three Rivers
New York . . .
Pittsburgh . . . Chicago

41			◄ Train Number ►			40
Daily			◄ Days of Operation ►			Daily
Read Down	Mile	▼		▲		Read Up
12 45P	0	Dp	New York, NY–Penn Sta. (ET)	Ar		7 25P
1 03P	10		Newark, NJ–Penn Sta.			6 53P
1 48P	58	▼	Trenton, NJ	▲		6 00P
2 20P	91	Ar	Philadelphia, PA–30th St. Sta.	Dp		5 25P
3 00P		Dp		Ar		4 52P
3 29P	110		Paoli, PA			4 13P
4 20P	159	▼	Lancaster, PA	▲		3 24P
5 05P	195	Ar	Harrisburg, PA (Scranton,	Dp		2 31P
5 25P		Dp	Reading)	Ar		2 16P
6 37P	256		Lewistown, PA			12 47P
7 17P	293		Huntingdon, PA	▲		12 06P
8 03P	327		Altoona, PA			11 20A
9 07P	366	▼	Johnstown, PA			10 14A
9 59P	413		Greensburg, PA			9 22A
10 55P	444	Ar	Pittsburgh, PA	Dp		8 38A
11 25P		Dp		Ar		8 23A
1 15A	518		Youngstown, OH	▲		5 58A
2 14A	571		Akron, OH (Canton)			4 50A
4 10A	682		Fostoria, OH (Lima)			3 05A
5 42A	817	▼	Nappanee, IN (Warsaw) (ET)			11 38P
7 00A	900	▼	Hammond-Whiting, IN (CT)			10 14P
8 25A	915	Ar	Chicago, IL–Union Sta. (CT)	Dp		9 20P

A	Time symbol for A.M.	Dp	Train Departs	(ET)	Eastern Time
P	Time symbol for P.M.	Ar	Train Arrives	(CT)	Central Time

Note: Times on schedules are local.
New York (Eastern Time) is 1 hour ahead of Chicago (Central Time).

Work

During the last 100 years, the fraction of the population doing certain kinds of work has greatly decreased. In other cases, it has greatly increased.

Working People Who Were ...		
	Farm Workers	**Engineers**
1900	1 in 3	1 in 764
1930	1 in 4	1 in 224
1960	1 in 16	1 in 78
2000	1 in 40	1 in 65

There is a third pattern. In some occupations, the number of people has grown at about the same rate as the total population. Why might this be?

Working People Who Were ...		
	Clergy	**Photographers**
1900	1 in 316	1 in 1,161
1930	1 in 327	1 in 1,475
1960	1 in 337	1 in 1,283
2000	1 in 350	1 in 775

Total Number of Working People

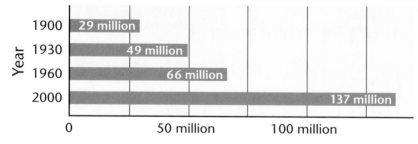

NOTE: Figures for 1900 and 1930 are based on workers 14 years and older. Figures for 1960 and 2000 are based on workers 16 years and older. All figures are for the civilian population only. Members of the Armed Forces are excluded.

For many kinds of jobs, more work is being done by fewer people. This is because technology keeps improving and workers are better educated. As a result, fewer workers are needed in some occupations (such as farming).

In 1900

There were about 10 million farm workers.

The farms fed a population of about 75 million people.

In 2000

There were about 3.5 million farm workers.

Key

 Farm worker population of 5 million people

 Population of 5 million people

The farms fed a population of about 280 million people.

Entertainment

During the first half of the twentieth century, going to movies was the most popular form of entertainment.

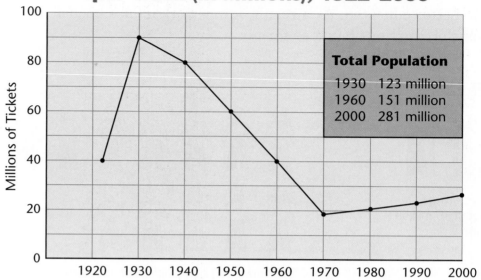

Average Number of Movie Tickets Sold per Week (in Millions), 1922–2000

Total Population	
1930	123 million
1960	151 million
2000	281 million

In the second half of the twentieth century, movies had to compete with other forms of entertainment.

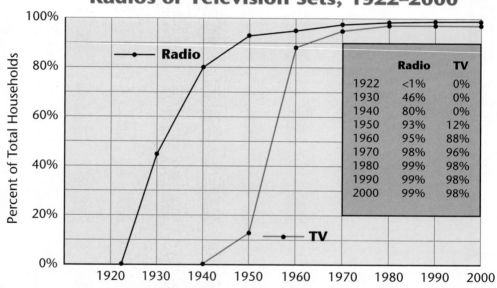

Estimated Percent of Households with Radios or Television Sets, 1922–2000

	Radio	TV
1922	<1%	0%
1930	46%	0%
1940	80%	0%
1950	93%	12%
1960	95%	88%
1970	98%	96%
1980	99%	98%
1990	99%	98%
2000	99%	98%

Today, watching television is the most popular form of entertainment.

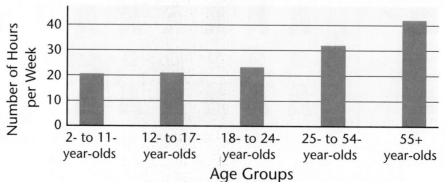

Average Number of Hours of TV Watched During a Typical Week, 2000

Play

What kinds of sports and outdoor activities do 11-year-olds participate in? The table below shows how popular different activities are among 11-year-olds, other age groups, all people, males, and females.

The data are from a sample of 15,000 households. Unless otherwise noted, the data estimates the percent of people who engaged in these activities more than once during the year.

Participation in Selected Activities during a Year							
Activity	Age 11 Years	Age 35–44 Years	Age 65+ Years	Male	Female	All People	Rank (for all people)
Aerobics or Step Aerobics[1]	6 %	15 %	4 %	6 %	18 %	12 %	8
Backpacking	9	7	0.5	7	4	6	15
Baseball	20	3	1	9	3	6	16
Basketball	32	8	0.7	16	7	11	9
Bicycle Riding[1]	39	15	5	18	13	16	7
Bowling	25	16	4	17	15	16	5
Camping	28	23	5	20	17	18	3
Exercise Walking[1]	14	33	27	22	35	28	1
Exercising with Equipment[1]	10	21	9	17	17	17	4
Fishing (Freshwater)	22	18	6	22	9	16	6
Football (Touch)	10	2	0.3	6	1	4	23
Golf	7	13	7	17	4	11	11
Hiking	13	13	4	11	10	10	13
Paintball Games	6	0.9	0.1	4	0.6	2	25
Roller or In-Line Skating	38	7	0.7	10	11	11	10
Running/Jogging[1]	14	10	2	11	9	10	14
Scooter Riding	25	1	0.4	6	4	5	19
Skateboarding	19	0.4	0.2	6	1	4	22
Skiing (Alpine/Downhill)	5	3	0.4	4	2	3	24
Soccer	17	2	0.6	7	5	6	17
Softball	11	5	0.9	6	5	5	18
Swimming[1]	41	22	10	21	23	22	2
Tennis	6	5	2	4	4	4	21
Volleyball	11	5	0.5	5	5	5	20
Work Out at Club	4	13	4	10	11	11	12

[1]Participant engaged in this activity at least six times in a year.

What Do Adults Do?

In the year 2000, there were approximately 200 million adults living in the United States. During the year, the following numbers of adults did these things at least once:

Activity	Number (million)	%	Activity	Number (million)	%
Barbecued	65	32.5	Read a book	80	40
Did charity work	58	29	Went to a movie	120	60
Dined out	100	50	Went to a sports event	70	35
Entertained friends at home	78	39	Went to an amusement park	84	42
Made home improvements	84	42	Went to the beach	50	25
Played a board game	32	16	Worked a crossword puzzle	32	16

School

Throughout the history of the United States, schooling has been important.

The Northwest Ordinance of 1787 created the rules for forming new states. It also showed the nation's belief in the importance of schooling. It stated —

> Being necessary to good government and the happiness of mankind, schools and the means of education shall forever be preserved.

Who Went to School in 1790?

In the northern states, most children between the ages of 4 and 14 went to school for part of the year. In the southern states, many, but not all, white children of these ages went to school. African Americans who were slaves did not receive formal schooling. Often they were not allowed to learn how to read. Some, however, found ways to learn secretly.

Most schools were in rural areas. Many children had to walk long distances to reach them. Schools were often in session for only two to three months in winter and then again in the summer. After age 10, many children attended school only in the winter, when farm work was light.

Who Went to School in 1900?

In 1900, parents reported to census takers that 80% of 10- to 14-year-olds had attended school at some point during the previous six months.

Almost all children who went to school in 1900 attended elementary school, which usually had eight grades. Approximately 15 million students attended public school in 1900. Only about 500,000 (roughly 3%) were in high school.

Three examples of early American schoolhouses

In rural schools, students were usually not separated by age. Five-and 6-year-olds were often in the same elementary classroom with 15- and 16-year-olds. The older students were not slow learners. They had to do farm work and could only attend school part time.

How Much Schooling Did Students Receive in 1900?

In 1900, the number of days students attended school was different from state to state. Students in North Carolina were in school only about 36 days per year, while students in Massachusetts were in school about 145 days per year.

The following tables and graph show three different ways to examine these data. In the first table, state averages are ranked from high to low for each region.

Average Number of Days in School per Student, 1900							
Northeast	**Days**	**South**	**Days**	**Midwest**	**Days**	**West**	**Days**
Massachusetts	145	Delaware	116	Illinois	123	California	121
Rhode Island	136	Maryland	110	Ohio	122	Nevada	108
Connecticut	135	Louisiana	89	Indiana	115	Utah	101
New York	131	Kentucky	72	Michigan	115	Colorado	93
Pennsylvania	123	Texas	71	Wisconsin	111	Oregon	84
New Jersey	119	Virginia	70	South Dakota	111	Washington	82
Vermont	111	West Virginia	69	Iowa	105	Arizona	77
New Hampshire	106	Georgia	69	Nebraska	102	Wyoming	73
Maine	105	Florida	69	Missouri	92	Montana	71
		Tennessee	67	Minnesota	91	Idaho	63
		South Carolina	63	North Dakota	87	New Mexico	59
		Oklahoma	61	Kansas	84		
		Alabama	61				
		Mississippi	59				
		Arkansas	48				
		North Carolina	36				

NOTE: No data available for Alaska and Hawaii.

The stem-and-leaf display[1] shows all the state averages from the table above.

Days in School, 1900: State Averages	
Stems (10s)	**Leaves (1s)**
3	6
4	8
5	9 9
6	1 1 3 3 7 9 9 9
7	0 1 1 2 3 7
8	2 4 4 7 9
9	1 2 3
10	1 2 5 5 6 8
11	0 1 1 1 5 5 6 9
12	1 2 3 3
13	1 5 6
14	5

[1]**How to read this stem-and-leaf display:** It shows that there was only one state where the number of days was in the 30s. That number was 36. Eight states had numbers in the 60s. Those numbers were 61, 61, 63, 63, 67, 69, 69, and 69. Other numbers are shown in a similar way.

The bar graph shows the median number of days in school for each region.

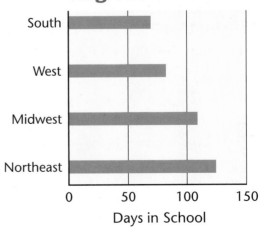

Days in School, 1900: Regional Medians

three hundred sixty-one

Elementary Schooling in the Twentieth Century

During the first half of the twentieth century, elementary schooling became a requirement for all children. The official school year was made longer. The number of student absences decreased. The time students spent in school rose. In 1900, students averaged 99 days per year in school. By 1960, they were in school an average of 160 days per year. After 1960, the average number of school days per year increased very little.

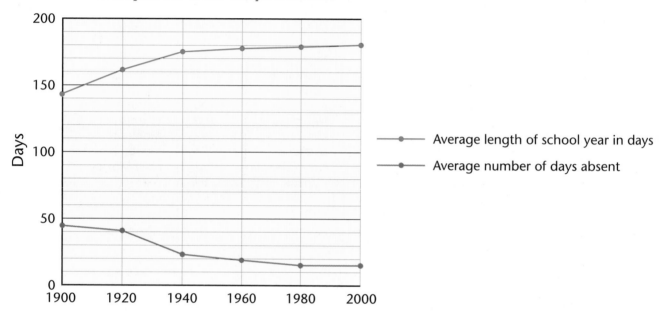

**Days in School Year
Compared with Days Absent**

— Average length of school year in days

— Average number of days absent

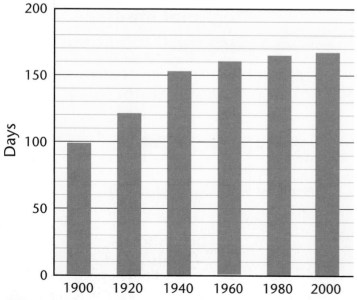

Days in School

Average number of days in school (per student)

Food

The American diet has changed over the last 30 years. The table below shows patterns of increased use and decreased use for many basic foods.

From 1970 to 2000, there was a 25% increase in the amount of bananas, apples, oranges, and grapes consumed. There was a 43% increase in the amount of lettuce, carrots, tomatoes, and broccoli consumed.

In 2005, the Department of Health and Human Services introduced a new guide to daily food choices. The new guidelines include these recommendations:

♦ Eat 2 cups of fruit and 2.5 cups of vegetables per day.

♦ Eat at least 3 ounces of whole-grain products per day.

♦ Consume 3 cups of fat-free or low-fat milk or milk products per day.

♦ When eating meat or poultry, look for lean or low-fat products.

As you study the table below, remember that *per capita* means "for each person."

Per Capita[1] Food Consumption, 1970–2000 (pounds per person per year)				
Foods	**1970**	**1980**	**1990**	**2000**
Red Meat	132	126	112	114
Poultry	34	41	56	67
Fish	12	12	15	15
Cheese	11	18	25	30
Ice Cream	18	18	16	17
Butter and Margarine	16	16	15	13
Wheat Flour	111	117	136	146
Sugar	102	84	64	66
Bananas	17	21	24	29
Apples	17	19	20	18
Oranges	16	14	12	12
Grapes	2.5	3.5	8	7
Lettuce	22	26	28	24
Carrots	6	6	8	10
Tomatoes	12	13	16	18
Broccoli	0.5	1.5	3.5	6

[1]*Per capita* means "by or for each individual person."

Age

The two graphs shown below are called **age-pyramid graphs.**
They show how the sizes of different age groups have changed
during the past 100 years. In each graph, data for males are
shown to the left of the center line; data for females are shown
to the right of the center line.

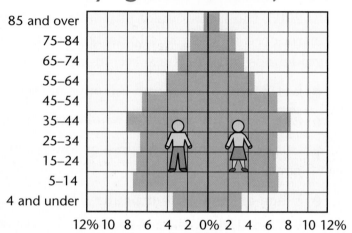

**U.S. Population
by Age and Gender, 1900**

85 and over	
75–84	
65–74	
55–64	
45–54	
35–44	
25–34	
15–24	
5–14	
4 and under	

12% 10 8 6 4 2 0% 2 4 6 8 10 12%

**U.S. Population
by Age and Gender, 2000**

85 and over	
75–84	
65–74	
55–64	
45–54	
35–44	
25–34	
15–24	
5–14	
4 and under	

12% 10 8 6 4 2 0% 2 4 6 8 10 12%

Example The bottom bar on each graph shows information for children
ages 4 and under.

- In 1900, 6.1% of the total U.S. population was male, ages 4 and under;
  and 6.0% was female, ages 4 and under.
- In 2000, 3.5% of the population was male, ages 4 and under; 3.3% was female,
  ages 4 and under.

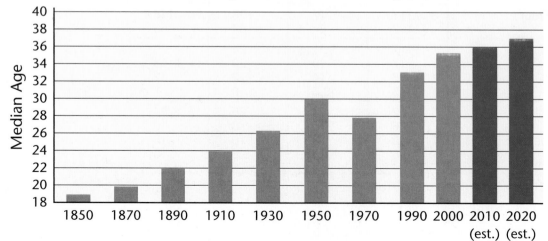

Median Age of the U.S. Population

Life Expectancy

The average lifetime for a person born in the United States in 1900 was about 47 years. By 2000, the average lifetime had increased to about 77 years.

Better health care is one of the major reasons for this increase. During the last 100 years, doctors have developed new and more efficient ways to treat illness. We have been able to control many infectious diseases. The development and use of vaccines has become commonplace.

Improved highway safety, safer workplaces, and better nutrition have also helped people live longer.

Years of Life Expected at Birth, 1900–2010 (males and females combined)												
	1900	1910	1920	1930	1940	1950	1960	1970	1980	1990	2000	2010 (estimated)
Years of Life Expected at Birth	47.3	50.0	54.1	59.7	62.9	68.2	69.7	70.8	73.7	75.4	77.0	78.5

Average lifetimes are different for males and females. The table above gives expected years of life for males and females combined. The graph below shows the data for males and females as separate lines.

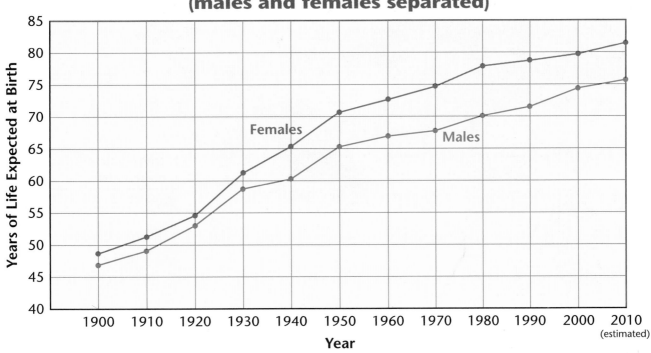

Years of Life Expected at Birth 1900–2010 (males and females separated)

Government

The laws that govern the United States are passed by Congress and signed by the president. The Congress is divided into two bodies.

House of Representatives	Senate
There are 435 representatives.	There are 100 senators.
Representatives are elected for two-year terms.	Senators are elected for six-year terms.
The entire House of Representatives is elected in each even-numbered year.	One-third of the Senate is elected in each even-numbered year.
A representative must be at least 25 years old and must have been a U.S. citizen for at least seven years.	A senator must be at least 30 years old and must have been a U.S. citizen for at least nine years.
The number of representatives from each state is based on population. Population is counted in the Census every ten years. The total population of the 50 states is divided by 435. In 2000, this resulted in each state receiving one representative for approximately every 650,000 people in the state. If a state's population was less than 650,000, it was allowed one representative.	The number of senators from each state is not based on population. Each state elects two senators.

Number of Representatives in the 2000s

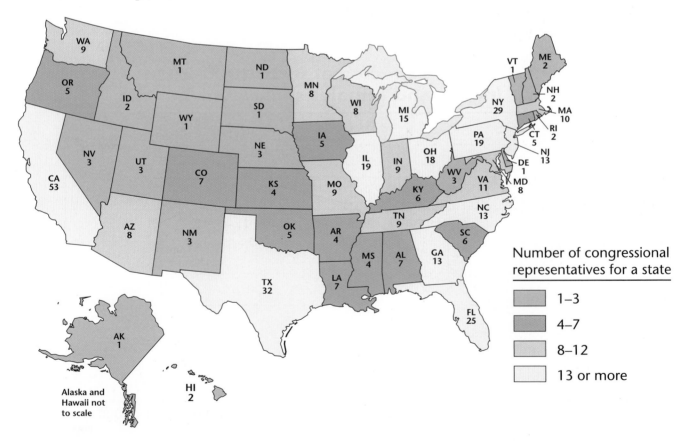

Number of congressional representatives for a state

- 1–3
- 4–7
- 8–12
- 13 or more

Alaska and Hawaii not to scale

Electing a President

The president of the United States is elected every four years. The president must be at least 35 years old. He or she must have been born in the United States.

When people vote for president, they are really voting to tell someone called an **elector** how to vote. Each state has as many electors as it has representatives and senators. In addition, Washington, D.C., has three electors. After the voting, the electors from all states meet to vote for president. In most states, the candidate who receives the greatest number of votes (called **popular votes**) wins *all* of the state's **electoral votes.**

To become president, a candidate must win more than half of all the electoral votes. In 1824, 1876, 1888, and 2000 the candidates with the most popular votes did not become president because they did not win more than half of the electoral votes.

	Electoral Vote	Popular Vote
1824[1]		
John Quincy Adams	84	108,740
Andrew Jackson	99	152,544
Henry Clay	37	47,136
W.H. Crawford	41	46,618
1876		
Rutherford B. Hayes	185	4,036,572
Samuel Tilden	184	4,284,020
1888		
Benjamin Harrison	233	5,447,129
Grover Cleveland	168	5,537,857
2000		
George W. Bush	271	50,459,211
Albert Gore	266	51,003,894

[1]No candidate received more than half of the electoral votes. The election was decided in the House of Representatives. It voted to elect John Quincy Adams.

Electoral Votes in the 2000s

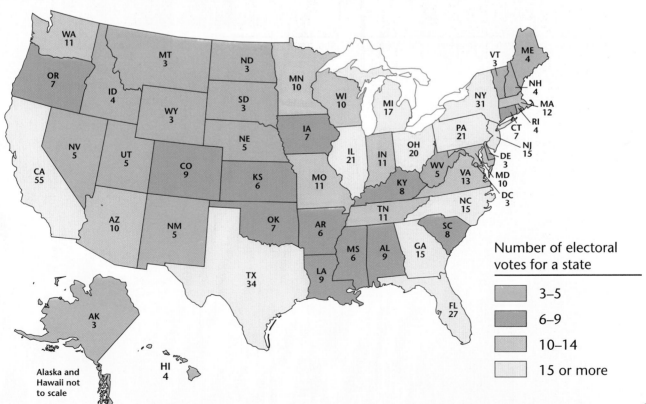

Number of electoral votes for a state

- 3–5
- 6–9
- 10–14
- 15 or more

Alaska and Hawaii not to scale

Voting in Presidential Elections

Who Can Vote for President?	
Year	**Changes in the Law that Extended the Right to Vote**
1828	Electors are chosen by direct popular vote in all but two states.
1870	The 15th Amendment to the Constitution forbids state and federal governments from denying citizens the right to vote due to "race, color, or previous condition of servitude."
1920	The 19th Amendment to the Constitution gives women the same voting rights as men.
1964	The 24th Amendment to the Constitution prohibits charging a poll tax[1] in federal elections.
1965	The Civil Rights Act of 1965 prohibits use of literacy tests for voters.
1971	The 26th Amendment to the Constitution lowers the voting age from 21 to 18.

[1]A tax that must be paid before a person is allowed to vote

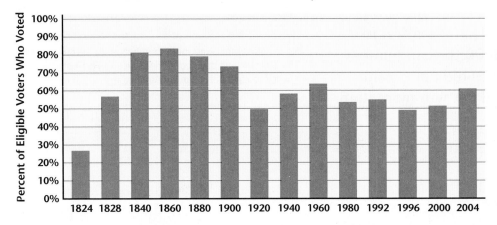

Percent of Eligible Voters Who Voted in Presidential Elections, 1824–2004

Punch ballot (top) and touch-screen ballot (bottom)

The U.S. Decennial Census

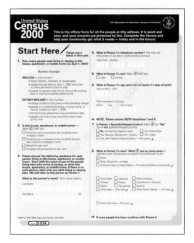

What Is It?

A census is a count of a nation's population. Other information is also collected at the same time as the people are counted.

The word *census* comes from the Latin word *censere,* meaning "to tax," or "to appraise." The U.S. Census is called decennial because it is taken every 10 years.

How Do We Take It?

Since 1970, most census forms have been sent out and returned by mail. Some people are hard to reach by mail or do not respond. Personal visits and phone calls are used to collect information from these people.

Why Do We Take It?

It's the law. Although many countries throughout history have taken censuses, the United States was the first nation in history to require a regular census in its Constitution. The following passages are taken from Article I, Section 2 of the U.S. Constitution:

> Representatives. . . shall be apportioned [divided up] among the several states which may be included within this union according to their respective numbers. . . .

> The actual enumeration shall be made within three years after the first meeting of the Congress of the United States, and within every subsequent term of ten years. . .

Note

The population information collected by the census has always been used to determine how many representatives each state will have in the House of Representatives. Population totals are also used to determine boundaries for congressional districts within each state. Many government offices and private businesses use the census information to plan and provide services.

1790 Census	2000 Census
Information was collected in person. It asked 5 questions.	Most information was collected by mail. It asked 53 questions.
Every household was asked the same set of questions.	Some questions were asked only of a sample group of 1 in 6 households.
It took 18 months to collect information.	Most information was collected in the first 3 months.
Results were tabulated by hand.	Results were processed by computer.
Most people lived in isolated rural areas; roads were scarce and of poor quality.	Many people were difficult to find or reach because they were traveling, homeless, or lived in remote locations. People living in the country illegally were also difficult to find and count.
Many people did not understand the reasons for the census. People would hide from enumerators—and sometimes attack them!	

The table below is a reproduction of the official report of the 1790 Census. This was the first official counting of Americans.

The last two areas in the "DISTRICTS" column are not state names.

♦ "S. Weft. Territory" means Southwest Territory. This area included what is now the state of Tennessee.

♦ "N. Do." means Northwest Territory. This area included what are now the states of Ohio, Indiana, Michigan, Illinois, Wisconsin, and part of Minnesota. The first census did not count people in the Northwest Territory.

SCHEDULE of the whole number of PERSONS within the several Districts of the United States, taken according to " An Act providing for the Enumeration of the Inhabitants of the United States ;" passed March the 1st, 1790.

DISTRICTS.	Free white Males of sixteen years & upwards including heads of families.	Free white Males under sixteen years.	Free white Females including heads of families.	All other free persons.	Slaves.	Total.
• Vermont	22,435	22,328	40,505	255	16	85,539
New Hampshire	36,086	34,851	70,160	630	158	141,885
Maine	24,384	24,748	46,870	538	•	96,540 }
Massachusetts	95,453	87,289	190,582	5,463	•	378,787 }
Rhode Island	16,019	15,799	32,652	3,407	948	68,825
Connecticut	60,523	54,403	117,448	2,808	2,764	237,946
New York	83,700	78,122	152,320	4,654	21,324	340,120
New Jersey	45,251	41,416	83,287	2,762	11,423	184,139
Pennsylvania	110,788	106,948	206,363	6,537	3,737	434,373
Delaware	11,783	12,143	22,384	3,899	8,887	59,096
Maryland	55,915	51,339	101,395	8,043	103,036	319,728
Virginia	110,936	116,135	215,046	12,866	292,627	747,610 }
Kentucky	15,154	17,057	28,922	114	12,430	73,677 }
North Carolina	69,988	77,506	140,710	4,975	100,572	393,751
South Carolina	35,576	37,722	66,880	1,801	107,094	249,073
Georgia	13,103	14,044	25,739	398	29,264	82,548
					Total,	3,893,635

DISTRICTS.	Free white males of twenty-one years and upwards, including heads of families.	Free males under twenty-one years of age.	Free white females, including heads of families.	All other persons.	Slaves.	Total.
S. Weft. Territory	6,271	10,277	15,365	361	3,417	35,691
N. Do.	-	•	•	•	•	•

Truly stated from the original returns deposited in the office of the Secretary of State.

TH: JEFFERSON.

October 24th, 1791.

* This return was not figned by the marfhal, but was enclofed and referred to in a letter written and figned by him.

Population Estimates for Colonial and Continental Periods, 1610–1790

Year	Vermont	New Hampshire[1]	Maine	Massachusetts[1]	Rhode Island[1]	Connecticut[1]	New York[1]	New Jersey[1]	Pennsylvania[1]	Delaware[1]	Maryland[1]	Virginia[1]	Kentucky	North Carolina[1]	South Carolina[1]	Georgia[1]	Tennessee	TOTAL
1610	—	—	—	—	—	—	—	—	—	—	—	210	—	—	—	—	—	210
1620	—	—	—	100	—	—	—	—	—	—	—	2,400	—	—	—	—	—	2,500
1630	—	500	400	1,300	—	—	500	—	—	—	—	3,000	—	—	—	—	—	5,700
1640	—	800	700	14,000	300	2,000	1,000	—	—	—	1,500	7,600	—	—	—	—	—	28,000
1650	—	1,400	1,000	18,000	800	6,000	3,000	—	—	—	4,500	17,000	—	—	—	—	—	52,000
1660	—	2,300	[3]	25,000[3]	1,500	8,000	6,000	—	—	—	8,000	33,000	—	1,000	—	—	—	85,000
1670	—	3,000	[3]	30,000[3]	2,500	10,000	9,000	2,500	—	—	15,000	40,000	—	2,500	—	—	—	115,000
1680	—	4,000	[3]	40,000[3]	4,000	13,000	14,000	6,000	—	500	20,000	49,000	—	4,000	1,100	—	—	156,000
1690	—	5,000	[3]	54,000[3]	5,000	18,000	20,000	9,000	12,000[4]	[4]	25,000	58,000	—	3,000	4,500	—	—	214,000
1700	—	6,000	[3]	70,000[3]	6,000	24,000	19,000	14,000	20,000[4]	[4]	31,000	72,000	—	5,000	8,000	—	—	275,000
1710	—	7,500	[3]	80,000[3]	8,000	31,000	26,000	20,000	35,000[4]	[4]	43,000	87,000	—	7,000	13,000	—	—	358,000
1720	—	9,500	[3]	92,000[3]	11,000	40,000	36,000	26,000	48,000[4]	[4]	62,000	116,000	—	13,000	21,000	—	—	474,000
1730	[2]	12,000	[3]	125,000[3]	17,000	55,000	49,000[2]	37,000	65,000[4]	[4]	82,000	153,000	—	30,000	30,000	—	—	655,000
1740	[2]	22,000	[3]	158,000[3]	24,000	70,000	63,000[2]	52,000	100,000[4]	[4]	105,000	200,000	—	50,000	45,000	—	—	889,000
1750	[2]	31,000	[3]	180,000[3]	35,000	100,000	80,000[2]	66,000	150,000[4]	[4]	137,000	275,000	—	80,000	68,000	5,000	—	1,207,000
1760	[2]	38,000	[3]	235,000[3]	44,000	142,000	113,000[2]	91,000	220,000[4]	[4]	162,000	346,000	—	115,000	95,000	9,000	—	1,610,000
1770	25,000	60,000	34,000[3]	265,000	55,000	175,000	160,000	110,000	250,000	25,000	200,000	450,000[5]	[5]	230,000	140,000	26,000	—	2,205,000
1780	40,000	85,000	56,000[3]	307,000	52,000	203,000	200,000	137,000	335,000	37,000	250,000	520,000	45,000	300,000	160,000	55,000	—	2,781,000
1790	86,000	142,000	97,000[3]	379,000	69,000	238,000	340,000	184,000	434,000	59,000	320,000	748,000	74,000	394,000	249,000	83,000	36,000	3,929,000

[1] Original colony
[2] Vermont was included with New York, 1730–1760. Vermont was admitted to statehood in 1791.
[3] Maine was included with Massachusetts, 1660–1760. Maine was admitted to statehood in 1820.
[4] Delaware was included with Pennsylvania, 1690–1760.
[5] Kentucky was included with Virginia in 1770. Kentucky became a state in 1792.

Most estimates in the table have been rounded to the nearest thousand.

The bottom line of the table shows the state totals given in the 1790 Census report. (See page 370.)
The census counts have been rounded to the nearest thousand.

2000 United States Census Questionnaire

In March of 2000, a census questionnaire was sent to all households in the United States. Every household was required to answer a small number of population and housing questions. A longer form was sent to a sample of 17% of all households.

The Bureau of the Census included this letter with each census questionnaire:

U.S. Department of Commerce
Bureau of the Census
Washington, D.C. 20233-2000

United States Census 2000

Office of the Director

March 13, 2000

To all households:

This is your official form for the United States Census 2000. It is used to count every person living in this house or apartment—people of all ages, citizens and non-citizens.

Your answers are important. First, the number of representatives each state has in Congress depends on the number of people living in the state.

The second reason may be more important to you and your community. The amount of government money your neighborhood receives depends on your answers. That money gets used for schools, employment services, housing assistance, roads, services for children and the elderly, and many other local needs.

Your privacy is protected by law (Title 13 of the United States Code), which also requires that you answer these questions. That law ensures that your information is only used for statistical purposes and that no unauthorized person can see your form or find out what you tell us—no other government agency, no court of law, NO ONE.

Please be as accurate and complete as you can in filling out your census form, and return it in the enclosed postage-paid envelope. Thank you.

Sincerely,

Kenneth Prewitt
Director, Bureau of the Census

Every household was required to answer a short list of questions. Some of these questions are shown below.

One out of six households was asked to answer a longer list of questions. Some of these questions are shown below.

---

3. What is Person 1's name? *Print name below.*

Last Name

| | | | | | | | | | | | | | | | |

First Name   MI

| | | | | | | | | | | | | | | | |

---

4. What is Person 1's telephone number? *We may call this person if we don't understand an answer.*

Area Code + Number

| | | | | | | | | | | |

---

5. What is Person 1's sex? Mark ☒ ONE box.

☐ Male   ☐ Female

---

6. What is Person 1's age and date of birth?

Age on April 1, 2000

| | | | |

Print numbers in boxes.

Month   Day   Year of Birth

| | | | | | | | | | |

→ **NOTE: Please answer BOTH Questions 7 and 8.**

7. Is Person 1 Spanish/Hispanic/Latino? *Mark ☒ the "No" box if **not** Spanish/Hispanic/Latino.*

☐ **No,** not Spanish/Hispanic/Latino   ☐ Yes, Puerto Rican
☐ Yes, Mexican, Mexican Am., Chicano   ☐ Yes, Cuban
☐ Yes, other Spanish/Hispanic/Latino—*Print group.*

| | | | | | | | | | | | | | |

---

8. What is Person 1's race? *Mark ☒ **one or more races** to indicate what this person considers himself/herself to be.*

☐ White
☐ Black, African Am. or Negro
☐ American Indian or Alaska Native—*Print name of enrolled or principal tribe.*

| | | | | | | | | | | | | |

☐ Asian Indian   ☐ Japanese   ☐ Native Hawaiian
☐ Chinese   ☐ Korean   ☐ Guamanian or Chamorro
☐ Filipino   ☐ Vietnamese   ☐ Samoan
☐ Other Asian—*Print race.*   ☐ Other Pacific Islander—*Print race.*

| | | | | | | | | | | | | |

☐ Some other race—*Print race.*

| | | | | | | | | | | | | |

→ **If more people live here, continue with Person 2.**

---

8. b. What grade or level of school was this person attending? Mark ☒ ONE box.

☐ Nursery school, preschool
☐ Kindergarten
☐ Grade 1 to grade 4
☐ Grade 5 to grade 8
☐ Grade 9 to grade 12
☐ College undergraduate (freshman to senior)
☐ Graduate or professional school *(for example: medical, dental, or law school)*

---

10. What is this person's ancestry or ethnic origin?

| | | | | | | | | | | | | | | | | |

(For example: African Am., Mexican, Polish)

---

11. a. Does this person speak a language other than English at home?

☐ Yes   ☐ No → *Skip to 12*

b. What is this language?

| | | | | | | | | | | | | | | | |

c. How well does this person speak English?

☐ Very well   ☐ Well
☐ Not well   ☐ Not at all

---

12. Where was this person born?

☐ In the United States—*Print name of state.*

| | | | | | | | | | | | | | | | |

☐ Outside the United States—*Print name of foreign country, or Puerto Rico, Guam, etc.*

| | | | | | | | | | | | | | | | |

---

21. LAST WEEK, did this person do ANY work for either pay or profit? Mark ☒ the "Yes" box *even if the person only worked 1 hour, or helped without pay in a family business or farm for 15 hours or more, or was on active duty in the Armed Forces.*

☐ Yes   ☐ No → *Skip to 25a*

---

41. Is there telephone service available in this house, apartment, or mobile home from which you can both make and receive calls?

☐ Yes   ☐ No

State Populations, 1790–2000

State	1790	1850	1900	1950	2000
NORTHEAST REGION	**1,968,000**	**8,627,000**	**21,047,000**	**39,478,000**	**53,594,000**
Maine	97,000	583,000	694,000	914,000	1,275,000
New Hampshire	142,000	318,000	412,000	533,000	1,236,000
Vermont	86,000	314,000	344,000	378,000	609,000
Massachusetts	379,000	995,000	2,805,000	4,691,000	6,349,000
Rhode Island	69,000	148,000	429,000	792,000	1,048,000
Connecticut	238,000	371,000	908,000	2,007,000	3,406,000
New York	340,000	3,097,000	7,269,000	14,830,000	18,976,000
New Jersey	184,000	490,000	1,884,000	4,835,000	8,414,000
Pennsylvania	434,000	2,312,000	6,302,000	10,498,000	12,281,000
SOUTH REGION	**1,961,000**	**8,983,000**	**24,524,000**	**47,197,000**	**100,235,000**
Delaware	59,000	92,000	185,000	318,000	784,000
Maryland	320,000	583,000	1,188,000	2,343,000	5,296,000
District of Columbia	–	52,000	279,000	802,000	572,000
Virginia	748,000	1,119,000	1,854,000	3,319,000	7,078,000
West Virginia	–	302,000	959,000	2,006,000	1,808,000
North Carolina	394,000	869,000	1,894,000	4,062,000	8,049,000
South Carolina	249,000	669,000	1,340,000	2,117,000	4,012,000
Georgia	83,000	906,000	2,216,000	3,445,000	8,186,000
Florida	–	87,000	529,000	2,771,000	15,982,000
Kentucky	74,000	982,000	2,147,000	2,945,000	4,042,000
Tennessee	36,000	1,003,000	2,021,000	3,292,000	5,689,000
Alabama	–	772,000	1,829,000	3,062,000	4,447,000
Mississippi	–	607,000	1,551,000	2,179,000	2,845,000
Arkansas	–	210,000	1,312,000	1,910,000	2,673,000
Louisiana	–	518,000	1,382,000	2,684,000	4,469,000
Oklahoma	–	–	790,000	2,233,000	3,451,000
Texas	–	213,000	3,049,000	7,711,000	20,852,000

NOTE: The state and region totals are taken from final census reports for 1790, 1850, 1900, 1950, and 2000.
All totals have been rounded to the nearest thousand.

State	1790	1850	1900	1950	2000
MIDWEST REGION	–	**5,404,000**	**26,333,000**	**44,461,000**	**64,390,000**
Ohio	–	1,980,000	4,158,000	7,947,000	11,353,000
Indiana	–	988,000	2,516,000	3,934,000	6,080,000
Illinois	–	851,000	4,822,000	8,712,000	12,419,000
Michigan	–	398,000	2,421,000	6,372,000	9,938,000
Wisconsin	–	305,000	2,069,000	3,435,000	5,364,000
Minnesota	–	6,000	1,751,000	2,982,000	4,919,000
Iowa	–	192,000	2,232,000	2,621,000	2,926,000
Missouri	–	682,000	3,107,000	3,955,000	5,595,000
North Dakota	–	–	319,000	620,000	642,000
South Dakota	–	–	402,000	653,000	755,000
Nebraska	–	–	1,066,000	1,326,000	1,711,000
Kansas	–	–	1,470,000	1,905,000	2,688,000
WEST REGION	–	**179,000**	**4,309,000**	**20,190,000**	**63,198,000**
Montana	–	–	243,000	591,000	902,000
Idaho	–	–	162,000	589,000	1,294,000
Wyoming	–	–	93,000	291,000	494,000
Colorado	–	–	540,000	1,325,000	4,301,000
New Mexico	–	62,000	195,000	681,000	1,819,000
Arizona	–	–	123,000	750,000	5,131,000
Utah	–	11,000	277,000	689,000	2,233,000
Nevada	–	–	42,000	160,000	1,998,000
Washington	–	1,000	518,000	2,379,000	5,894,000
Oregon	–	12,000	414,000	1,521,000	3,421,000
California	–	93,000	1,485,000	10,586,000	33,872,000
Alaska	–	–	64,000	129,000	627,000
Hawaii	–	–	154,000	500,000	1,212,000
UNITED STATES TOTAL	**3,929,000**	**23,192,000**	**76,212,000**	**151,326,000**	**281,422,000**

State Populations, 1790-2000 (continued)

The U.S. Census Bureau has estimated the United States total population for years to come.

	2010	2020	2030	2040	2050
United States Total Population	309,000,000	336,000,000	364,000,000	392,000,000	420,000,000

The United States in 1790

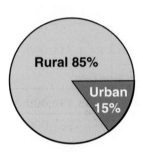
Rural 95% — Urban 5%

Travel Time, New York to Chicago:
about 4–6 weeks by horse, foot, and canoe

Household Size

12%	39%	49%
1 or 2 people	3–5 people	6 or more people

The United States in 1850

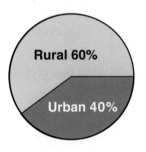
Rural 85% — Urban 15%

Travel Time, New York to Chicago:
about 2–3 weeks by stagecoach

Household Size

14%	42%	44%
1 or 2 people	3–5 people	6 or more people

The United States in 1900

Rural 60% — Urban 40%

Travel Time, New York to Chicago:
about 18 hours by train

Household Size

20%	49%	31%
1 or 2 people	3–5 people	6 or more people

The United States in 2000

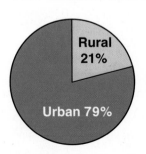
Rural 21% — Urban 79%

Travel Time, New York to Chicago:
about 2 hours by airliner

Household Size

59%	37%	4%
1 or 2 people	3–5 people	6 or more people

Urban means communities with 2,500 or more people.
Rural means communities with fewer than 2,500 people.

Population Density in 2000, by State

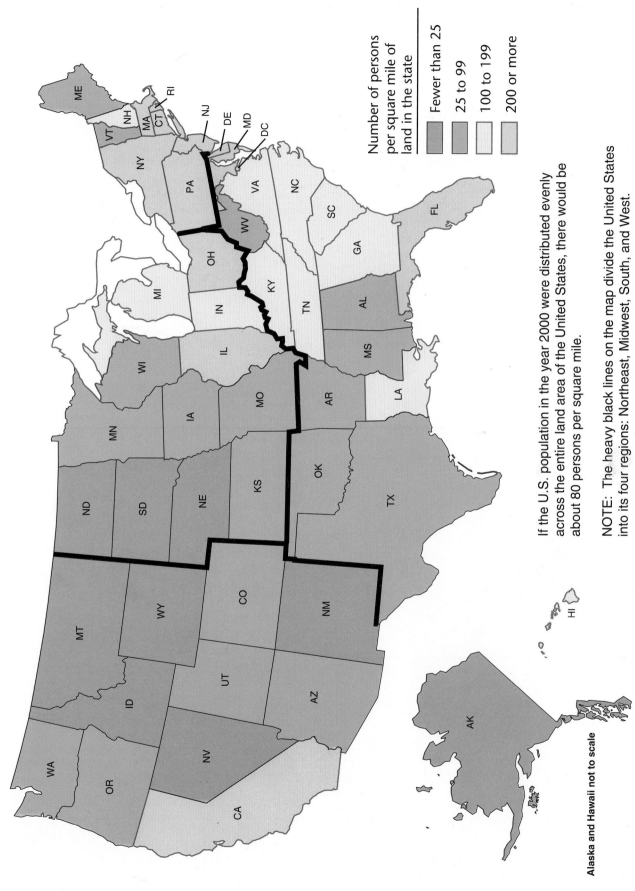

Number of persons per square mile of land in the state

Fewer than 25
25 to 99
100 to 199
200 or more

If the U.S. population in the year 2000 were distributed evenly across the entire land area of the United States, there would be about 80 persons per square mile.

NOTE: The heavy black lines on the map divide the United States into its four regions: Northeast, Midwest, South, and West.

Alaska and Hawaii not to scale

Climate

Average Temperature in. . .

January

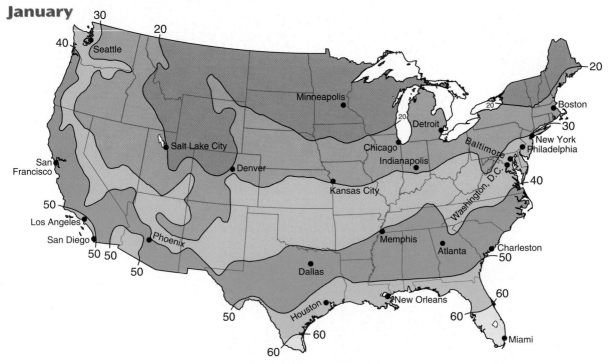

April

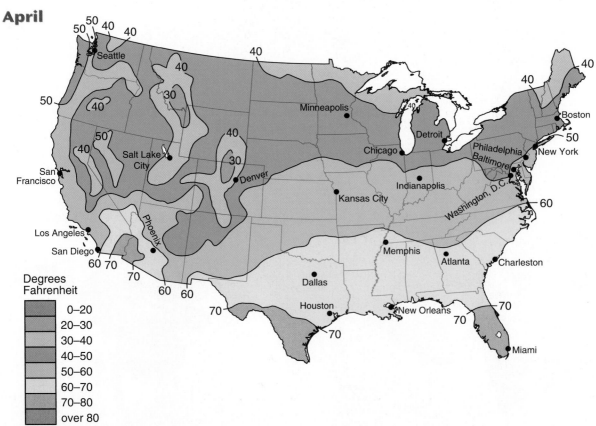

Degrees
Fahrenheit

- 0–20
- 20–30
- 30–40
- 40–50
- 50–60
- 60–70
- 70–80
- over 80

Average Temperature in. . .

July

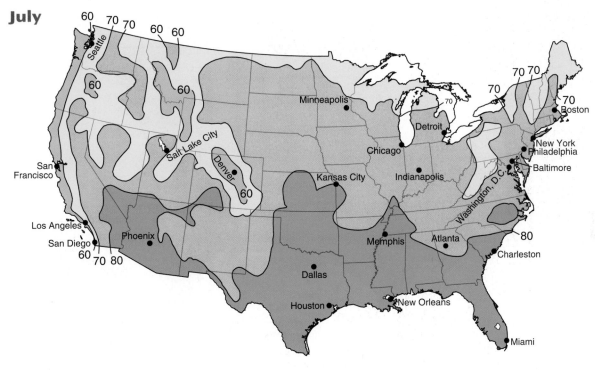

October

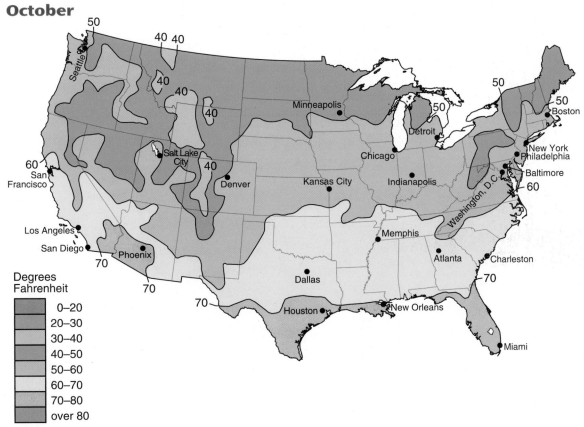

Degrees
Fahrenheit

	0–20
	20–30
	30–40
	40–50
	50–60
	60–70
	70–80
	over 80

Growing Seasons in the United States

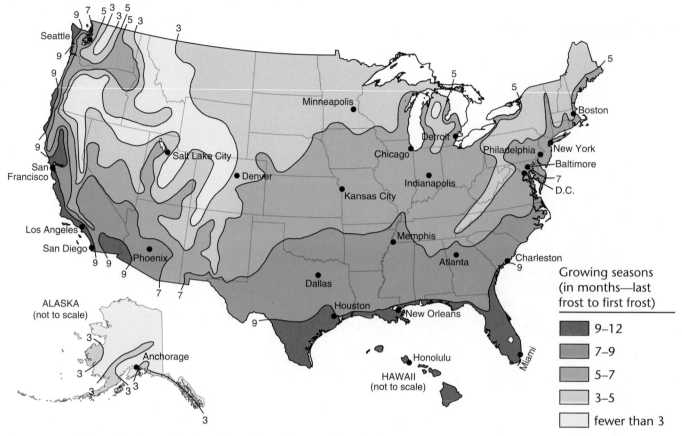

Growing seasons
(in months—last
frost to first frost)

■	9–12
■	7–9
■	5–7
□	3–5
□	fewer than 3

Average Yearly Precipitation in the United States

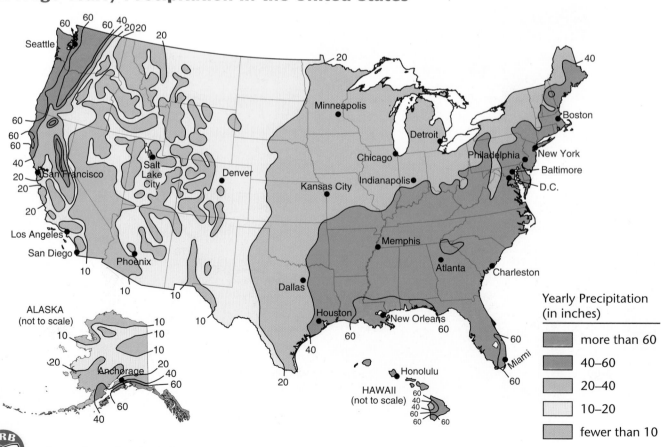

Yearly Precipitation
(in inches)

■	more than 60
■	40–60
□	20–40
□	10–20
■	fewer than 10

Geography

Landform Map of the United States

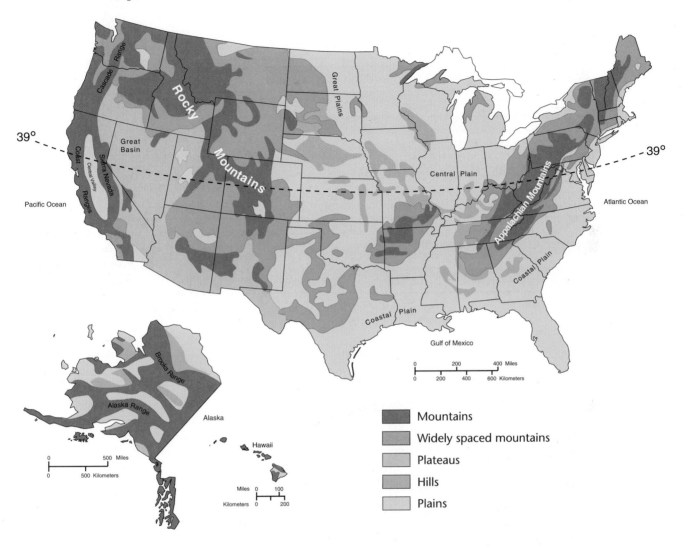

Mountains

Widely spaced mountains

Plateaus

Hills

Plains

Elevation along the 39th Parallel

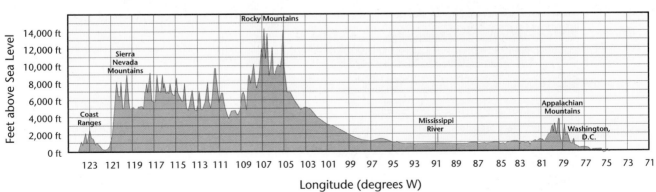

Estimated Percent of Land That Is Farmland and Forest

■ Forest   ■ Farmland   ■ Other

Note

Farmland includes cropland, rangeland, and pastureland.

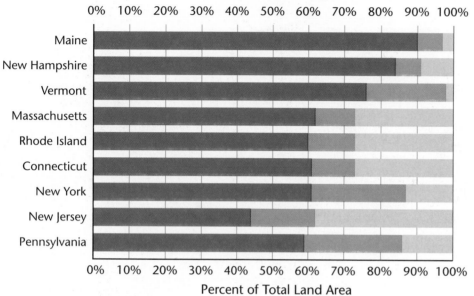

Northeast Region

Percent of Total Land Area

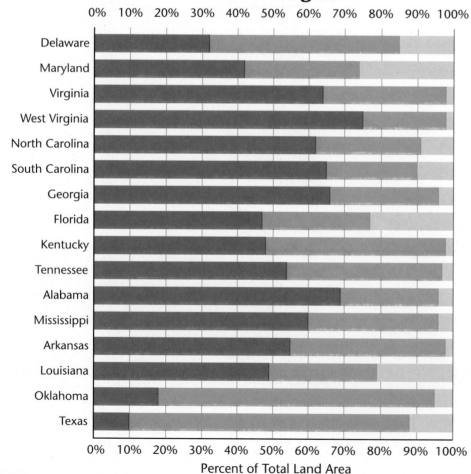

South Region

Percent of Total Land Area

■ Forest   ■ Farmland   ☐ Other

Midwest Region

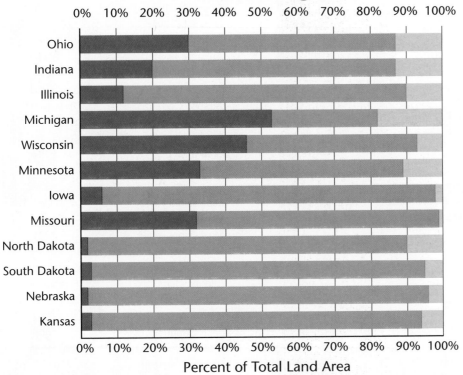

Percent of Total Land Area

West Region

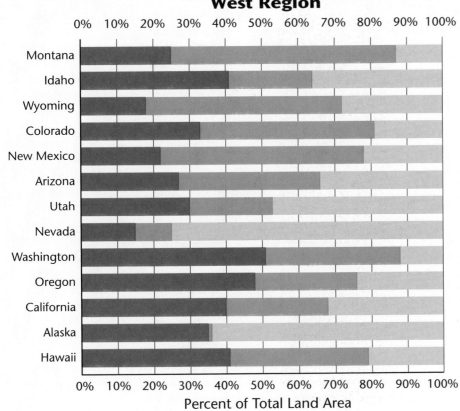

Percent of Total Land Area

Highest and Lowest Elevations in the United States

State	Highest Point	Altitude (ft)	Lowest Point	Elevation (ft)
Alabama	Cheaha Mountain	2,405	Gulf of Mexico	Sea level
Alaska	Mount McKinley	20,320	Pacific Ocean	Sea level
Arizona	Humphreys Peak	12,633	Colorado River	70
Arkansas	Magazine Mountain	2,753	Ouachita River	55
California	Mount Whitney	14,494	Death Valley	−282
Colorado	Mount Elbert	14,433	Arikaree River	3,315
Connecticut	Mount Frissell	2,380	Long Island Sound	Sea level
Delaware	Ebright Road (New Castle Co)	448	Atlantic Ocean	Sea level
Florida	Sec. 30, T6N, R20W (Walton Co)[1]	345	Atlantic Ocean	Sea level
Georgia	Brasstown Bald	4,784	Atlantic Ocean	Sea level
Hawaii	Mauna Kea	13,796	Pacific Ocean	Sea level
Idaho	Borah Peak	12,662	Snake River	710
Illinois	Charles Mound	1,235	Mississippi River	279
Indiana	Franklin Township (Wayne Co)	1,257	Ohio River	320
Iowa	Sec. 29, T100N, R41W (Osceola Co)[1]	1,670	Mississippi River	480
Kansas	Mount Sunflower	4,039	Verdigris River	679
Kentucky	Black Mountain	4,139	Mississippi River	257
Louisiana	Driskill Mountain	535	New Orleans	−8
Maine	Mount Katahdin	5,267	Atlantic Ocean	Sea level
Maryland	Backbone Mountain	3,360	Atlantic Ocean	Sea level
Massachusetts	Mount Greylock	3,487	Atlantic Ocean	Sea level
Michigan	Mount Arvon	1,979	Lake Erie	571
Minnesota	Eagle Mountain	2,301	Lake Superior	600
Mississippi	Woodall Mountain	806	Gulf of Mexico	Sea level
Missouri	Taum Sauk Mountain	1,772	St. Francis River	230
Montana	Granite Peak	12,799	Kootenai River	1,800
Nebraska	Johnson Township (Kimball Co)	5,424	Missouri River	840
Nevada	Boundary Peak	13,143	Colorado River	479
New Hampshire	Mount Washington	6,288	Atlantic Ocean	Sea level
New Jersey	High Point	1,803	Atlantic Ocean	Sea level
New Mexico	Wheeler Peak	13,161	Red Bluff Reservoir	2,842
New York	Mount Marcy	5,344	Atlantic Ocean	Sea level
North Carolina	Mount Mitchell	6,684	Atlantic Ocean	Sea level
North Dakota	White Butte	3,506	Red River	750
Ohio	Campbell Hill	1,549	Ohio River	455
Oklahoma	Black Mesa	4,973	Little River	289
Oregon	Mount Hood	11,239	Pacific Ocean	Sea level
Pennsylvania	Mount Davis	3,213	Delaware River	Sea level
Rhode Island	Jerimoth Hill	812	Atlantic Ocean	Sea level
South Carolina	Sassafras Mountain	3,560	Atlantic Ocean	Sea level
South Dakota	Harney Peak	7,242	Big Stone Lake	966
Tennessee	Clingmans Dome	6,643	Mississippi River	178
Texas	Guadalupe Peak	8,749	Gulf of Mexico	Sea level
Utah	Kings Peak	13,528	Beaverdam Wash	2,000
Vermont	Mount Mansfield	4,393	Lake Champlain	95
Virginia	Mount Rogers	5,729	Atlantic Ocean	Sea level
Washington	Mount Rainier	14,410	Pacific Ocean	Sea level
West Virginia	Spruce Knob	4,861	Potomac River	240
Wisconsin	Timms Hill	1,951	Lake Michigan	579
Wyoming	Gannett Peak	13,804	Belle Fourche River	3,099

[1]"Sec." means Section; "T" means Township; "R" means Range; "N" means North; and "W" means West.

Latitude and Longitude of State Capitals

Postal Abbreviation	State	Capital	Latitude	Longitude
AL	Alabama	Montgomery	32° 22' N	86° 18' W
AK	Alaska	Juneau	58° 18' N	134° 25' W
AZ	Arizona	Phoenix	33° 27' N	112° 04' W
AR	Arkansas	Little Rock	34° 45' N	92° 17' W
CA	California	Sacramento	38° 35' N	121° 30' W
CO	Colorado	Denver	39° 44' N	104° 59' W
CT	Connecticut	Hartford	41° 46' N	72° 41' W
DE	Delaware	Dover	39° 10' N	75° 31' W
FL	Florida	Tallahassee	30° 26' N	84° 17' W
GA	Georgia	Atlanta	33° 45' N	84° 23' W
HI	Hawaii	Honolulu	21° 18' N	157° 52' W
ID	Idaho	Boise	43° 37' N	116° 12' W
IL	Illinois	Springfield	39° 48' N	89° 39' W
IN	Indiana	Indianapolis	39° 46' N	86° 09' W
IA	Iowa	Des Moines	41° 36' N	93° 37' W
KS	Kansas	Topeka	39° 03' N	95° 41' W
KY	Kentucky	Frankfort	38° 11' N	84° 52' W
LA	Louisiana	Baton Rouge	30° 27' N	91° 09' W
ME	Maine	Augusta	44° 19' N	69° 47' W
MD	Maryland	Annapolis	38° 58' N	76° 30' W
MA	Massachusetts	Boston	42° 21' N	71° 04' W
MI	Michigan	Lansing	42° 44' N	84° 33' W
MN	Minnesota	St. Paul	44° 57' N	93° 06' W
MS	Mississippi	Jackson	32° 18' N	90° 11' W
MO	Missouri	Jefferson City	38° 34' N	92° 11' W
MT	Montana	Helena	46° 36' N	112° 02' W
NE	Nebraska	Lincoln	40° 48' N	96° 40' W
NV	Nevada	Carson City	39° 10' N	119° 45' W
NH	New Hampshire	Concord	43° 12' N	71° 32' W
NJ	New Jersey	Trenton	40° 13' N	74° 46' W
NM	New Mexico	Santa Fe	35° 41' N	105° 56' W
NY	New York	Albany	42° 39' N	73° 45' W
NC	North Carolina	Raleigh	35° 46' N	78° 38' W
ND	North Dakota	Bismarck	46° 48' N	100° 47' W
OH	Ohio	Columbus	39° 58' N	83° 00' W
OK	Oklahoma	Oklahoma City	35° 28' N	97° 31' W
OR	Oregon	Salem	44° 57' N	123° 02' W
PA	Pennsylvania	Harrisburg	40° 16' N	76° 53' W
RI	Rhode Island	Providence	41° 49' N	71° 25' W
SC	South Carolina	Columbia	34° 00' N	81° 02' W
SD	South Dakota	Pierre	44° 22' N	100° 21' W
TN	Tennessee	Nashville	36° 10' N	86° 47' W
TX	Texas	Austin	30° 16' N	97° 45' W
UT	Utah	Salt Lake City	40° 46' N	111° 53' W
VT	Vermont	Montpelier	44° 16' N	72° 35' W
VA	Virginia	Richmond	37° 33' N	77° 28' W
WA	Washington	Olympia	47° 03' N	122° 54' W
WV	West Virginia	Charleston	38° 21' N	81° 38' W
WI	Wisconsin	Madison	43° 04' N	89° 24' W
WY	Wyoming	Cheyenne	41° 08' N	104° 49' W

National Capital
State Capital
City

0   100   200
1 inch represents
200 miles.

U.S. Highway Distances (miles)

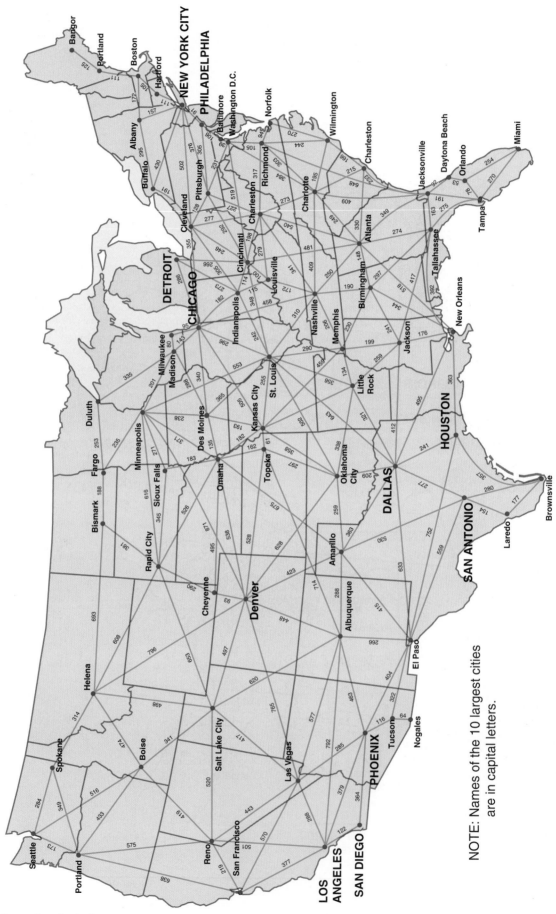

NOTE: Names of the 10 largest cities are in capital letters.

U.S. Air Distances (miles)

	Atlanta	Boston	Chicago	Dallas	Denver	Detroit	Houston	Kansas City	Los Angeles	Miami	Minneapolis	New Orleans	New York	Omaha	Philadelphia	Phoenix	Pittsburgh	Portland	St. Louis	Salt Lake City	San Francisco	Seattle
Boston	940																					
Chicago	600	860																				
Dallas	720	1550	790																			
Denver	1200	1760	900	650																		
Detroit	590	630	230	980	1130																	
Houston	680	1600	920	210	860	1090																
Kansas City	680	1250	400	450	540	630	640															
Los Angeles	1940	2610	1740	1240	840	1970	1370	1360														
Miami	590	1250	1190	1110	1710	1140	960	1230	2340													
Minneapolis	900	1120	330	850	690	520	1040	390	1530	1500												
New Orleans	420	1360	830	430	1060	930	300	690	1670	670	1040											
New York	760	180	740	1380	1630	500	1410	1110	2470	1090	1020	1180										
Omaha	820	1280	410	580	480	650	790	150	1330	1390	280	840	1150									
Philadelphia	660	280	670	1290	1560	450	1320	1030	2400	1010	980	1090	90	1040								
Phoenix	1580	2300	1440	870	580	1680	1010	1040	370	1970	1270	1300	2140	1030	2080							
Pittsburgh	520	490	410	1060	1300	200	1120	760	2130	1010	720	910	340	820	260	1810						
Portland	2170	2530	1730	1630	980	1950	1830	1490	830	2700	1420	2050	2450	1360	2410	1000	2140					
St. Louis	480	1040	250	540	780	440	660	220	1590	1060	440	600	890	340	810	1260	550	1700				
Salt Lake City	1580	2100	1240	1010	380	1480	1200	920	590	2580	990	1420	1980	830	1930	500	1650	630	1730			
San Francisco	2130	2700	1840	1670	950	2070	1630	1490	330	2580	1580	1910	2580	1430	2520	650	2250	550	1730	590		
Seattle	2180	2490	1720	1670	1010	1930	1870	1480	950	2720	1390	2080	2420	1360	2380	1100	2120	130	1700	680	670	
Washington, D.C.	530	410	590	1160	1460	380	1180	920	2280	910	900	960	220	1000	130	1950	180	2330	690	1830	2410	2300

Area, Length, and Width of States

State	Land Area (sq mi)	Inland Water Area (sq mi)	Total Area (sq mi)	Length[1] (mi)	Width[1] (mi)
Alabama	50,750	968	51,718	330	190
Alaska	571,951	17,243	589,194	1,480	810
Arizona	113,642	364	114,006	400	310
Arkansas	52,075	1,107	53,182	260	240
California	155,973	2,674	158,647	770	250
Colorado	103,729	371	104,100	380	280
Connecticut	4,845	161	5,006	110	70
Delaware	1,955	71	2,026	100	30
Florida	53,937	4,683	58,620	500	160
Georgia	57,919	1,011	58,930	300	230
Hawaii	6,423	36	6,459	—	—
Idaho	82,751	823	83,574	570	300
Illinois	55,593	750	56,343	390	210
Indiana	35,870	315	36,185	270	140
Iowa	55,875	401	56,276	310	200
Kansas	81,823	459	82,282	400	210
Kentucky	39,732	679	40,411	380	140
Louisiana	43,566	4,153	47,719	380	130
Maine	30,865	2,263	33,128	320	190
Maryland	9,775	680	10,455	250	90
Massachusetts	7,838	424	8,262	190	50
Michigan	56,809	1,704	58,513	490	240
Minnesota	79,617	4,780	84,397	400	250
Mississippi	46,914	781	47,695	340	170
Missouri	68,898	811	69,709	300	240
Montana	145,556	1,490	147,046	630	280
Nebraska	76,878	481	77,359	430	210
Nevada	109,806	761	110,567	490	320
New Hampshire	8,969	314	9,283	190	70
New Jersey	7,419	371	7,790	150	70
New Mexico	121,364	234	121,598	370	343
New York	47,224	1,888	49,112	330	283
North Carolina	48,718	3,954	52,672	500	150
North Dakota	68,994	1,710	70,704	340	211
Ohio	40,953	376	41,329	220	220
Oklahoma	68,679	1,224	69,903	400	220
Oregon	96,002	1,050	97,052	360	261
Pennsylvania	44,820	490	45,310	283	160
Rhode Island	1,045	178	1,223	40	30
South Carolina	30,111	1,006	31,117	260	200
South Dakota	75,896	1,225	77,121	380	210
Tennessee	41,219	926	42,145	440	120
Texas	261,914	4,959	266,873	790	660
Utah	82,168	2,736	84,904	350	270
Vermont	9,249	366	9,615	160	80
Virginia	39,598	1,000	40,598	430	200
Washington	66,581	1,545	68,126	360	240
West Virginia	24,087	145	24,232	240	130
Wisconsin	54,314	1,831	56,145	310	260
Wyoming	97,105	714	97,819	360	280

[1]Length and width are approximate averages for each state.

Land area is for dry land and land temporarily or partially covered by water, such as marshland and swamps.

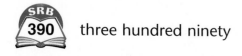

National Facts

Fact or Feature	Location	Data Recorded as of 2004
Largest State	Alaska	589,194 sq mi
Smallest State	Rhode Island	1,223 sq mi
Northernmost Point	Point Barrow, Alaska	71° 23' N
Southernmost Point	Ka Lae (South Cape), Hawaii	18° 55' N
Easternmost Point	Semisopochnoi Island, Alaska[1]	179° 46' E
Westernmost Point	Amatignak Island, Alaska	179° 06' W
Highest Settlement	Climax, Colorado	11,360 ft above sea level
Lowest Settlement	Calipatria, California	184 ft below sea level
Oldest National Park	Yellowstone National Park Wyoming, Montana, Idaho	Established 1872
Largest National Park	Wrangell–St. Elias, Alaska	13,005 sq mi
Smallest National Park	Hot Springs, Arkansas	9 sq mi
Highest Waterfall	Yosemite Falls:	
	Total of three sections	2,425 ft
	Upper Yosemite Fall	1,430 ft
	Cascades in middle section	675 ft
	Lower Yosemite Fall	320 ft
Longest River	Mississippi–Missouri	3,710 mi
Highest Mountain	Mount McKinley, Alaska	20,320 ft above sea level
Lowest Point	Death Valley, California	282 ft below sea level
Deepest Lake	Crater Lake, Oregon	1,932 ft
Largest Gorge	Grand Canyon, Colorado River, Arizona	277 mi long, 600 ft to 18 mi wide, 1 mi deep
Deepest Gorge	Hell's Canyon, Snake River, Idaho–Oregon	7,900 ft deep
Rainiest Spot	Mt. Waialeale, Hawaii	Annual average rainfall: 460 inches
Strongest Surface Wind	Mount Washington, New Hampshire, recorded 1934	231 mph
Biggest Dam	New Cornelia Tailings, Tenmile Wash, Arizona	274,026,000 cu yds of material used
Tallest Building	Willis Tower, Chicago, Illinois	1,450 ft
Largest Building	Boeing Manufacturing Plant, Everett, Washington	472,000,000 cu ft; covers 98 acres
Largest Office Building	Pentagon, Arlington, Virginia	77,025,000 cu ft; covers 29 acres
Tallest Structure	TV Tower, Blanchard, North Dakota	2,063 ft
Longest Bridge Span	Verrazano–Narrows, New York	4,260 ft
Highest Bridge	Royal Gorge, Colorado	1,053 ft above water
Deepest Well	Gas Well, Washita County, Oklahoma	31,441 ft

[1] Alaska's Aleutian Islands extend into the Eastern Hemisphere. The Aleutian Islands technically contain the easternmost point in the United States. If Alaska is excluded, the easternmost point in the United States is West Quoddy Head, Maine (66° 57' W).

Abbreviations: ft = foot   mi = mile   mph = miles per hour
sq mi = square mile   cu ft = cubic foot
cu yd = cubic yard

Explore More

Here are some questions that can be answered with information in the American Tour. Think of more questions on your own.

♦ How does your state compare with other states? Use data displays to compare your state to the average or median state, or to show where your state falls in the distribution of all states. Place your results in the American Tour corner of your classroom.

♦ What are some interesting features and facts about your state or another state? Use the information in the American Tour to create a State Almanac. For example, in what year did your state become a state? What are the highest and lowest points in your state? How does your state's population in 2000 compare with its population in 1900? In 1850? In 1790?

Here are some questions that can be answered by looking in other almanacs, atlases, and reference books.

♦ What are the highest and lowest recorded temperatures in different states? Make a display that shows which states had the highest and lowest temperatures in each decade of the twentieth century. Look for patterns.

♦ How has technology spread? How many cars, telephones, computers, and other devices were there in each decade of the twentieth century? How many of each were there per person? Make a line graph or bar graph that shows this information.

♦ In the last U.S. presidential election, how many electoral and popular votes did each candidate receive? What percent of the total electoral or popular vote did each receive? What percent of the total popular vote did each candidate receive in your state? Make circle graphs that display this information.

♦ When did important historical events occur? When were things such as the telephone, polio vaccine, or the computer invented, discovered, or first used? When did famous people such as artists, actors, scientists, or sports figures live? Make timelines to show your findings.

References

Bibliography

Crystal, David, Ed. *The Cambridge Factfinder.* New York: Cambridge University Press, 1997.

Dillehay, Thomas D. *The Settlement of the Americas.* New York: Basic Books, 2000.

Graves, William, Ed. *Historical Atlas of the United States.* Washington, D.C.: National Geographic Society, 1993.

Hakim, Joy. *Colonies to Country,* Book 3, *A History of US* and *Liberty for All,* Book 5, *A History of US* New York: Oxford University Press, 1994.

Kaestle, Carl. *Pillars of the Republic: Common Schools and American Society, 1780–1860.* New York: Hill and Wang, 1983.

Larkin, Jack. *The Reshaping of Everyday Life, 1790–1840.* New York: HarperCollins, 1989.

Rand McNally, *The Road Atlas, 2005.* Chicago: Rand McNally, 2004.

Snipp, C. Matthew. *American Indians: The First of This Land.* New York: Russell Sage Foundation, 1989.

Tankersley, Kenneth. *In Search of Ice Age Americans.* Layton, Utah: Gibbs Smith, 2002.

U.S. Department of Commerce, National Oceanic and Atmospheric Administration (NOAA). Washington, D.C.: http://www.noaa.gov and the National Weather Service: http://www.nws.noaa.gov.

U.S. Department of Commerce, U.S. Census Bureau. Washington, D.C.: http://www.census.gov.

U.S. Department of Education, National Center for Education Statistics. Washington, D.C.: http://nces.ed.gov.

U.S. Department of the Interior. U.S. Geological Survey (USGS). Reston, VA: http://www.usgs.gov.

U.S. Government Printing Office, National Education Commission on Time and Learning. *Prisoners of Time.* Washington, D.C., 1994.

U.S. Government Printing Office, reprint of 1909 publication. *A Century of Population Growth, 1790–1900.* Baltimore: Genealogical Publishing Co., 1989.

U.S. Government Printing Office, *Statistical Atlas of the United States.* Washington, D.C., 1914.

U.S. Government Printing Office, U.S. Bureau of Education. *Statistics of State School Systems.* Washington, D.C., 1901.

U.S. Government Printing Office, U.S. Department of Commerce, U.S. Census Bureau. *Historical Statistics of the United States: Colonial Times to 1970.* Washington, D.C., 1975.

U.S. Government Printing Office, U.S. Department of Commerce, U.S. Census Bureau. *Statistical Abstract of the United States: 2004* (and earlier editions). Washington, D.C., 2004.

Wetterau, Bruce, Ed. *The New York Public Library Book of Chronologies.* New York: Prentice Hall, 1990.

The World Almanac® *and Book of Facts, 2005* (and earlier editions). New York: World Almanac Books, 2005.

Sources

339 Information and migration map: Information from Dillehay and Tankersley; Migration map based on Dillehay and Tankersley.

340 Information from Dillehay and Tankersley; Artifact photographs: Tankersley.

341 Map: *Historical Atlas of the United States.*

342 Map: based on 2000 U.S. Census data.

343 Graph: *Historical Atlas of the United States;* Table: data from *Statistical Abstract of the United States.*

344 Immigration Graph and Table: *Historical Statistics of the United States* and *Statistical Abstract of the United States.*

345 Foreign-Born Population Graph and Table: *TheWorld Almanac.*

346 Map: based on 2000 U.S. Census data.

347 Map: *Historical Atlas of United States;* Percent bar and table: information from *Historical Statistics of the United States.*

348 "19th Century Settlement Patterns" Map: based on information from *Statistical Atlas of the United States;* "The Center of Population Moving West" Map: *Statistical Abstract of the United States.*

349 Map: information from *The New York Public Library Book of Chronologies* and USCMP research.

350–351 Data for Area from *Historical Statistics of the United States;* Data for Population from *Statistical Abstract of the United States, The World Almanac,* and 2000 U.S. Census.

352 Table: data based on *A Century of Population Growth 1790–1900.*

353 Table: information from *Historical Atlas of the United States;* Larkin; and Hakim, *Liberty for All;* Graph: based on *Historical Atlas of the United States.*

354 Map: information from *Historical Atlas of the United States.*

355 Information from *The New York Public Library Book of Chronologies;* Train schedule information from Amtrak; Airline schedule information from American Airlines.

356–357 Data from *Historical Statistics of the United States, Statistical Abstract of the United States,* and *The World Almanac.*

358 Data from *Historical Statistics of the United States, The World Almanac, Statistical Abstract of the United States,* and Nielsen Research.

359 Data from *Statistical Abstract of the United States.*

360 "Who Went to School in 1790?" information from Kaestle; "Who Went to School in 1900?" information from 1900 U.S. Census and *Statistics of State School Systems.*

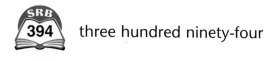

361 Data from *Statistics of State School Systems.*

362 Graphs: data from the National Center for Education Statistics.

363 Data from *Statistical Abstract of the United States.*

364 Graph "U.S. Population by Age and Gender, 1900": data from 1900 U.S. Census; Graph "U.S. Population by Age and Gender, 2000": data from 2000 U.S. Census; Graph "Median Age of the U.S. Population": data from *Statistical Abstract of the United States* and *The World Almanac.*

365 Table and Graph: data from *The World Almanac* and *Statistical Abstract of the United States.*

366–367 Information from the U.S. Constitution, *Historical Statistics of the United States,* and 2000 U.S. Census.

368 Information from *Historical Statistics of the United States;* Graph: data from *Historical Statistics of the United States* and UCSMP research.

369–370 Information from the U.S. Census Bureau.

371 Data from *A Century of Population Growth 1790–1900.*

372–373 Derived from 2000 U.S. Census Forms D-61A and D-61B.

374–375 Data from *Statistical Abstract of the United States* and U.S. Census Bureau projections.

376 Urban/Rural Population data from *Historical Statistics of the United States* and United Nations Statistics Division; Household Size data from *Statistical Abstract of the United States.*

377 Map: information from *The World Almanac.*

378–379 Temperature Maps: based on data from the National Oceanic and Atmospheric Administration.

380 Growing Seasons Map: based on data from the National Weather Service; Precipitation Map: based on data from the National Oceanic and Atmospheric Administration.

381 Landform Map: based on data from the U.S. Geological Survey; Graph: based on *Historical Atlas of the United States.*

382–383 Based on data from *The World Almanac* and *Statistical Abstract of the United States.*

384 Data from *The World Almanac.*

385 Data from *The World Almanac.*

386–387 Map: based on data from Arc World 1:3m and World 25m by ESRI Data & Maps.

388 Highway distances: based on *The Rand McNally Road Atlas, 2005.*

389 Air distances: based on *The Cambridge Factfinder.*

390 Land and Inland Water Area: data from *Statistical Abstract of the United States;* Length and Width: data from *The World Almanac.*

391 Information from *The World Almanac.*

Place-Value Chart

billions	100 millions	10 millions	millions	100 thousands	10 thousands	thousands	hundreds	tens	ones	.	tenths	hundredths	thousandths
1,000 millions	100,000,000s	10,000,000s	1,000,000s	100,000s	10,000s	1,000s	100s	10s	1s	.	0.1s	0.01s	0.001s
10^9	10^8	10^7	10^6	10^5	10^4	10^3	10^2	10^1	10^0	.	10^{-1}	10^{-2}	10^{-3}

Prefixes

uni-	one	tera-	trillion (10^{12})
bi-	two	giga-	billion (10^9)
tri-	three	mega-	million (10^6)
quad-	four	kilo-	thousand (10^3)
penta-	five	hecto-	hundred (10^2)
hexa-	six	deca-	ten (10^1)
hepta-	seven	uni-	one (10^0)
octa-	eight	deci-	tenth (10^{-1})
nona-	nine	centi-	hundredth (10^{-2})
deca-	ten	milli-	thousandth (10^{-3})
dodeca-	twelve	micro-	millionth (10^{-6})
icosa-	twenty	nano-	billionth (10^{-9})

Multiplication and Division Table

*,/	1	2	3	4	5	6	7	8	9	10	11	12
1	1	2	3	4	5	6	7	8	9	10	11	12
2	2	4	6	8	10	12	14	16	18	20	22	24
3	3	6	9	12	15	18	21	24	27	30	33	36
4	4	8	12	16	20	24	28	32	36	40	44	48
5	5	10	15	20	25	30	35	40	45	50	55	60
6	6	12	18	24	30	36	42	48	54	60	66	72
7	7	14	21	28	35	42	49	56	63	70	77	84
8	8	16	24	32	40	48	56	64	72	80	88	96
9	9	18	27	36	45	54	63	72	81	90	99	108
10	10	20	30	40	50	60	70	80	90	100	110	120
11	11	22	33	44	55	66	77	88	99	110	121	132
12	12	24	36	48	60	72	84	96	108	120	132	144

The numbers on the diagonal are square numbers.

Rules for Order of Operations

1. Do operations within parentheses or other grouping symbols before doing anything else.
2. Calculate all powers.
3. Do multiplications and divisions in order, from left to right.
4. Then do additions and subtractions in order, from left to right.

Metric System

Units of Length

1 kilometer (km) = 1,000 meters (m)

1 meter = 10 decimeters (dm)
= 100 centimeters (cm)
= 1,000 millimeters (mm)

1 decimeter = 10 centimeters

1 centimeter = 10 millimeters

Units of Area

1 square meter (m^2) = 100 square decimeters (dm^2)
= 10,000 square centimeters (cm^2)

1 square decimeter = 100 square centimeters

1 square kilometer = 1,000,000 square meters

Units of Volume

1 cubic meter (m^3) = 1,000 cubic decimeters (dm^3)
= 1,000,000 cubic centimeters (cm^3)

1 cubic decimeter = 1,000 cubic centimeters

Units of Capacity

1 kiloliter (kL) = 1,000 liters (L)

1 liter = 1,000 milliliters (mL)

1 cubic centimeter = 1 milliliter

Units of Weight

1 metric ton (t) = 1,000 kilograms (kg)

1 kilogram = 1,000 grams (g)

1 gram = 1,000 milligrams (mg)

U.S. Customary System

Units of Length

1 mile (mi) = 1,760 yards (yd)
= 5,280 feet (ft)

1 yard = 3 feet
= 36 inches (in.)

1 foot = 12 inches

Units of Area

1 square yard (yd^2) = 9 square feet (ft^2)
= 1,296 square inches ($in.^2$)

1 square foot = 144 square inches

1 acre = 43,560 square feet

1 square mile (mi^2) = 640 acres

Units of Volume

1 cubic yard (yd^3) = 27 cubic feet (ft^3)

1 cubic foot = 1,728 cubic inches ($in.^3$)

Units of Capacity

1 gallon (gal) = 4 quarts (qt)

1 quart = 2 pints (pt)

1 pint = 2 cups (c)

1 cup = 8 fluid ounces (fl oz)

1 fluid ounce = 2 tablespoons (tbs)

1 tablespoon = 3 teaspoons (tsp)

Units of Weight

1 ton (T) = 2,000 pounds (lb)

1 pound = 16 ounces (oz)

System Equivalents

1 inch is about 2.5 cm (2.54)

1 kilometer is about 0.6 mile (0.621)

1 mile is about 1.6 kilometers (1.609)

1 meter is about 39 inches (39.37)

1 liter is about 1.1 quarts (1.057)

1 ounce is about 28 grams (28.350)

1 kilogram is about 2.2 pounds (2.205)

Units of Time

1 century = 100 years

1 decade = 10 years

1 year (yr) = 12 months
= 52 weeks (plus one or two days)
= 365 days (366 days in a leap year)

1 month (mo) = 28, 29, 30, or 31 days

1 week (wk) = 7 days

1 day (d) = 24 hours

1 hour (hr) = 60 minutes

1 minute (min) = 60 seconds (sec)

Decimal and Percent Equivalents for "Easy" Fractions

"Easy" Fractions	Decimals	Percents
$\frac{1}{2}$	0.50	50%
$\frac{1}{3}$	$0.\overline{3}$	$33\frac{1}{3}\%$
$\frac{2}{3}$	$0.\overline{6}$	$66\frac{2}{3}\%$
$\frac{1}{4}$	0.25	25%
$\frac{3}{4}$	0.75	75%
$\frac{1}{5}$	0.20	20%
$\frac{2}{5}$	0.40	40%
$\frac{3}{5}$	0.60	60%
$\frac{4}{5}$	0.80	80%
$\frac{1}{6}$	$0.1\overline{6}$	$16\frac{2}{3}\%$
$\frac{5}{6}$	$0.8\overline{3}$	$83\frac{1}{3}\%$
$\frac{1}{8}$	0.125	$12\frac{1}{2}\%$
$\frac{3}{8}$	0.375	$37\frac{1}{2}\%$
$\frac{5}{8}$	0.625	$62\frac{1}{2}\%$
$\frac{7}{8}$	0.875	$87\frac{1}{2}\%$
$\frac{1}{10}$	0.10	10%
$\frac{3}{10}$	0.30	30%
$\frac{7}{10}$	0.70	70%
$\frac{9}{10}$	0.90	90%

The Global Grid

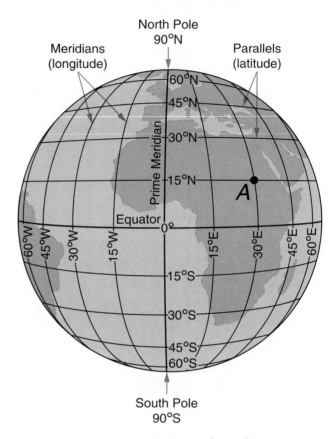

Point *A* is located at 15°N, 30°E.

Fraction-Decimal Number Line

0		0.125		0.25		0.375		0.5		0.625		0.75		0.875		1
$\frac{0}{2}$	$\frac{1}{16}$	$\frac{1}{8}$	$\frac{3}{16}$	$\frac{1}{4}$	$\frac{5}{16}$	$\frac{3}{8}$	$\frac{7}{16}$	$\frac{1}{2}$	$\frac{9}{16}$	$\frac{5}{8}$	$\frac{11}{16}$	$\frac{3}{4}$	$\frac{13}{16}$	$\frac{7}{8}$	$\frac{15}{16}$	$\frac{2}{2}$
$\frac{0}{4}$		$\frac{2}{16}$		$\frac{2}{8}$		$\frac{6}{16}$		$\frac{2}{4}$		$\frac{10}{16}$		$\frac{6}{8}$		$\frac{14}{16}$		$\frac{4}{4}$
$\frac{0}{8}$				$\frac{4}{16}$				$\frac{4}{8}$				$\frac{12}{16}$				$\frac{8}{8}$
$\frac{0}{16}$								$\frac{8}{16}$								$\frac{16}{16}$

Fraction-Stick and Decimal Number-Line Chart

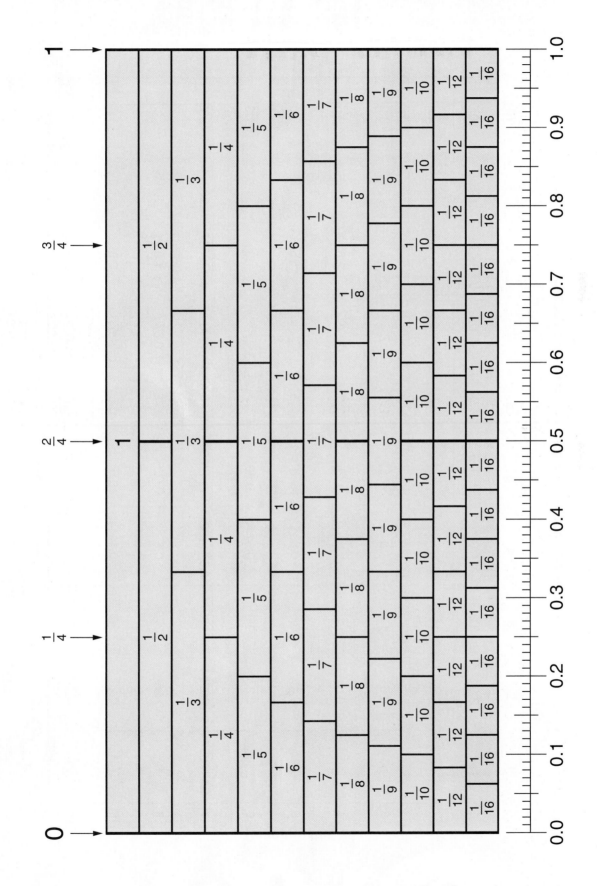

Table of Decimal Equivalents for Fractions

Numerator

Denominator \ Numerator	1	2	3	4	5	6	7	8	9	10
1	1.0	2.0	3.0	4.0	5.0	6.0	7.0	8.0	9.0	10.0
2	0.5	1.0	1.5	2.0	2.5	3.0	3.5	4.0	4.5	5.0
3	$0.\overline{3}$	$0.\overline{6}$	1.0	$1.\overline{3}$	$1.\overline{6}$	2.0	$2.\overline{3}$	$2.\overline{6}$	3.0	$3.\overline{3}$
4	0.25	0.5	0.75	1.0	1.25	1.5	1.75	2.0	2.25	2.5
5	0.2	0.4	0.6	0.8	1.0	1.2	1.4	1.6	1.8	2.0
6	$0.1\overline{6}$	$0.\overline{3}$	0.5	$0.\overline{6}$	$0.8\overline{3}$	1.0	$1.1\overline{6}$	$1.\overline{3}$	1.5	$1.\overline{6}$
7	$0.\overline{142857}$	$0.\overline{285714}$	$0.\overline{428571}$	$0.\overline{571428}$	$0.\overline{714285}$	$0.\overline{857142}$	1.0	$1.\overline{142857}$	$1.\overline{285714}$	$1.\overline{428571}$
8	0.125	0.25	.375	0.5	0.625	0.75	0.875	1.0	1.125	1.25
9	$0.\overline{1}$	$0.\overline{2}$	$0.\overline{3}$	$0.\overline{4}$	$0.\overline{5}$	$0.\overline{6}$	$0.\overline{7}$	$0.\overline{8}$	1.0	$1.\overline{1}$
10	0.1	0.2	0.3	0.4	0.5	0.6	0.7	0.8	0.9	1.0

Denominator

Equivalent Fractions, Decimals, and Percents

															Decimal	Percent
$\frac{1}{2}$	$\frac{2}{4}$	$\frac{3}{6}$	$\frac{4}{8}$	$\frac{5}{10}$	$\frac{6}{12}$	$\frac{7}{14}$	$\frac{8}{16}$	$\frac{9}{18}$	$\frac{10}{20}$	$\frac{11}{22}$	$\frac{12}{24}$	$\frac{13}{26}$	$\frac{14}{28}$	$\frac{15}{30}$	0.5	50%
$\frac{1}{3}$	$\frac{2}{6}$	$\frac{3}{9}$	$\frac{4}{12}$	$\frac{5}{15}$	$\frac{6}{18}$	$\frac{7}{21}$	$\frac{8}{24}$	$\frac{9}{27}$	$\frac{10}{30}$	$\frac{11}{33}$	$\frac{12}{36}$	$\frac{13}{39}$	$\frac{14}{42}$	$\frac{15}{45}$	$0.\overline{3}$	$33\frac{1}{3}\%$
$\frac{2}{3}$	$\frac{4}{6}$	$\frac{6}{9}$	$\frac{8}{12}$	$\frac{10}{15}$	$\frac{12}{18}$	$\frac{14}{21}$	$\frac{16}{24}$	$\frac{18}{27}$	$\frac{20}{30}$	$\frac{22}{33}$	$\frac{24}{36}$	$\frac{26}{39}$	$\frac{28}{42}$	$\frac{30}{45}$	$0.\overline{6}$	$66\frac{2}{3}\%$
$\frac{1}{4}$	$\frac{2}{8}$	$\frac{3}{12}$	$\frac{4}{16}$	$\frac{5}{20}$	$\frac{6}{24}$	$\frac{7}{28}$	$\frac{8}{32}$	$\frac{9}{36}$	$\frac{10}{40}$	$\frac{11}{44}$	$\frac{12}{48}$	$\frac{13}{52}$	$\frac{14}{56}$	$\frac{15}{60}$	0.25	25%
$\frac{3}{4}$	$\frac{6}{8}$	$\frac{9}{12}$	$\frac{12}{16}$	$\frac{15}{20}$	$\frac{18}{24}$	$\frac{21}{28}$	$\frac{24}{32}$	$\frac{27}{36}$	$\frac{30}{40}$	$\frac{33}{44}$	$\frac{36}{48}$	$\frac{39}{52}$	$\frac{42}{56}$	$\frac{45}{60}$	0.75	75%
$\frac{1}{5}$	$\frac{2}{10}$	$\frac{3}{15}$	$\frac{4}{20}$	$\frac{5}{25}$	$\frac{6}{30}$	$\frac{7}{35}$	$\frac{8}{40}$	$\frac{9}{45}$	$\frac{10}{50}$	$\frac{11}{55}$	$\frac{12}{60}$	$\frac{13}{65}$	$\frac{14}{70}$	$\frac{15}{75}$	0.2	20%
$\frac{2}{5}$	$\frac{4}{10}$	$\frac{6}{15}$	$\frac{8}{20}$	$\frac{10}{25}$	$\frac{12}{30}$	$\frac{14}{35}$	$\frac{16}{40}$	$\frac{18}{45}$	$\frac{20}{50}$	$\frac{22}{55}$	$\frac{24}{60}$	$\frac{26}{65}$	$\frac{28}{70}$	$\frac{30}{75}$	0.4	40%
$\frac{3}{5}$	$\frac{6}{10}$	$\frac{9}{15}$	$\frac{12}{20}$	$\frac{15}{25}$	$\frac{18}{30}$	$\frac{21}{35}$	$\frac{24}{40}$	$\frac{27}{45}$	$\frac{30}{50}$	$\frac{33}{55}$	$\frac{36}{60}$	$\frac{39}{65}$	$\frac{42}{70}$	$\frac{45}{75}$	0.6	60%
$\frac{4}{5}$	$\frac{8}{10}$	$\frac{12}{15}$	$\frac{16}{20}$	$\frac{20}{25}$	$\frac{24}{30}$	$\frac{28}{35}$	$\frac{32}{40}$	$\frac{36}{45}$	$\frac{40}{50}$	$\frac{44}{55}$	$\frac{48}{60}$	$\frac{52}{65}$	$\frac{56}{70}$	$\frac{60}{75}$	0.8	80%
$\frac{1}{6}$	$\frac{2}{12}$	$\frac{3}{18}$	$\frac{4}{24}$	$\frac{5}{30}$	$\frac{6}{36}$	$\frac{7}{42}$	$\frac{8}{48}$	$\frac{9}{54}$	$\frac{10}{60}$	$\frac{11}{66}$	$\frac{12}{72}$	$\frac{13}{78}$	$\frac{14}{84}$	$\frac{15}{90}$	$0.1\overline{6}$	$16\frac{2}{3}\%$
$\frac{5}{6}$	$\frac{10}{12}$	$\frac{15}{18}$	$\frac{20}{24}$	$\frac{25}{30}$	$\frac{30}{36}$	$\frac{35}{42}$	$\frac{40}{48}$	$\frac{45}{54}$	$\frac{50}{60}$	$\frac{55}{66}$	$\frac{60}{72}$	$\frac{65}{78}$	$\frac{70}{84}$	$\frac{75}{90}$	$0.8\overline{3}$	$83\frac{1}{3}\%$
$\frac{1}{7}$	$\frac{2}{14}$	$\frac{3}{21}$	$\frac{4}{28}$	$\frac{5}{35}$	$\frac{6}{42}$	$\frac{7}{49}$	$\frac{8}{56}$	$\frac{9}{63}$	$\frac{10}{70}$	$\frac{11}{77}$	$\frac{12}{84}$	$\frac{13}{91}$	$\frac{14}{98}$	$\frac{15}{105}$	0.143	14.3%
$\frac{2}{7}$	$\frac{4}{14}$	$\frac{6}{21}$	$\frac{8}{28}$	$\frac{10}{35}$	$\frac{12}{42}$	$\frac{14}{49}$	$\frac{16}{56}$	$\frac{18}{63}$	$\frac{20}{70}$	$\frac{22}{77}$	$\frac{24}{84}$	$\frac{26}{91}$	$\frac{28}{98}$	$\frac{30}{105}$	0.286	28.6%
$\frac{3}{7}$	$\frac{6}{14}$	$\frac{9}{21}$	$\frac{12}{28}$	$\frac{15}{35}$	$\frac{18}{42}$	$\frac{21}{49}$	$\frac{24}{56}$	$\frac{27}{63}$	$\frac{30}{70}$	$\frac{33}{77}$	$\frac{36}{84}$	$\frac{39}{91}$	$\frac{42}{98}$	$\frac{45}{105}$	0.429	42.9%
$\frac{4}{7}$	$\frac{8}{14}$	$\frac{12}{21}$	$\frac{16}{28}$	$\frac{20}{35}$	$\frac{24}{42}$	$\frac{28}{49}$	$\frac{32}{56}$	$\frac{36}{63}$	$\frac{40}{70}$	$\frac{44}{77}$	$\frac{48}{84}$	$\frac{52}{91}$	$\frac{56}{98}$	$\frac{60}{105}$	0.571	57.1%
$\frac{5}{7}$	$\frac{10}{14}$	$\frac{15}{21}$	$\frac{20}{28}$	$\frac{25}{35}$	$\frac{30}{42}$	$\frac{35}{49}$	$\frac{40}{56}$	$\frac{45}{63}$	$\frac{50}{70}$	$\frac{55}{77}$	$\frac{60}{84}$	$\frac{65}{91}$	$\frac{70}{98}$	$\frac{75}{105}$	0.714	71.4%
$\frac{6}{7}$	$\frac{12}{14}$	$\frac{18}{21}$	$\frac{24}{28}$	$\frac{30}{35}$	$\frac{36}{42}$	$\frac{42}{49}$	$\frac{48}{56}$	$\frac{54}{63}$	$\frac{60}{70}$	$\frac{66}{77}$	$\frac{72}{84}$	$\frac{78}{91}$	$\frac{84}{98}$	$\frac{90}{105}$	0.857	85.7%
$\frac{1}{8}$	$\frac{2}{16}$	$\frac{3}{24}$	$\frac{4}{32}$	$\frac{5}{40}$	$\frac{6}{48}$	$\frac{7}{56}$	$\frac{8}{64}$	$\frac{9}{72}$	$\frac{10}{80}$	$\frac{11}{88}$	$\frac{12}{96}$	$\frac{13}{104}$	$\frac{14}{112}$	$\frac{15}{120}$	0.125	$12\frac{1}{2}\%$
$\frac{3}{8}$	$\frac{6}{16}$	$\frac{9}{24}$	$\frac{12}{32}$	$\frac{15}{40}$	$\frac{18}{48}$	$\frac{21}{56}$	$\frac{24}{64}$	$\frac{27}{72}$	$\frac{30}{80}$	$\frac{33}{88}$	$\frac{36}{96}$	$\frac{39}{104}$	$\frac{42}{112}$	$\frac{45}{120}$	0.375	$37\frac{1}{2}\%$
$\frac{5}{8}$	$\frac{10}{16}$	$\frac{15}{24}$	$\frac{20}{32}$	$\frac{25}{40}$	$\frac{30}{48}$	$\frac{35}{56}$	$\frac{40}{64}$	$\frac{45}{72}$	$\frac{50}{80}$	$\frac{55}{88}$	$\frac{60}{96}$	$\frac{65}{104}$	$\frac{70}{112}$	$\frac{75}{120}$	0.625	$62\frac{1}{2}\%$
$\frac{7}{8}$	$\frac{14}{16}$	$\frac{21}{24}$	$\frac{28}{32}$	$\frac{35}{40}$	$\frac{42}{48}$	$\frac{49}{56}$	$\frac{56}{64}$	$\frac{63}{72}$	$\frac{70}{80}$	$\frac{77}{88}$	$\frac{84}{96}$	$\frac{91}{104}$	$\frac{98}{112}$	$\frac{105}{120}$	0.875	$87\frac{1}{2}\%$
$\frac{1}{9}$	$\frac{2}{18}$	$\frac{3}{27}$	$\frac{4}{36}$	$\frac{5}{45}$	$\frac{6}{54}$	$\frac{7}{63}$	$\frac{8}{72}$	$\frac{9}{81}$	$\frac{10}{90}$	$\frac{11}{99}$	$\frac{12}{108}$	$\frac{13}{117}$	$\frac{14}{126}$	$\frac{15}{135}$	$0.\overline{1}$	$11\frac{1}{9}\%$
$\frac{2}{9}$	$\frac{4}{18}$	$\frac{6}{27}$	$\frac{8}{36}$	$\frac{10}{45}$	$\frac{12}{54}$	$\frac{14}{63}$	$\frac{16}{72}$	$\frac{18}{81}$	$\frac{20}{90}$	$\frac{22}{99}$	$\frac{24}{108}$	$\frac{26}{117}$	$\frac{28}{126}$	$\frac{30}{135}$	$0.\overline{2}$	$22\frac{2}{9}\%$
$\frac{4}{9}$	$\frac{8}{18}$	$\frac{12}{27}$	$\frac{16}{36}$	$\frac{20}{45}$	$\frac{24}{54}$	$\frac{28}{63}$	$\frac{32}{72}$	$\frac{36}{81}$	$\frac{40}{90}$	$\frac{44}{99}$	$\frac{48}{108}$	$\frac{52}{117}$	$\frac{56}{126}$	$\frac{60}{135}$	$0.\overline{4}$	$44\frac{4}{9}\%$
$\frac{5}{9}$	$\frac{10}{18}$	$\frac{15}{27}$	$\frac{20}{36}$	$\frac{25}{45}$	$\frac{30}{54}$	$\frac{35}{63}$	$\frac{40}{72}$	$\frac{45}{81}$	$\frac{50}{90}$	$\frac{55}{99}$	$\frac{60}{108}$	$\frac{65}{117}$	$\frac{70}{126}$	$\frac{75}{135}$	$0.\overline{5}$	$55\frac{5}{9}\%$
$\frac{7}{9}$	$\frac{14}{18}$	$\frac{21}{27}$	$\frac{28}{36}$	$\frac{35}{45}$	$\frac{42}{54}$	$\frac{49}{63}$	$\frac{56}{72}$	$\frac{63}{81}$	$\frac{70}{90}$	$\frac{77}{99}$	$\frac{84}{108}$	$\frac{91}{117}$	$\frac{98}{126}$	$\frac{105}{135}$	$0.\overline{7}$	$77\frac{7}{9}\%$
$\frac{8}{9}$	$\frac{16}{18}$	$\frac{24}{27}$	$\frac{32}{36}$	$\frac{40}{45}$	$\frac{48}{54}$	$\frac{56}{63}$	$\frac{64}{72}$	$\frac{72}{81}$	$\frac{80}{90}$	$\frac{88}{99}$	$\frac{96}{108}$	$\frac{104}{117}$	$\frac{112}{126}$	$\frac{120}{135}$	$0.\overline{8}$	$88\frac{8}{9}\%$

Note: The decimals for sevenths have been rounded to the nearest thousandth.

Probability Meter

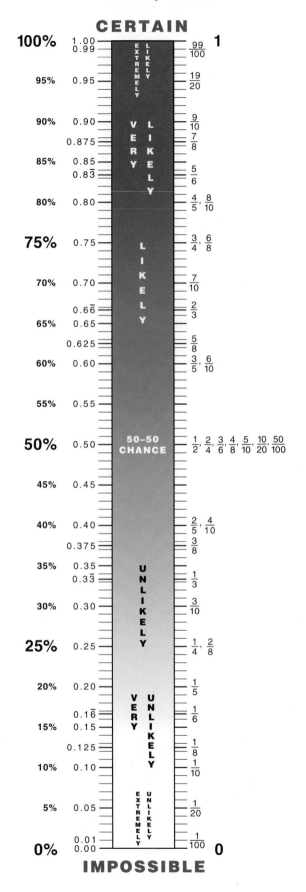

Formulas	Meaning of Variables
Rectangles • Perimeter: $p = (2 * l) + (2 * w)$ • Area: $A = b * h$	p = perimeter; l = length; w = width A = area; b = length of base; h = height
Squares • Perimeter: $p = 4 * s$ • Area: $A = s^2$	p = perimeter; s = length of side A = area
Parallelograms • Area: $A = b * h$	A = area; b = length of base; h = height
Triangles • Area: $A = \frac{1}{2} * b * h$	A = area; b = length of base; h = height
Regular Polygons • Perimeter: $p = n * s$	p = perimeter; n = number of sides; s = length of side
Circles • Circumference: $c = \pi * d$, or $c = 2 * \pi * r$ • Area: $A = \pi * r^2$	c = circumference; d = diameter; r = radius A = area
Rectangular Prisms • Volume: $V = B * h$, or $V = l * w * h$ • Surface area: $S = 2 * ((l * w) + (l * h) + (w * h))$ The surface area formula is true only when all of the faces of the prism are rectangles.	V = volume; B = area of base; l = length; w = width; h = height S = surface area
Cubes • Volume: $V = e^3$ • Surface area: $S = 6 * e^2$	V = volume; e = length of edge S = surface area
Cylinders • Volume: $V = B * h$, or $V = \pi * r^2 * h$ • Surface area: $S = (2 * \pi * r^2) + ((2 * \pi * r) * h)$ The surface area formula is true only when the line through the centers of the circular bases is perpendicular to the bases.	V = volume; B = area of base; h = height; r = radius of base S = surface area
Pyramids • Volume: $V = \frac{1}{3} * B * h$	V = volume; B = area of base; h = height
Cones • Volume: $V = \frac{1}{3} * B * h$, or $V = \frac{1}{3} * \pi * r^2 * h$	V = volume; B = area of base; h = height; r = radius of base
Distances • $d = r * t$	d = distance traveled; r = rate of speed; t = time of travel

Roman Numerals

A **numeral** is a symbol used to represent a number. **Roman numerals,** developed about 500 B.C., use letters to represent numbers.

Seven different letters are used in Roman numerals. Each letter stands for a different number.

A string of letters means that their values should be added together. For example, CCC = 100 + 100 + 100 = 300, and CLXII = 100 + 50 + 10 + 1 + 1 = 162.

Roman Numeral	Number
I	1
V	5
X	10
L	50
C	100
D	500
M	1,000

If a smaller value is placed *before* a larger value, the smaller value is subtracted instead of added. For example, IV = 5 − 1 = 4, and CDX = 500 − 100 + 10 = 410.

There are several **rules for subtracting letters.**

◆ The letters I (1), X (10), C (100), and M (1,000) represent powers of ten. These are the only letters that may be subtracted. For example, 95 in Roman numerals is XCV (VC for 95 is incorrect because V is not a power of ten).

◆ One letter may not be subtracted from a second letter if the value of the second letter is more than 10 times the value of the first. The letter I may be subtracted only from V or X. The letter X may be subtracted only from L or C. For example, 49 in Roman numerals is XLIX (IL for 49 is incorrect). And 1990 in Roman numerals is MCMXC (MXM for 1990 is incorrect).

◆ Only a *single* letter may be subtracted from another letter that follows. For example, 7 in Roman numerals is VII (IIIX for 7 is incorrect). And 300 in Roman numerals is CCC (CCD for 300 is incorrect).

The largest Roman numeral, M, stands for 1,000. One way to write large numbers is to write a string of Ms. For example, MMMM stands for 4,000. Another way to write large numbers is to write a bar above a numeral. The bar means that the numeral beneath should be multiplied by 1,000. So, $\overline{\text{IV}}$ also stands for 4,000. And $\overline{\text{M}}$ stands for 1,000 ∗ 1,000 = 1 million.

A

Abundant number
A counting number whose *proper factors* add up to more than the number itself. For example, 12 is an abundant number because the sum of its proper factors is $1 + 2 + 3 + 4 + 6 = 16$, and 16 is greater than 12. See also *deficient number* and *perfect number*.

Acre
A unit of *area* equal to 43,560 square feet in the U.S. customary system of measurement. An acre is roughly the size of a football field. A square mile equals 640 acres.

Addend
Any one of a set of numbers that are added. For example, in $5 + 3 + 1 = 9$, the addends are 5, 3, and 1.

Adjacent angles
Angles that are next to each other; adjacent angles have a common vertex and common side, but no other overlap. In the diagram, angles 1 and 2 are adjacent angles; so are angles 2 and 3, angles 3 and 4, and angles 4 and 1.

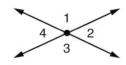

Algebraic expression
An *expression* that contains a variable. For example, if Maria is 2 inches taller than Joe and if the variable M represents Maria's height, then the algebraic expression $M - 2$ represents Joe's height.

Algorithm
A set of step-by-step instructions for doing something, such as carrying out a computation or solving a problem.

Angle
A figure that is formed by two rays or two line segments with a common endpoint. The rays or segments are called the *sides* of the angle. The common endpoint is called the *vertex* of the angle. Angles are measured in *degrees* (°). An *acute angle* has a measure greater than 0° and less than 90°. An *obtuse angle* has a measure greater than 90° and less than 180°. A *reflex angle* has a measure greater than 180° and less than 360°. A *right angle* measures 90°. A *straight angle* measures 180°.

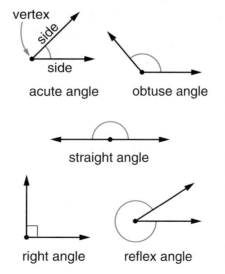

acute angle   obtuse angle

straight angle

right angle   reflex angle

Apex
In a pyramid or a cone, the vertex opposite the base. In a pyramid, all the faces except the base meet at the apex. See also *base of a pyramid or a cone*.

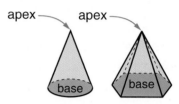

apex   apex

base   base

Arc
Part of a circle, from one point on the circle to another. For example, a *semicircle* is an arc whose endpoints are the endpoints of a diameter of the circle.

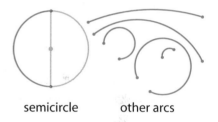

semicircle   other arcs

Area
The amount of surface inside a closed boundary. Area is measured in square units, such as square inches or square centimeters.

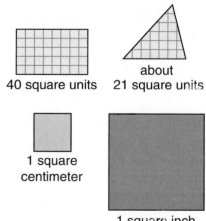

40 square units   about 21 square units

1 square centimeter

1 square inch

Area model (1) A model for multiplication problems in which the length and width of a rectangle represent the factors, and the area of the rectangle represents the product. (2) A model for showing fractions as parts of circles, rectangles, or other geometric figures.

Area model for 3 * 5 = 15

Area model for $\frac{2}{3}$

Array (1) An arrangement of objects in a regular pattern, usually in rows and columns. (2) A *rectangular array*. In *Everyday Mathematics,* an array is a rectangular array unless specified otherwise.

Associative Property A property of addition and multiplication (but not of subtraction or division) that says that when you add or multiply three numbers, it does not matter which two you add or multiply first. For example:

$(4 + 3) + 7 = 4 + (3 + 7)$
and $(5 * 8) * 9 = 5 * (8 * 9)$.

Average A typical value for a set of numbers. The word *average* usually refers to the *mean* of a set of numbers.

Axis (plural: **axes**) (1) Either of the two number lines that intersect to form a *coordinate grid*.

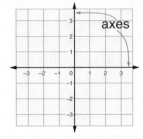

axes

(2) A line about which a solid figure rotates.

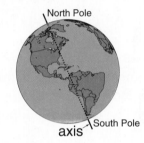

North Pole

South Pole

axis

B

Bar graph A graph that uses horizontal or vertical bars to represent data.

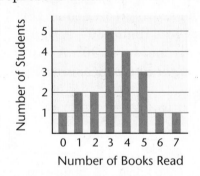

Number of Students

Number of Books Read

Vegetables
Bakery Goods
Fruit
Fast Food
Red Meat

0 10 20 30 40
Percent Wasted

Base (in exponential notation) The number that is raised to a power. For example, in 5^3, the base is 5. See also *exponential notation* and *power of a number*.

Base of a polygon A side on which a polygon "sits." The height of a polygon may depend on which side is called the base. See also *height of a parallelogram* and *height of a triangle*.

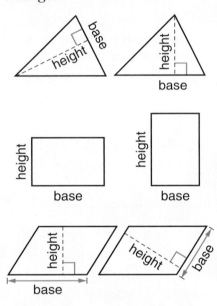

Base of a prism or a cylinder Either of the two parallel and congruent faces that define the shape of a prism or a cylinder.

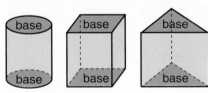

Base of a pyramid or a cone The face of a pyramid or a cone that is opposite its *apex*. The base of a pyramid is the only face that does not include the apex.

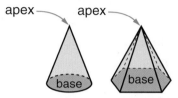

apex — apex —

base   base

Base-ten Our system for writing numbers that uses only 10 symbols, called *digits*. The digits are 0, 1, 2, 3, 4, 5, 6, 7, 8, and 9. You can write any number using only these 10 digits. Each digit has a value that depends on its place in the number. In this system, moving a digit one place to the left makes that digit worth 10 times as much. And moving a digit one place to the right makes that digit worth one-tenth as much. See also *place value*.

Bisect To divide a segment, an angle, or another figure into two equal parts.

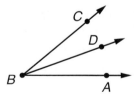

Ray *BD* bisects angle *ABC*.

Bisector A line, segment, or ray that divides a segment, an angle, or a figure into two equal parts. See also *bisect*.

Broken-line graph A graph in which data points are connected by line segments. Broken-line graphs are often used to show how something has changed over a period of time. Same as *line graph*.

Attendance for the First Week of School

Number of Students (y-axis: 0, 5, 10, 15, 20, 25)

Day of the Week (x-axis: Mon, Tue, Wed, Thu, Fri)

C

Capacity (1) The amount a container can hold. The *volume* of a container. Capacity is usually measured in units such as gallons, pints, cups, fluid ounces, liters, and milliliters. (2) The heaviest weight a scale can measure.

Census An official count of a country's population. A census is taken every 10 years in the United States.

Change diagram A diagram used in *Everyday Mathematics* to represent situations in which quantities are increased or decreased.

Change

| Start | | End |
| 14 | −5 | ? |

Circle The set of all points in a plane that are the same distance from a fixed point in the plane. The fixed point is the *center* of the circle, and the distance is the *radius*. The center and *interior* of a circle are not part of the circle. A circle together with its interior is called a *disk* or a *circular region*. See also *diameter*.

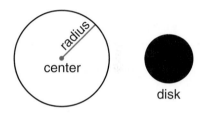

radius

center

disk

Circle graph A graph in which a circle and its interior are divided by radii into parts (*sectors*) to show the parts of a set of data. The whole circle represents the whole set of data. Same as *pie graph*.

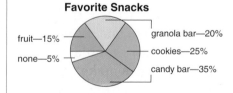

Favorite Snacks

fruit—15%
none—5%
granola bar—20%
cookies—25%
candy bar—35%

Circumference The distance around a circle; the perimeter of a circle.

circumference

Glossary

Column-addition method
A method for adding numbers in which the addends' digits are first added in each place-value column separately, and then 10-for-1 trades are made until each column has only one digit. Lines are drawn to separate the place-value columns.

	100s	10s	1s
	2	4	8
+	1	8	7
	3	12	15
	3	13	5
	4	3	5

$$248 + 187 = 435$$

Column-division method
A division procedure in which vertical lines are drawn between the digits of the dividend. As needed, trades are made from one column into the next column at the right. The lines make the procedure easier to carry out.

	1	7	2
5⟌	8	6̸	3̸
	−5	36	13
	3̸	−35	−10
		1̸	3

$$863 / 5 \rightarrow 172 \text{ R}3$$

Common denominator
(1) If two fractions have the same denominator, that denominator is called a common denominator. (2) For two or more fractions, any number that is a *common multiple* of their denominators. For example, the fractions $\frac{1}{2}$ and $\frac{2}{3}$ have the common denominators 6, 12, 18, and so on. See also *quick common denominator*.

Common factor A counting number is a common factor of two or more counting numbers if it is a *factor* of each of those numbers. For example, 4 is a common factor of 8 and 12. See also *factor of a counting number* n.

Common multiple A number is a common multiple of two or more numbers if it is a *multiple* of each of those numbers. For example, the multiples of 2 are 2, 4, 6, 8, 10, 12, and so on; the multiples of 3 are 3, 6, 9, 12, and so on; and the common multiples of 2 and 3 are 6, 12, 18, and so on.

Commutative Property
A property of addition and multiplication (but not of subtraction or division) that says that changing the order of the numbers being added or multiplied does not change the answer. These properties are often called *turn-around facts* in *Everyday Mathematics*. For example: $5 + 10 = 10 + 5$ and $3 * 8 = 8 * 3$.

Comparison diagram
A diagram used in *Everyday Mathematics* to represent situations in which two quantities are compared.

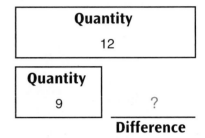

Complementary angles
Two angles whose measures total 90°.

∠1 and ∠2 are complementary angles

Composite number
A counting number that has more than 2 different factors. For example, 4 is a composite number because it has three factors: 1, 2, and 4.

Concave polygon A polygon in which at least one vertex is "pushed in." At least one inside angle of a concave polygon is a *reflex angle* (has a measure greater than 180°). Same as *nonconvex polygon*.

Concentric circles Circles that have the same center but radii of different lengths.

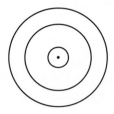

Cone A 3-dimensional shape that has a circular *base*, a curved surface, and one vertex, which is called the *apex*. The points on the curved surface of a cone are on straight lines connecting the apex and the boundary of the base.

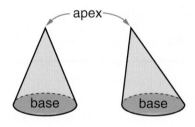

apex

base   base

Congruent Having the same shape and size. Two 2-dimensional figures are congruent if they match exactly when one is placed on top of the other. (It may be necessary to flip one of the figures over.)

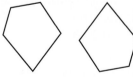

congruent pentagons

congruent prisms

Constant A quantity that does not change.

Contour line A curve on a map through places where a certain measurement (such as temperature or elevation) is the same. Often contour lines separate regions that have been colored differently to show a range of conditions.

Contour map A map that uses *contour lines* to show a particular feature (such as elevation or climate).

Convex polygon A polygon in which all vertices are "pushed outward." Each inside angle of a convex polygon has a measure less than 180°.

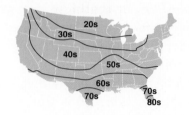

Coordinate (1) A number used to locate a point on a number line. (2) One of the two numbers in an ordered number pair. The number pair is used to locate a point on a *coordinate grid*.

Coordinate grid See *rectangular coordinate grid*.

Corresponding Having the same relative position in *similar* or *congruent figures*. In the diagram, pairs of *corresponding sides* are marked with the same number of slash marks, and *corresponding angles* are marked with the same number of arcs.

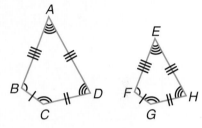

corresponding sides and angles

Counting numbers The numbers used to count things. The set of counting numbers is {1, 2, 3, 4, ...}. Compare to *whole numbers*.

Cube A polyhedron with 6 square faces. A cube has 8 vertices and 12 edges.

Cubic centimeter A metric unit of volume equal to the volume of a cube with 1 cm edges. $1 \text{ cm}^3 = 1$ mL.

Cubic unit A unit used in measuring volume, such as a cubic centimeter or a cubic foot.

Cubit An ancient unit of length, measured from the point of the elbow to the end of the middle finger. A cubit is about 18 inches.

Curved surface A surface that is rounded rather than flat. Spheres, cylinders, and cones each have one curved surface.

Cylinder A 3-dimensional shape that has two circular bases that are parallel and congruent and are connected by a curved surface. A soup can is shaped like a cylinder.

Data Information that is gathered by counting, measuring, questioning, or observing.

Decimal A number written in *standard, base-10 notation* that contains a decimal point, such as 2.54. A whole number is a decimal, but is usually written without a decimal point.

Decimal point A dot used to separate the ones and tenths places in decimal numbers.

Deficient number A counting number whose *proper factors* add up to less than the number itself. For example, 10 is a deficient number because the sum of its proper factors is $1 + 2 + 5 = 8$, and 8 is less than 10. See also *abundant number* and *perfect number*.

Degree (°) (1) A unit of measure for angles based on dividing a circle into 360 equal parts. Latitude and longitude are measured in degrees, and these degrees are based on angle measures. (2) A unit of measure for temperature. In all cases, a small raised circle (°) is used to show degrees.

Denominator The number below the line in a fraction. A fraction may be used to name part of a whole. If the *whole* (the *ONE,* or the *unit*) is divided into equal parts, the denominator represents the number of equal parts into which the whole is divided. In the fraction $\frac{a}{b}$, b is the denominator.

Density A *rate* that compares the *weight* of an object with its *volume.* For example, suppose a ball has a weight of 20 grams and a volume of 10 cubic centimeters. To find its density, divide its weight by its volume: $20 \text{ g} / 10 \text{ cm}^3 = 2 \text{ g} / \text{cm}^3$, or 2 grams per cubic centimeter.

Diameter (1) A line segment that passes through the center of a circle or sphere and has endpoints on the circle or sphere. (2) The length of this line segment. The diameter of a circle or sphere is twice the length of its *radius.*

Difference The result of subtracting one number from another. See also *minuend* and *subtrahend.*

Digit One of the number symbols 0, 1, 2, 3, 4, 5, 6, 7, 8, and 9 in the standard, *base-ten* system.

Discount The amount by which the regular price of an item is reduced.

Distributive Property A property that relates multiplication and addition or subtraction. This property gets its name because it "distributes" a factor over terms inside parentheses.

Distributive property of multiplication over addition:
$a * (b + c) = (a * b) + (a * c)$,
so $2 * (5 + 3) = (2 * 5) + (2 * 3)$
$= 10 + 6 = 16.$

Distributive property of multiplication over subtraction:
$a * (b - c) = (a * b) - (a * c)$,
so $2 * (5 - 3) = (2 * 5) - (2 * 3)$
$= 10 - 6 = 4.$

Dividend The number in division that is being divided. For example, in $35 \div 5 = 7$, the dividend is 35.

Divisible by If one counting number can be divided by a second counting number with a remainder of 0, then the first number is divisible by the second number. For example, 28 is divisible by 7 because 28 divided by 7 is 4, with a remainder of 0.

Divisibility test A test to find out whether one counting number is *divisible by* another counting number without actually doing the division. A divisibility test for 5, for example, is to check the digit in the 1s place: if that digit is 0 or 5, then the number is divisible by 5.

Divisor In division, the number that divides another number. For example, in $35 \div 5 = 7$, the divisor is 5.

Dodecahedron A polyhedron with 12 faces.

Edge A line segment or curve where two surfaces meet.

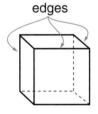

edges

edge

Endpoint A point at the end of a *line segment* or *ray*. A line segment is named using the letter labels of its endpoints. A ray is named using the letter labels of its endpoint and another point on the ray.

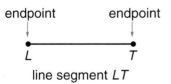

endpoint   endpoint

L   T

line segment *LT*

Enlarge To increase the size of an object or a figure without changing its shape. See also *size-change factor*.

Equally likely outcomes If all of the possible *outcomes* for an experiment or situation have the same *probability,* they are called equally likely outcomes. In the case of equally likely outcomes, the probability of an *event* is equal to this fraction:

$$\frac{\text{number of favorable outcomes}}{\text{number of possible outcomes}}$$

See also *favorable outcome*.

Equation A number sentence that contains an equal sign. For example, $15 = 10 + 5$ is an equation.

Equilateral triangle A triangle with all three sides equal in length. In an equilateral triangle, all three angles have the same measure.

Equivalent Equal in value but possibly in a different form. For example, $\frac{1}{2}$, 0.5, and 50% are all equivalent.

Equivalent fractions Fractions with different denominators that name the same number. For example, $\frac{1}{2}$ and $\frac{4}{8}$ are equivalent fractions.

Equivalent rates *Rates* that make the same comparison. For example, the rates $\frac{60 \text{ miles}}{1 \text{ hour}}$ and $\frac{1 \text{ mile}}{1 \text{ minute}}$ are equivalent. Two rates named as fractions using the *same units* are equivalent if the fractions (ignoring the units) are equivalent. For example, $\frac{12 \text{ pages}}{4 \text{ minutes}}$ and $\frac{6 \text{ pages}}{2 \text{ minutes}}$ are equivalent rates because $\frac{12}{4}$ and $\frac{6}{2}$ are equivalent.

Equivalent ratios *Ratios* that make the same comparison. Two or more ratios are equivalent if they can be named as equivalent fractions. For example, the ratios 12 to 20, 6 to 10, and 3 to 5 are equivalent ratios because $\frac{12}{20} = \frac{6}{10} = \frac{3}{5}$.

Estimate An answer that should be close to an exact answer. *To estimate* means to give an answer that should be close to an exact answer.

Evaluate To find a value for. To evaluate a mathematical *expression,* carry out the operations. If there are variables, first replace them with numbers. To evaluate a *formula,* find the value of one variable in the formula when the values of the other variables are given.

Even number A counting number that can be divided by 2 with no remainder. The even numbers are 2, 4, 6, 8, and so on. 0, −2, −4, −6, and so on are also usually considered even.

Event Something that happens. The *probability* of an event is the chance that the event will happen. For example, rolling a number smaller than 4 with a die is an event. The possible *outcomes* of rolling a die are 1, 2, 3, 4, 5, and 6. The event "roll a number smaller than 4" will happen if the outcome is 1 or 2 or 3. And the chance that this will happen is $\frac{3}{6}$. If the probability of an event is 0, the event is *impossible*. If the probability is 1, the event is *certain*.

Expanded notation A way of writing a number as the sum of the values of each digit. For example, in expanded notation, 356 is written 300 + 50 + 6. See also *standard notation, scientific notation,* and *number-and-word notation*.

Exponent A small, raised number used in *exponential notation* to tell how many times the *base* is used as a *factor*. For example, in 5^3, the base is 5, the exponent is 3, and $5^3 = 5 * 5 * 5 = 125$. See also *power of a number*.

Exponential notation A way to show repeated multiplication by the same factor. For example, 2^3 is exponential notation for $2 * 2 * 2$. The small raised 3 is the *exponent*. It tells how many times the number 2, called the *base,* is used as a factor.

$$2^3 \leftarrow \text{exponent}$$
$$\uparrow$$
$$\text{base}$$

Expression A group of mathematical symbols that represents a number—or can represent a number if values are assigned to any variables in the expression. An expression may include numbers, variables, operation symbols, and grouping symbols – but *not* relation symbols (=, >, <, and so on). Any expression that contains one or more variables is called an *algebraic expression*.

2π	$3 + 4$	$5 * (7 - 3)$

expressions

x	$\pi * r$ (or πr)	$a^2 + (a/5)$

algebraic expressions

Extended multiplication fact A multiplication fact involving multiples of 10, 100, and so on. For example, $6 * 70$, $60 * 7$, and $60 * 70$ are extended multiplication facts.

Face A flat surface on a 3-dimensional shape.

Fact family A set of related addition and subtraction facts, or related multiplication and division facts. For example, $5 + 6 = 11$, $6 + 5 = 11$, $11 - 5 = 6$, and $11 - 6 = 5$ are a fact family. $5 * 7 = 35$, $7 * 5 = 35$, $35 \div 5 = 7$, and $35 \div 7 = 5$ are another fact family.

Factor (in a product) Whenever two or more numbers are multiplied to give a product, each of the numbers that is multiplied is called a factor. For example, in $4 * 1.5 = 6$, 6 is the product and 4 and 1.5 are called factors. See also *factor of a counting number* n.

$$4 * 1.5 = 6$$
$$\uparrow \quad \uparrow \qquad \uparrow$$
$$\text{factors} \quad \text{product}$$

NOTE: This definition of *factor* is much less important than the definition below.

Factor of a counting number *n* A counting number whose product with some other counting number equals *n*. For example, 2 and 3 are factors of 6 because $2 * 3 = 6$. But 4 is not a factor of 6 because $4 * 1.5 = 6$ and 1.5 is not a counting number.

$$2 * 3 = 6$$
$$\uparrow \quad \uparrow \qquad \uparrow$$
$$\text{factors} \quad \text{product}$$

NOTE: This definition of *factor* is much more important than the definition above.

Factor pair Two factors of a counting number whose product is the number. A number may have more than one factor pair. For example, the factor pairs for 18 are 1 and 18, 2 and 9, 3 and 6.

Factor rainbow A way to show factor pairs in a list of all the factors of a counting number. A factor rainbow can be used to check whether a list of factors is correct.

Factor rainbow for 24

Factor string A counting number written as a product of two or more of its factors. The number 1 is never part of a factor string. For example, a factor string for 24 is 2 * 3 * 4. This factor string has three factors, so its length is 3. Another factor string for 24 is 2 * 3 * 2 * 2 (length 4).

Factor tree A way to get the *prime factorization* of a counting number. Write the original number as a product of counting-number factors. Then write each of these factors as a product of factors, and so on, until the factors are all prime numbers. A factor tree looks like an upside-down tree, with the root (the original number) at the top and the leaves (the factors) beneath it.

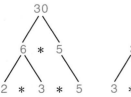

Fair Free from bias. Each side of a fair die or coin will come up about equally often.

Fair game A game in which every player has the same chance of winning.

False number sentence A number sentence that is not true. For example, 8 = 5 + 5 is a false number sentence.

Fathom A unit used by people who work with boats and ships to measure depths underwater and lengths of cables. A fathom is now defined as 6 feet.

Favorable outcome An *outcome* that satisfies the conditions of an event of interest. For example, suppose a 6-sided die is rolled and the event of interest is rolling an even number. There are 6 possible outcomes: 1, 2, 3, 4, 5, or 6. There are 3 favorable outcomes: 2, 4, or 6. See also *equally likely outcomes.*

Figurate numbers Numbers that can be shown by specific geometric patterns. Square numbers and triangular numbers are examples of figurate numbers.

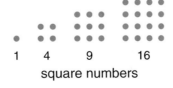

triangular numbers

square numbers

Formula A general rule for finding the value of something. A formula is often written using letters, called *variables,* which stand for the quantities involved. For example, the formula for the area of a rectangle may be written as $A = l * w$, where A represents the area of the rectangle, l represents its length, and w represents its width.

Fraction (primary definition) A number in the form $\frac{a}{b}$ where a and b are whole numbers and b is not 0. A fraction may be used to name part of a whole, or to compare two quantities. A fraction may also be used to represent division. For example, $\frac{2}{3}$ can be thought of as 2 divided by 3. See also *numerator* and *denominator.*

Fraction (other definitions) (1) A fraction that satisfies the definition above, but includes a unit in both the numerator and denominator. This definition of fraction includes any rate that is written as a fraction. For example, $\frac{50 \text{ miles}}{1 \text{ gallon}}$ and $\frac{40 \text{ pages}}{10 \text{ minutes}}$. (2) Any number written using a fraction bar, where the fraction bar is used to indicate division. For example, $\frac{2.3}{6.5}$, $\frac{1\frac{4}{5}}{12}$, and $\frac{\frac{3}{4}}{\frac{5}{8}}$.

Fraction-Stick Chart A diagram used in *Everyday Mathematics* to represent simple fractions.

G

Geometric solid
A 3-dimensional shape, such as a prism, pyramid, cylinder, cone, or sphere. Despite its name, a geometric solid is hollow; it does not contain the points in its interior.

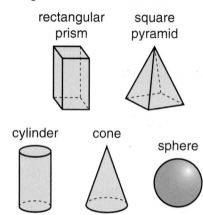

rectangular prism   square pyramid

cylinder   cone   sphere

Geometry Template An *Everyday Mathematics* tool that includes a millimeter ruler, a ruler with sixteenth-inch intervals, half-circle and full-circle protractors, a percent circle, pattern-block shapes, and other geometric figures. The template can also be used as a compass.

Great span The distance from the tip of the thumb to the tip of the little finger (pinkie) when the hand is stretched as far as possible.

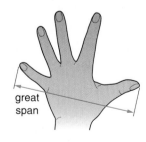

great span

Greatest common factor (GCF) The largest factor that two or more counting numbers have in common. For example, the common factors of 24 and 36 are 1, 2, 3, 4, 6, and 12. The greatest common factor of 24 and 36 is 12.

Grouping symbols Symbols such as parentheses ( ), brackets [ ], and braces { } that tell the order in which operations in an expression are to be done. For example, in the expression $(3 + 4) * 5$, the operation in the parentheses should be done first. The expression then becomes $7 * 5 = 35$.

H

Height of a parallelogram The length of the shortest line segment between the *base* of a parallelogram and the line containing the opposite side. That shortest segment is perpendicular to the base and is also called the *height*. See also *base of a polygon*.

height
base

height
base

Height of a prism or a cylinder The length of the shortest line segment between the base of a prism or a cylinder and the plane containing the opposite base. That shortest segment is perpendicular to the base and is also called the *height*. See also *base of a prism or a cylinder*.

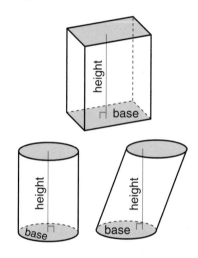

height
base

height
base

height
base

Height of a pyramid or a cone The length of the shortest line segment between the vertex of a pyramid or a cone and the plane containing its base. That shortest segment is also perpendicular to the plane containing the base and is called the *height*. See also *base of a pyramid or a cone*.

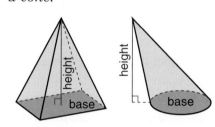

height
base

height
base

Height of a triangle
The length of the shortest line segment between the line containing a base of a triangle and the vertex opposite that base. That shortest segment is perpendicular to the line containing the base and is also called the *height*. See also *base of a polygon*.

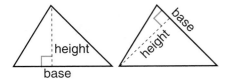

Hemisphere Half of Earth's surface. Also, half of a sphere.

Heptagon A polygon with seven sides.

Hexagon A polygon with six sides.

Hexagram A 6-pointed star formed by extending the sides of a regular hexagon.

Horizontal In a left-right orientation; parallel to the horizon.

Icosahedron A polyhedron with 20 faces.

Image The reflection of an object that you see when you look in a mirror. Also, a figure that is produced by a *transformation* (a *reflection*, *translation*, or *rotation*, for example) of another figure. See also *preimage*.

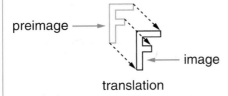
translation

Improper fraction A fraction whose numerator is greater than or equal to its denominator. For example, $\frac{4}{3}$, $\frac{5}{2}$, $\frac{4}{4}$, and $\frac{24}{12}$ are improper fractions. In *Everyday Mathematics,* improper fractions are sometimes called "top-heavy" fractions.

Inequality A number sentence with $>$, $<$, $\geq$, $\leq$, or $\neq$. For example, the sentence $8 < 15$ is an inequality.

Inscribed polygon
A polygon whose vertices are all on the same circle.

inscribed square

Integer A number in the set $\{\ldots, -4, -3, -2, -1, 0, 1, 2, 3, 4, \ldots\}$; a *whole number* or the *opposite* of a whole number, where 0 is its own opposite.

Interior The inside of a closed 2-dimensional or 3-dimensional figure. The interior is usually not considered to be part of the figure.

Intersect To meet or cross.

Intersecting Meeting or crossing one another. For example, lines, segments, rays, and planes can intersect.

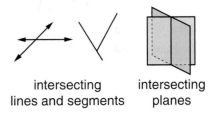
intersecting lines and segments   intersecting planes

Interval (1) The set of all numbers between two numbers, *a* and *b,* which may include *a* or *b* or both. (2) A part of a line, including all points between two specific points.

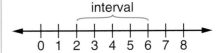
interval

Interval estimate
An estimate that places an unknown quantity in a range. For example, an interval estimate of a person's weight might be "between 100 and 110 pounds."

Irrational number
A number that cannot be written as a fraction, where both the numerator and the denominator are *integers* and the denominator is not zero. For example, π (pi) is an irrational number.

Isosceles triangle
A triangle with at least two sides equal in length. In an isosceles triangle, at least two angles have the same measure. A triangle with all three sides the same length is an isosceles triangle, but is usually called an *equilateral triangle*.

Kite A quadrilateral with two pairs of adjacent equal sides. The four sides cannot all have the same length, so a rhombus is not a kite.

Landmark A notable feature of a data set. Landmarks include the *median, mode, maximum, minimum,* and *range*. The *mean* can also be thought of as a landmark.

Latitude A measure, in degrees, that tells how far north or south of the equator a place is.

Lattice method A very old way to multiply multidigit numbers.

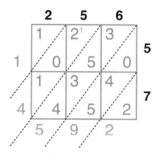

$$256 * 57 = 14{,}592$$

Leading-digit estimation A way to estimate in which the left-most, nonzero digit in a number is not changed, but all other digits are replaced by zeros. For example, to estimate $432 + 76$, use the leading-digit estimates 400 and 70: $400 + 70 = 470$.

Least common denominator (LCD) The *least common multiple* of the denominators of every fraction in a given collection. For example, the least common denominator of $\frac{1}{2}$, $\frac{4}{5}$, and $\frac{3}{8}$ is 40.

Least common multiple (LCM) The smallest number that is a multiple of two or more numbers. For example, while some common multiples of 6 and 8 are 24, 48, and 72, the least common multiple of 6 and 8 is 24.

Left-to-right subtraction A subtraction method in which you start at the left and subtract column by column. For example, to subtract $932 - 356$:

		9	3	2
Subtract the 100s.	−	3	0	0
		6	3	2
Subtract the 10s.	−		5	0
		5	8	2
Subtract the 1s.	−			6
		5	7	6

$$932 - 356 = 576$$

Like In some situations, like means *the same*. The fractions $\frac{2}{5}$ and $\frac{3}{5}$ have like denominators. The measurements 23 cm and 52 cm have like units.

Line A straight path that extends infinitely in opposite directions.

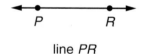

line PR

Line graph See *broken-line graph*.

Line of reflection (mirror line) A line halfway between a figure (preimage) and its reflected image. In a *reflection,* a figure is "flipped over" the line of reflection.

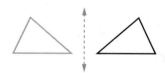

line of reflection

Line of symmetry A line drawn through a figure so that it is divided into two parts that are mirror images of each other. The two parts look alike but face in opposite directions. See also *line symmetry*.

Line plot A sketch of data in which check marks, Xs, or other marks above a labeled line show the frequency of each value.

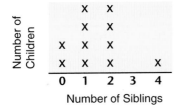

Line segment A straight path joining two points. The two points are called *endpoints* of the segment.

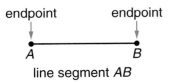

line segment *AB*

Line symmetry A figure has line symmetry if a line can be drawn through it so that it is divided into two parts that are mirror images of each other. The two parts look alike but face in opposite directions. See also *line of symmetry*.

Lines of latitude Lines that run east-west on a map or globe and locate a place with reference to the equator, which is also a line of latitude. On a globe, lines of latitude are circles, and are called *parallels* because the planes containing these circles are parallel to the plane containing the equator.

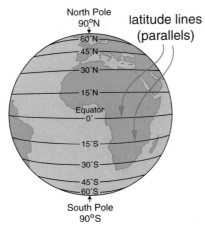

Lines of longitude Lines that run north-south on a map or globe and locate a place with reference to the *prime meridian,* which is also a line of longitude. On a globe, lines of longitude are *semicircles* that meet at the North and South Poles. They are also called *meridians*.

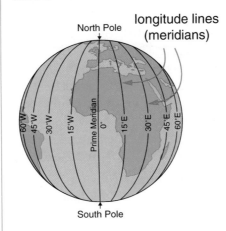

Longitude A measure, in degrees, that tells how far east or west of the *prime meridian* a place is.

Lowest terms See *simplest form.*

M

Magnitude estimate A rough *estimate*. A magnitude estimate tells whether an answer should be in the tens, hundreds, thousands, ten-thousands, and so on.

Map legend (map key) A diagram that explains the symbols, markings, and colors on a map.

Map scale A tool that helps you estimate real distances between places shown on a map. It relates distances on the map to distances in the real world. For example, a map scale may show that 1 inch on a map represents 100 miles in the real world. See also *scale.*

Maximum The largest amount; the greatest number in a set of data.

Mean The sum of a set of numbers divided by the number of numbers in the set. The mean is often referred to simply as the *average.*

Median The middle value in a set of data when the data are listed in order from smallest to largest, or from largest to smallest. If there are an even number of data points, the median is the *mean* of the two middle values.

Metric system of measurement A measurement system based on the *base-ten* numeration system. It is used in most countries around the world.

Minimum The smallest amount; the smallest number in a set of data.

Minuend In subtraction, the number from which another number is subtracted. For example, in $19 - 5 = 14$, the minuend is 19. See also *subtrahend*.

Mixed number A number that is written using both a whole number and a fraction. For example, $2\frac{1}{4}$ is a mixed number equal to $2 + \frac{1}{4}$.

Mode The value or values that occur most often in a set of data.

Multiple of a number n (1) A product of n and a counting number. For example, the multiples of 7 are 7, 14, 21, 28, … . (2) a product of n and an integer. The multiples of 7 are …, -21, -14, -7, 0, 7, 14, 21, … .

Multiplication counting principle A way of determining the total number of possible outcomes for two or more separate choices. For example, suppose you roll a die and then flip a coin. There are 6 choices for which face of the die shows and 2 choices for which side of the coin shows. So there are $6 * 2$, or 12 possible outcomes in all: (1,H), (1,T), (2,H), (2,T), (3,H), (3,T), (4,H), (4,T), (5,H), (5,T), (6,H), (6,T).

Multiplication diagram A diagram used for problems in which there are several equal groups. The diagram has three parts: a number of groups, a number in each group, and a total number. Also called *multiplication/division diagram*. See also *rate diagram*.

rows	chairs per row	total chairs
15	25	?

Name-collection box A diagram that is used for writing equivalent names for a number.

25	$37 - 12$	$20 + 5$
//// //// //// //// ////		5^2
twenty-five		*veinticinco*

Negative number A number that is less than zero; a number to the left of zero on a horizontal number line or below zero on a vertical number line. The symbol $-$ may be used to write a negative number. For example, "negative 5" is usually written as -5.

n-gon A polygon with n sides. For example, a 5-gon is a pentagon, and an 8-gon is an octagon.

Nonagon A polygon with nine sides.

Nonconvex polygon See *concave polygon*.

n-to-1 ratio A ratio with 1 in the denominator.

Number-and-word notation A way of writing a number using a combination of numbers and words. For example, 27 billion is number-and-word notation for 27,000,000,000.

Number model A *number sentence* or *expression* that models or fits a number story or situation. For example, the story *Sally had $5, and then she earned $8,* can be modeled as the number sentence $5 + 8 = 13$, or as the expression $5 + 8$.

Number sentence At least two *numbers* or *expressions* separated by a *relation symbol* ($=, >, <, \geq, \leq, \neq$). Most number sentences contain at least one *operation symbol* ($+$, $-$, $\times$, $*$, $\div$, or $/$). Number sentences may also have *grouping symbols,* such as parentheses and brackets.

Number story A story with a problem that can be solved using arithmetic.

Numeral A word, symbol, or figure that represents a number. For example, six, VI, and 6 are numerals that represent the same number.

Numerator The number above the line in a fraction. A fraction may be used to name part of a whole. If the *whole* (the *ONE,* or the *unit*) is divided into equal parts, the numerator represents the number of equal parts being considered. In the fraction $\frac{a}{b}$, a is the numerator.

Octagon A polygon with eight sides.

Octahedron A polyhedron with 8 faces.

Odd number A counting number that cannot be evenly divided by 2. When an odd number is divided by 2, there is a remainder of 1. The odd numbers are 1, 3, 5, and so on.

ONE See *whole* and *unit*.

Open sentence A *number sentence* which has *variables* in place of one or more missing numbers. An open sentence is usually neither true nor false. For example, $5 + x = 13$ is an open sentence. The sentence is true if 8 is substituted for x. The sentence is false if 4 is substituted for x.

Operation symbol A symbol used to stand for a mathematical operation. Common operation symbols are $+$, $-$, $\times$, $*$, $\div$, and $/$.

Opposite of a number A number that is the same distance from 0 on the number line as a given number, but on the opposite side of 0. For example, the opposite of $+3$ is -3, and the opposite of -5 is $+5$.

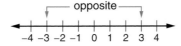
opposite

Order of operations Rules that tell in what order to perform operations in arithmetic and algebra. The order of operations is as follows:
1. Do the operations in parentheses first. (Use rules 2–4 inside the parentheses.)
2. Calculate all the expressions with exponents.
3. Multiply and divide in order from left to right.
4. Add and subtract in order from left to right.

Ordered number pair (ordered pair) Two numbers that are used to locate a point on a *rectangular coordinate grid.* The first number gives the position along the horizontal axis, and the second number gives the position along the vertical axis. The numbers in an ordered pair are called *coordinates.* Ordered pairs are usually written inside parentheses: (5,3). See *rectangular coordinate grid* for an illustration.

Origin (1) The 0 point on a number line. (2) The point (0,0) where the two axes of a coordinate grid meet.

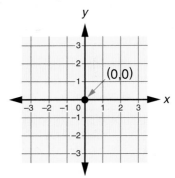

The ordered pair (0,0) names the origin.

Outcome A possible result of an experiment or situation. For example, heads and tails are the two possible outcomes of tossing a coin. See also *event* and *equally likely outcomes.*

Pan balance A tool used to weigh objects or compare weights. The pan balance is also used as a model in balancing and solving equations.

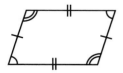

Parallel Lines, line segments, or rays in the same plane are parallel if they never cross or meet, no matter how far they are extended. Two planes are parallel if they never cross or meet. A line and a plane are parallel if they never cross or meet. The symbol ∥ means *is parallel to*.

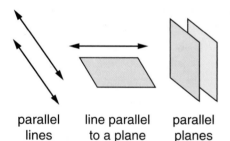

parallel   line parallel   parallel
lines   to a plane   planes

Parallelogram
A quadrilateral with two pairs of parallel sides. Opposite sides of a parallelogram are congruent. Opposite angles in a parallelogram have the same measure.

Parentheses Grouping symbols, ( ), used to tell which parts of an expression should be calculated first.

Partial-differences method A way to subtract in which differences are computed for each place (ones, tens, hundreds, and so on) separately. The partial differences are then combined to give the final answer.

```
                      9 3 2
                  –   3 5 6
900 – 300    →       6 0 0
30 – 50      → –       2 0
2 – 6        → –         4
600 – 20 – 4 →       5 7 6
```

932 – 356 = 576

Partial-products method A way to multiply in which the value of each digit in one factor is multiplied by the value of each digit in the other factor. The final product is the sum of these partial products.

```
                    6 7
               ×    5 3
50 × 60 → 3 0 0 0
50 × 7  →   3 5 0
3 × 60  →   1 8 0
3 × 7   →     2 1
Add.      3, 5 5 1
```

67 * 53 = 3,551

Partial-quotients method A way to divide in which the dividend is divided in a series of steps. The quotients for each step (called partial quotients) are added to give the final answer.

```
6)1010
 –  600 | 100
    410
 –  300 |  50
    110
 –   60 |  10
     50
 –   48 |   8
      2   168
```
↑ remainder   ↑ quotient

1,010 ÷ 6 → 168 R2

Partial-sums method A way to add in which sums are computed for each place (ones, tens, hundreds, and so on) separately. The partial-sums are then added to give the final answer.

```
                            2 6 8
                        +   4 8 3
Add the 100s.        →      6 0 0
Add the 10s.         →      1 4 0
Add the 1s.          →        1 1
Add the partial sums. →     7 5 1
```

268 + 483 = 751

Parts-and-total diagram
A diagram used in *Everyday Mathematics* to represent situations in which two or more quantities are combined to form a total quantity.

Total
13

Part	Part
8	?

Part-to-part ratio A *ratio* that compares a part of a whole to another part of the same whole. For example, the statement "There are 8 boys for every 12 girls" expresses a part-to-part ratio. See also *part-to-whole ratio*.

Part-to-whole ratio
A *ratio* that compares a part of a whole to the whole. For example, the statements "8 out of 20 students are boys" and "12 out of 20 students are girls," both express part-to-whole ratios. See also *part-to-part ratio*.

Pentagon A polygon with five sides.

Percent (%) Per hundred or out of a hundred. For example, "48% of the students in the school are boys" means that 48 out of every 100 students in the school are boys; $48\% = \frac{48}{100} = 0.48$.

Percent Circle A tool on the *Geometry Template* that is used to measure and draw figures that involve percents (such as circle graphs).

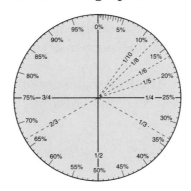

Perfect number A counting number whose *proper factors* add up to the number itself. For example, 6 is a perfect number because the sum of its proper factors is $1 + 2 + 3 = 6$. See also *abundant number* and *deficient number*.

Perimeter The distance around a 2-dimensional shape, along the boundary of the shape. The perimeter of a circle is called its *circumference*. A formula for the perimeter P of a rectangle with length l and width w is $P = 2 * (l + w)$.

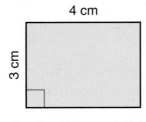

4 cm

3 cm

$P = 2 * (4 \text{ cm} + 3 \text{ cm})$
$= 2 * 7 \text{ cm} = 14 \text{ cm}$

Perpendicular Crossing or meeting at *right angles*. Lines, rays, line segments, or planes that cross or meet at right angles are perpendicular. The symbol $\perp$ means "is perpendicular to."

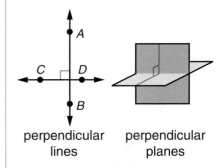

perpendicular lines   perpendicular planes

Per-unit rate A *rate* with 1 in the denominator. Per-unit rates tell how many of one thing there are for one of another thing. For example, "2 dollars per gallon" is a per-unit rate. "12 miles per hour" and "4 words per minute" are also examples of per-unit rates.

Pi (π) The ratio of the *circumference* of a circle to its *diameter*. Pi is also the ratio of the area of a circle to the square of its radius. Pi is the same for every circle and is an irrational number that is approximately equal to 3.14. Pi is the 16th letter of the Greek alphabet and is written π.

Pictograph A graph constructed with pictures or symbols. The *key* for a pictograph tells what each picture or symbol is worth.

Number of Cars Washed

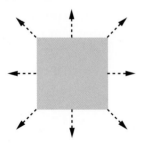

Friday	🚗 🚗
Saturday	🚗 🚗 🚗 🚗 🚗
Sunday	🚗 🚗 🚗 🚗

KEY: 🚗 = 6 cars

Pie graph See *circle graph*.

Place value A system that gives a digit a value according to its position in a number. In our *base-ten* system for writing numbers, moving a digit one place to the left makes that digit worth 10 times as much. And moving a digit one place to the right makes that digit worth one-tenth as much. For example, in the number 456, the 4 in the hundreds place is worth 400; but in the number 45.6, the 4 in the tens place is worth 40.

Plane A flat surface that extends forever.

Point An exact location in space. The center of a circle is a point. Lines have infinitely many points on them.

Polygon A 2-dimensional figure that is made up of three or more line segments joined end to end to make one closed path. The line segments of a polygon may not cross.

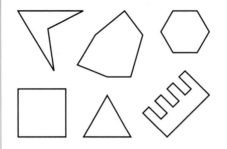

Polyhedron A geometric solid whose surfaces *(faces)* are all flat and formed by polygons. Each face consists of a polygon and the interior of that polygon. A polyhedron does not have any curved surface.

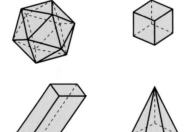

Population In data collection, the collection of people or objects that is the focus of study.

Positive number A number that is greater than zero; a number to the right of zero on a horizontal number line, or above zero on a vertical number line. A positive number may be written using the + symbol, but is usually written without it. For example, $+10 = 10$ and $\pi = +\pi$.

Power of a number The product of factors that are all the same. For example, $5 * 5 * 5$ (or 125) is called "5 to the third power" or "the third power of 5" because 5 is a factor three times. $5 * 5 * 5$ can also be written as 5^3. See also *exponent*.

Power of 10 A whole number that can be written as a *product of 10s*. For example, 100 is equal to $10 * 10$, or 10^2. 100 is called "the second power of 10" or "10 to the second power." A number that can be written as a *product of $\frac{1}{10}s$* is also a power of 10. For example, $10^{-2} = \frac{1}{10^2} = \frac{1}{10 * 10} = \frac{1}{10} * \frac{1}{10}$ is a power of 10.

Precise Exact or accurate. The smaller the unit or fraction of a unit used in measuring, the more precise the measurement is. For example, a measurement to the nearest inch is more precise than a measurement to the nearest foot. A ruler with $\frac{1}{16}$-inch markings is more precise than a ruler with $\frac{1}{4}$-inch markings.

Preimage A geometric figure that is changed (by a *reflection, rotation,* or *translation,* for example) to produce another figure. See also *image.*

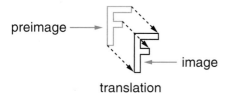

preimage →
image ←

translation

Prime factorization A counting number expressed as a product of prime factors. Every counting number greater than 1 can be written as a product of prime factors in only one way. For example, the prime factorization of 24 is 2 * 2 * 2 * 3. (The order of the factors does not matter; 2 * 3 * 2 * 2 is also the prime factorization of 24.) The prime factorization of a prime number is that number. For example, the prime factorization of 13 is 13.

Prime meridian An imaginary semicircle on Earth that connects the North and South Poles and passes through Greenwich, England.

Prime number A counting number that has exactly two different *factors:* itself and 1. For example, 5 is a prime number because its only factors are 5 and 1. The number 1 is not a prime number because that number has only a single factor, the number 1 itself.

Prism A polyhedron with two parallel *faces,* called *bases* that are the same size and shape. All of the other faces connect the bases and are shaped like parallelograms. The *edges* that connect the bases are parallel to each other. Prisms get their names from the shape of their bases.

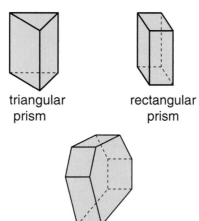

triangular prism
rectangular prism

hexagonal prism

Probability A number from 0 through 1 that tells the chance that an event will happen. The closer a probability is to 1, the more likely the event is to happen. See also *equally likely outcomes.*

Product The result of multiplying two numbers, called *factors.* For example, in 4 * 3 = 12, the product is 12.

Proper factor Any *factor of a counting number* except the number itself. For example, the *factors* of 10 are 1, 2, 5, and 10, and the *proper factors* of 10 are 1, 2, and 5.

Proper fraction A fraction in which the numerator is less than the denominator; a proper fraction names a number that is less than 1. For example, $\frac{3}{4}, \frac{2}{5},$ and $\frac{12}{24}$ are proper fractions.

Proportion A number model that states that two fractions are equal. Often the fractions in a proportion represent rates or ratios. For example, the problem *Alan's speed is 12 miles per hour. At the same speed, how far can he travel in 3 hours?* can be modeled by the proportion

$$\frac{12 \text{ miles}}{1 \text{ hour}} = \frac{n \text{ miles}}{3 \text{ hours}}$$

Protractor A tool on the *Geometry Template* that is used to measure and draw angles. The half-circle protractor can be used to measure and draw angles up to 180°; the full-circle protractor, to measure angles up to 360°.

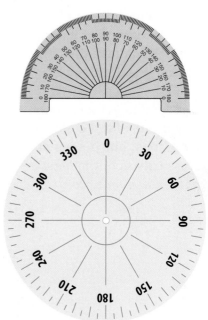

Pyramid A polyhedron in which one face, the *base,* may have any polygon shape. All of the other faces have triangle shapes and come together at a vertex called the *apex.* A pyramid takes its name from the shape of its base.

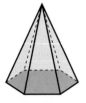 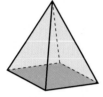

hexagonal   square
pyramid   pyramid

Quadrangle A polygon that has four angles. Same as *quadrilateral.*

Quadrilateral A polygon that has four sides. Same as *quadrangle.*

Quick common denominator The product of the denominators of two or more fractions. For example, the quick common denominator of $\frac{1}{4}$ and $\frac{3}{6}$ is 4 * 6, or 24. As the name suggests, this is a quick way to get a *common denominator* for a collection of fractions, but it does not necessarily give the *least common denominator.*

Quotient The result of dividing one number by another number. For example, in $35 \div 5 = 7$, the quotient is 7.

Radius (plural: **radii**) (1) A line segment from the center of a circle (or sphere) to any point on the circle (or sphere). (2) The length of this line segment.

Random numbers Numbers produced by an experiment, such as rolling a die or spinning a spinner, in which all *outcomes* are *equally likely.* For example, rolling a *fair* die produces random numbers because each of the six possible numbers 1, 2, 3, 4, 5 and 6 has the same chance of coming up.

Random sample A *sample* that gives all members of the *population* the same chance of being selected.

Range The difference between the *maximum* and the *minimum* in a set of data.

Rate A comparison by division of two quantities with *unlike units.* For example, a speed such as 55 miles per hour is a rate that compares distance with time. See also *ratio.*

Rate diagram A diagram used to model *rate* situations. See also *multiplication diagram.*

number of pounds	cost per pound	total cost
3	79¢	$2.37

Rate table A way of displaying *rate* information. In a rate table, the fractions formed by the two numbers in each column are equivalent fractions.

miles	35	70	105
gallons	1	2	3

Ratio A comparison by division of two quantities with *like units.* Ratios can be expressed with fractions, decimals, percents, or words. Sometimes they are written with a colon between the two numbers that are being compared. For example, if a team wins 3 games out of 5 games played, the ratio of wins to total games can be written as $\frac{3}{5}$, 0.6, 60%, 3 to 5, or 3:5. See also *rate.*

Rational number Any number that can be written or renamed as a *fraction* or the *opposite* of a fraction. Most of the numbers you have used are rational numbers. For example, $\frac{2}{3}$, $-\frac{2}{3}$, $60\% = \frac{60}{100}$, and $-1.25 = -\frac{5}{4}$ are all rational numbers.

Ray A straight path that starts at one point (called the *endpoint*) and continues forever in one direction.

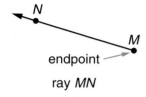

ray *MN*

Real number Any *rational* or *irrational* number.

Reciprocal numbers Two numbers whose product is 1. For example, the reciprocal of 5 is $\frac{1}{5}$, and the reciprocal of $\frac{1}{5}$ is 5; the reciprocal of 0.4 $(\frac{4}{10})$ is $\frac{10}{4}$, or 2.5, and the reciprocal of 2.5 is 0.4.

Rectangle A parallelogram with four right angles.

Rectangle method A method for finding area in which rectangles are drawn around a figure or parts of a figure. The rectangles form regions with boundaries that are rectangles or triangular halves of rectangles. The area of the original figure can be found by adding or subtracting the areas of these regions.

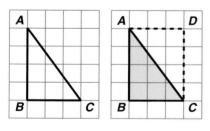

Rectangular array An arrangement of objects into rows and columns that form a rectangle. All rows and columns must be filled. Each row has the same number of objects. And each column has the same number of objects.

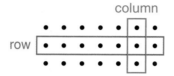

Rectangular coordinate grid A device for locating points in a plane using *ordered number pairs,* or *coordinates.* A rectangular coordinate grid is formed by two number lines that intersect at their zero points and form right angles. Also called a *coordinate grid.*

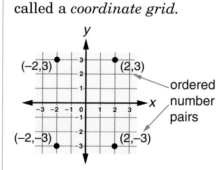

Rectangular prism A *prism* with rectangular *bases.* The four faces that are not bases are either rectangles or other parallelograms.

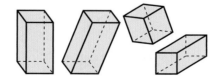

Reduce (1) To decrease the size of an object or figure, without changing its shape. See also *size-change factor.* (2) To put a fraction in *simpler form.*

Reflection The "flipping" of a figure over a line (the *line of reflection*) so that its *image* is the mirror image of the original figure *(preimage).* A reflection of a solid figure is a mirror-image "flip" over a plane.

Regular polygon A polygon whose sides are all the same length and whose interior angles are all equal.

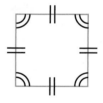

Regular polyhedron
A polyhedron whose faces are congruent and formed by *regular polygons,* and whose *vertices* all look the same. There are five regular polyhedrons:

| regular tetrahedron | 4 faces, each formed by an equilateral triangle |
regular icosahedron	
cube	6 faces, each formed by a square
regular octahedron	8 faces, each formed by an equilateral triangle
regular dodecahedron	12 faces, each formed by a regular pentagon
regular icosahedron	20 faces, each formed by an equilateral triangle

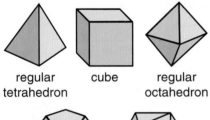

regular tetrahedron   cube   regular octahedron

regular dodecahedron   regular icosahedron

Regular tessellation
A *tessellation* made by repeating *congruent* copies of one *regular polygon.* (Each *vertex point* must be a vertex of *every* polygon around it.) There are three regular tessellations.

Relation symbol A symbol used to express a relationship between two quantities.

Symbol	Meaning
=	"is equal to"
≠	"is no equal to"
>	"is greater than"
<	"is less than"
≥	"is greater than or equal to"
≤	"is less than or equal to"

Remainder An amount left over when one number is divided by another number. For example, if 7 children share 38 cookies, each child gets 5 cookies and 3 are left over. We may write $38 \div 7 \rightarrow 5$ R3, where R3 stands for the remainder.

Repeating decimal
A *decimal* in which one digit or a group of digits is repeated without end. For example, 0.3333… and $23.\overline{147}$ = 23.147147… are repeating decimals. See also *terminating decimal.*

Rhombus A quadrilateral whose sides are all the same length. All rhombuses are parallelograms. Every square is a rhombus, but not all rhombuses are squares.

Right angle A 90° angle.

Right cone A cone whose base is perpendicular to the line joining the apex and the center of the base.

Right cylinder A cylinder whose bases are perpendicular to the line joining the centers of the bases.

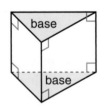

Right prism A prism whose bases are perpendicular to all of the edges that connect the two bases.

Right triangle A triangle that has a right angle (90°).

Rotation A movement of a figure around a fixed point, or axis; a *turn*.

Rotation symmetry A figure has rotation symmetry if it can be rotated less than a full turn around a point or an axis so that the resulting figure (the *image*) exactly matches the original figure (the *preimage*).

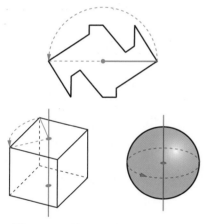

Shapes with rotation symmetry

Round To adjust a number to make it easier to work with or to make it better reflect the level of precision of the data. Often numbers are rounded to the nearest multiple of 10, 100, 1,000, and so on. For example, 12,964 rounded to the nearest thousand is 13,000.

S

Sample A part of a group chosen to represent the whole group. See also *population* and *random sample*.

Scale (1) The *ratio* of a distance on a map, globe, or drawing to an actual distance. (2) A system of ordered marks at fixed intervals used in measurement; or any instrument that has such marks. For example, a ruler with scales in inches and centimeters, and a thermometer with scales in °F and °C. See also *map scale* and *scale drawing*.

Scale drawing A drawing of an object or a region in which all parts are drawn to the same *scale*. Architects and builders use scale drawings.

Scale factor The *ratio* of the size of a drawing or model of an object to the actual size of the object. See also *scale model* and *scale drawing*.

Scale model A model of an object in which all parts are in the same proportions as in the actual object. For example, many model trains and airplanes are scale models of actual vehicles.

Scalene triangle A triangle with sides of three different lengths. In a scalene triangle, all three angles have different measures.

Scientific notation A system for writing numbers in which a number is written as the product of a *power of 10* and a number that is at least 1 and less than 10. Scientific notation allows you to write big and small numbers with only a few symbols. For example, $4 * 10^{12}$ is scientific notation for 4,000,000,000,000.

Sector A region bounded by an *arc* and two *radii* of a circle. The arc and 2 radii are part of the sector. A sector resembles a slice of pizza. The word *wedge* is sometimes used instead of sector.

sector

Semicircle Half of a circle. Sometimes the diameter joining the endpoints of the circle's arc is included. And sometimes the interior of this closed figure is also included.

Semiregular tessellation A *tessellation* made with *congruent* copies of two or more different *regular polygons*. The same combination of polygons must meet in the same order at each *vertex point*. (Each vertex point must be a vertex of *every* polygon around it.) There are eight semiregular tessellations. See also *regular tessellation*.

Side (1) One of the rays or segments that form an angle. (2) One of the line segments of a polygon. (3) One of the faces of a polyhedron.

Significant digits The *digits* in a number that convey useful and reliable information. A number with more significant digits is more *precise* than a number with fewer significant digits.

Similar Figures that have the same shape, but not necessarily the same size.

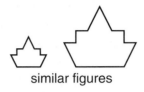
similar figures

Simpler form An equivalent fraction with a smaller numerator and smaller denominator. A fraction can be put in simpler form by dividing its numerator and denominator by a common factor greater than one. For example, dividing the numerator and denominator of $\frac{18}{24}$ by 2 gives the simpler form $\frac{9}{12}$.

Simplest form A fraction that cannot be renamed in simpler form. Also known as *lowest terms*. A *mixed number* is in simplest form if its fractional part is in simplest form.

Simplify To express a fraction in *simpler form*.

Size-change factor A number that tells the amount of enlargement or reduction. See also *enlarge, reduce, scale,* and *scale factor*.

Slanted prism or cone or cylinder A prism (or cone, or cylinder) that is *not* a right prism (or cone, or cylinder).

Slide See *translation*.

Slide rule An *Everyday Mathematics* tool used for adding and subtracting integers and fractions.

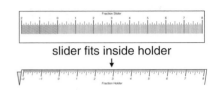
slider fits inside holder

Solution of an open sentence A value that makes an open sentence *true* when it is substituted for the variable. For example, 7 is a solution of $5 + n = 12$.

Span The distance from the tip of the thumb to the tip of the first (index) finger of an outstretched hand. Also called *normal span*.

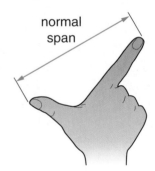
normal span

Speed A *rate* that compares a distance traveled with the time taken to travel that distance. For example, if a car travels 100 miles in 2 hours, then its speed is $\frac{100 \text{ mi}}{2 \text{ hr}}$, or 50 miles per hour.

Sphere The set of all points in space that are the same distance from a fixed point. The fixed point is the *center* of the sphere, and the distance is the *radius*.

Square A rectangle whose sides are all the same length.

Square number A number that is the product of a counting number with itself. For example, 25 is a square number because $25 = 5 * 5$. The square numbers are 1, 4, 9, 16, 25, and so on.

Square of a number The product of a number with itself. For example, 81 is the square of 9 because $81 = 9 * 9$. And 0.64 is the square of 0.8 because $0.64 = 0.8 * 0.8$.

Square root of a number The square root of a number n is a number that, when multiplied by itself, gives n. For example, 4 is the square root of 16 because $4 * 4 = 16$.

Square unit A unit used in measuring area, such as a square centimeter or a square foot.

Standard notation The most familiar way of representing whole numbers, integers, and decimals. In standard notation, numbers are written using the *base-ten place-value* system. For example, standard notation for three hundred fifty-six is 356. See also *expanded notation, scientific notation,* and *number-and-word notation.*

Stem-and-leaf plot A display of data in which digits with larger *place values* are "stems" and digits with smaller place values are "leaves."

Data List: 24, 24, 25, 26, 27, 27, 31, 31, 32, 32, 36, 36, 41, 41, 43, 45, 48, 50, 52.

Stems (10s)	Leaves (1s)
2	4 4 5 6 7 7
3	1 1 2 2 6 6
4	1 1 3 5 8
5	0 2

Straightedge A tool used to draw line segments. A straightedge does not need to have ruler marks on it; if you use a ruler as a straightedge, ignore the ruler marks.

Substitute To replace one thing with another. In a formula, to replace variables with numerical values.

Subtrahend In subtraction, the number being subtracted. For example, in $19 - 5 = 14$, the subtrahend is 5. See also *minuend.*

Sum The result of adding two or more numbers. For example, in $5 + 3 = 8$, the sum is 8. See also *addend.*

Supplementary angles Two angles whose measures total 180°.

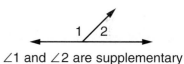

∠1 and ∠2 are supplementary

Surface (1) The boundary of a 3-dimensional object. The part of an object that is next to the air. Common surfaces include the top of a body of water, the outermost part of a ball, and the topmost layer of ground that covers Earth. (2) Any 2-dimensional layer, such as a *plane* or any one of the faces of a *polyhedron.*

Surface area The total area of all the surfaces of a 3-dimensional object. The surface area of a *rectangular prism* is the sum of the areas of its six faces. The surface area of a *cylinder* is the sum of the area of its curved surface and the areas of its two circular bases.

Survey A study that collects data.

Symmetric (1) Having two parts that are mirror images of each other. (2) Looking the same when turned by some amount less than 360°. See also *line symmetry,* and *rotation symmetry.*

Tally chart A table that uses marks, called *tallies,* to show how many times each value appears in a set of data.

Number of Pull-ups	Number of Children
0	𝍬𝍬 /
1	𝍬𝍬
2	𝍬𝍬
3	//

Terminating decimal A *decimal* that ends. For example, 0.5 and 2.125 are terminating decimals. See also *repeating decimal.*

Tessellate To make a *tessellation;* to tile.

Tessellation An arrangement of shapes that covers a surface completely without overlaps or gaps. Also called a *tiling.*

Tetrahedron A polyhedron with 4 faces. A tetrahedron is a triangular pyramid.

Theorem A mathematical statement that can be proved to be true.

3-dimensional (3-D) Having length, width, and thickness. Solid objects take up volume and are 3-dimensional. A figure whose points are not all in a single plane is 3-dimensional.

Time graph A graph that is constructed from a story that takes place over time. A time graph shows what has happened during a period of time.

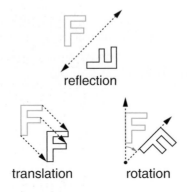

Growth of an Amaryllis

Trade-first subtraction method A subtraction method in which all trades are done before any subtractions are carried out.

Transformation Something done to a geometric figure that produces a new figure. The most common transformations are *translations* (slides), *reflections* (flips), and *rotations* (turns).

reflection

translation   rotation

Translation A movement of a figure along a straight line; a *slide.* In a translation, each point of the figure slides the same distance in the same direction.

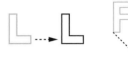

Trapezoid A quadrilateral that has exactly one pair of parallel sides.

Tree diagram A tree diagram is a network of points connected by line segments. A *factor tree* is a tree diagram that is used to factor numbers. Some tree diagrams are used to represent situations that consist of two or more choices or stages.

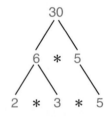

Prime factorization of 30

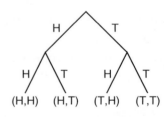

Flipping a coin twice

Triangle A polygon with three sides.

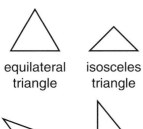

equilateral triangle   isosceles triangle

scalene triangle   right triangle

Triangular numbers
Counting numbers that can be shown by triangular arrangements of dots. The triangular numbers are 1, 3, 6, 10, 15, 21, 28, 36, 45, and so on.

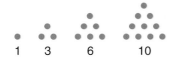

1   3   6   10

Triangular prism A prism whose bases are triangles.

Triangular pyramid
A pyramid in which all of the faces are triangles; also called a *tetrahedron*. Any one of the four faces of a triangular pyramid can be called the base. If all of the faces are equilateral triangles, the pyramid is a *regular tetrahedron*.

regular tetrahedron

True number sentence
A number sentence in which the relation symbol accurately connects the two sides. For example, $15 = 5 + 10$ and $25 > 20 + 3$ are both true number sentences.

Turn See *rotation*.

Turn-around facts A pair of multiplication or addition facts in which the order of the factors (or addends) is reversed. For example, $3 * 9 = 27$ and $9 * 3 = 27$ are turn-around multiplication facts. And $4 + 5 = 9$ and $5 + 4 = 9$ are turn-around addition facts. There are no turn-around facts for division or subtraction. See also *commutative property*.

Turn-around rule A rule for solving addition and multiplication problems based on the *commutative property*. For example, if you know that $6 * 8 = 48$, then, by the turn-around rule, you also know that $8 * 6 = 48$.

Twin primes Two *prime numbers* that have a *difference* of 2. For example, 3 and 5 are twin primes, and 11 and 13 are twin primes.

2-dimensional (2-D)
Having length and width but not thickness. A figure whose points are all in one plane is 2-dimensional. Circles and polygons are 2-dimensional. 2-dimensional shapes have area but not volume.

Unit A label used to put a number in context. The *ONE*. In measuring length, for example, the inch and the centimeter are units. In a problem about 5 apples, *apple* is the unit. See also *whole*.

Unit fraction A fraction whose numerator is 1. For example, $\frac{1}{2}$, $\frac{1}{3}$, $\frac{1}{8}$, and $\frac{1}{20}$ are unit fractions.

Unit percent
One percent (1%).

Unit price The cost for one item or for one unit of measure.

Unlike denominators
Denominators that are different, as in $\frac{1}{2}$ and $\frac{1}{3}$.

"Unsquaring" a number
Finding the *square root* of a number.

U.S. customary system of measurement
The measuring system most frequently used in the United States.

V

Variable A letter or other symbol that represents a number. In the number sentence $5 + n = 9$, any number may be substituted for the variable n, but only 4 makes the sentence true. In the inequality $x + 2 < 10$, any number may be substituted for the variable x, but only numbers less than 8 make the sentence true. In the equation $a + 3 = 3 + a$, any number may be substituted for the variable a, and every number makes the sentence true.

Venn diagram A picture that uses circles or rings to show relationships between sets.

Girls on Sports Teams

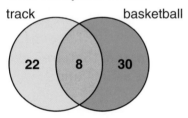

Vertex (plural: **vertices**) The point where the sides of an angle, the sides of a polygon, or the edges of a polyhedron meet.

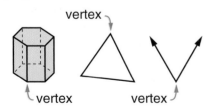

Vertex point A point where corners of shapes in a *tessellation* meet.

Vertical Upright; perpendicular to the horizon.

Vertical (opposite) angles When two lines intersect, the angles that do not share a common side. Vertical angles have equal measures.

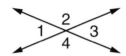

Angles 1 and 3 and angles 2 and 4 are pairs of vertical angles.

Volume A measure of how much space a solid object takes up. Volume is measured in cubic units, such as cubic centimeters or cubic inches. The volume or *capacity* of a container is a measure of how much the container will hold. Capacity is measured in units such as gallons or liters.

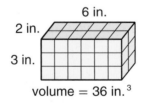

volume = 36 in.^{3}

1 cubic centimeter (actual size)

If the cubic centimeter were hollow it would hold exactly 1 milliliter.
1 milliliter (mL) = 1 cm^3.

W

"What's My Rule?" Problem A type of problem that asks for a rule connecting two sets of numbers. Also, a type of problem that asks for one of the sets of numbers, given a rule and the other set of numbers.

Whole (or ONE or unit) The entire object, collection of objects, or quantity being considered. The ONE, the *unit,* 100%.

Whole numbers The *counting numbers,* together with 0. The set of whole numbers is {0, 1, 2, 3, ...}

Page 4
1. 5,000 **2.** 500,000 **3.** 50 **4.** 50,000

Page 6
1. 49 **2.** 27 **3.** 1,000,000
4. 6 **5.** 376,996 **6.** 50,625

Page 7
1. $\frac{1}{25}$ **2.** $\frac{1}{1,000}$ **3.** 1
4. $\frac{1}{4}$ **5.** 32 **6.** 1

Page 8
1. 16 **2.** 27
3. 7 **4.** 8,000,000
5. 760,000 **6.** $5 * 10^2$
7. $4.4 * 10^4$ **8.** $6 * 10^8$

Page 9
1. false **2.** false **3.** true **4.** true

Page 10
1. 1, 3, 5, 15 **2.** 1, 2, 4, 8
3. 1, 2, 4, 7, 14, 28
4. 1, 2, 3, 4, 6, 9, 12, 18, 36
5. 1, 11
6. 1, 2, 4, 5, 10, 20, 25, 50, 100

Page 11
1. by 3 and by 5 **2.** by 2, 3, 5, 6, and 10
3. by 2 **4.** by 3 and 9
5. by 2, 3, 5, 6, 9, and 10

Page 12
1. $3 * 5$ **2.** $2 * 2 * 5$, or $2^2 * 5$
3. $2 * 2 * 2 * 5$, or $2^3 * 5$
4. $2 * 2 * 3 * 3$, or $2^2 * 3^2$
5. $2 * 2 * 2 * 2 * 2$, or 2^5
6. $2 * 2 * 5 * 5$, or $2^2 * 5^2$

Page 14
1. 887 **2.** 133 **3.** 321
4. 1,023 **5.** 863 **6.** 830

Page 15
1. 38 **2.** 382 **3.** 366
4. 262 **5.** 4,279

Page 16
1. 363 **2.** 159 **3.** 216 **4.** 243

Page 17
1. 456 **2.** 517 **3.** 283 **4.** 2,708

Page 18
1. 900 **2.** 36,000 **3.** 2,500
4. 32,000 **5.** 4,000 **6.** 56,000

Page 19
1. $3 * 200 = 600$
$3 * 60 = 180$
$3 * 5 = 15$
$3 * 265 = 795$
2. $40 * 60 = 2,400$
$40 * 7 = 280$
$2 * 60 = 120$
$2 * 7 = 14$
$42 * 67 = 2,814$
3. $40 * 50 = 2,000$
$40 * 8 = 320$
$0 * 50 = 0$
$0 * 8 = 0$
$40 * 58 = 2,320$
4. $80 * 50 = 4,000$
$80 * 4 = 320$
$3 * 50 = 150$
$3 * 4 = 12$
$83 * 54 = 4,482$
5. $50 * 300 = 15,000$
$50 * 70 = 3,500$
$50 * 2 = 100$
$50 * 372 = 18,600$

Page 20
1.

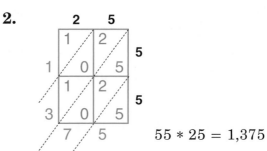

$6 * 78 = 468$

2.

$55 * 25 = 1,375$

3.

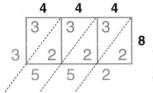

$77 * 89 = 6,853$

4.

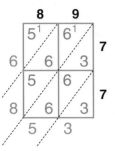

$8 * 444 = 3,552$

5.

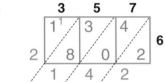

$357 * 6 = 2,142$

Page 21

1. 45　　　　**2.** 8,000　　　**3.** 6,000
4. 530　　　**5.** 70　　　　**6.** 900

Page 23

1. 17 R3　**2.** 147　**3.** 85 R2　**4.** 224 R2

Page 27

1. 0.80　　　　　　　**2.** 0.08

3. $\frac{70}{100}$ or $\frac{7}{10}$　　　　**4.** $\frac{4,506}{1,000}$ or $4\frac{506}{1,000}$

5. $\frac{2,468}{100}$ or $24\frac{68}{100}$　　**6.** $\frac{14}{1,000}$

Page 30

1. a. 20,000　**b.** $\frac{2}{100}$　　　**c.** $\frac{2}{1,000}$
2. a. 0.359　**b.** 0.953　　**c.** 0.539

Page 33

1. $0.59 > 0.059$　　　　**2.** $0.099 < 0.1$
3. $\frac{1}{4} < 0.30$　　　　　　**4.** $0.99 > 0.100$

Page 36

1. 16.02　　　**2.** 1.69　　　**3.** 0.023

Page 37

1. 456　**2.** 2,800　**3.** $4,500　**4.** 10.4

Page 39

1. 9.69　**2.** 19.572　**3.** 2.4644　**4.** 0.0063

Page 40

1.

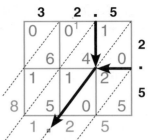

$32.5 * 2.5 = 81.25$

2.

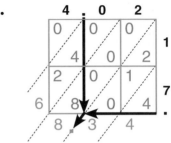

$4.02 * 17 = 68.34$

3.

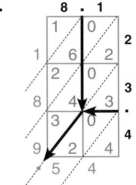

$8.1 * 23.4 = 189.54$

Page 41

1. 5.67　　**2.** 0.0047　**3.** $0.29　**4.** 0.006

Page 42

1. 24.8　　　**2.** 2.11　　　**3.** 1.3

Page 43

1. 2.9　　　**2.** 3.6　　　**3.** 15.1

Page 46

1. a. 1.6　　**b.** 36.5　　**c.** 1.9
2. a. 1.7　　**b.** 36.6　　**c.** 2.0
3. a. 1.6　　**b.** 36.6　　**c.** 11.0

Page 50

1. $48　　　**2.** $4　　　**3.** $600

Page 51

1. $3　　　　　　　**2.** $2.25

Page 53

1. $220 2. $150

3. 31,578,947, or about 31.6 million

Page 59

1. Sample answers: $\frac{2}{4}, \frac{3}{6}, \frac{4}{8}, \frac{5}{10}, \frac{6}{12}, \frac{8}{16}$

2. Sample answers: $\frac{4}{6}, \frac{6}{9}, \frac{8}{12}$

3. $\frac{2}{12}$

4. Sample answers: $\frac{1}{4}, \frac{2}{8}, \frac{3}{12}$

5. The chart shows no equivalent fraction for $\frac{5}{7}$.

Page 60

Sample answers:

1. a. $\frac{9}{12}, \frac{12}{16}$ b. $\frac{6}{16}, \frac{15}{40}$ c. $\frac{6}{15}, \frac{8}{20}$

   d. $\frac{8}{14}, \frac{12}{21}$ e. $\frac{16}{6}, \frac{24}{9}$ f. $\frac{22}{24}, \frac{33}{36}$

2. a. $\frac{2}{3}, \frac{4}{6}$ b. $\frac{2}{5}$ c. $\frac{2}{3}, \frac{4}{6}, \frac{8}{12}$

   d. $\frac{3}{4}, \frac{9}{12}, \frac{15}{20}$ e. $\frac{6}{4}, \frac{3}{2}$ f. $\frac{50}{4}, \frac{25}{2}$

Page 61

1. a. true b. true
   c. true d. false

2. Sample answer: $\frac{4}{6}, \frac{6}{9}, \frac{8}{12}, \frac{10}{15}, \frac{12}{18}$

Page 63

1. $\frac{19}{4}$ 2. $\frac{11}{3}$ 3. $\frac{28}{5}$

4. $\frac{29}{6}$ 5. $\frac{7}{3}$ 6. $12\frac{3}{4}$

7. $8\frac{2}{3}$ 8. $6\frac{4}{5}$ 9. $4\frac{6}{8}$

10. $3\frac{12}{16}$

Page 64

1. 12 2. 30 3. 45 4. 42

Page 65

Sample answers:

1. $\frac{2}{6}$ and $\frac{5}{6}$, $\frac{6}{18}$ and $\frac{15}{18}$

2. $\frac{5}{10}$ and $\frac{8}{10}$, $\frac{10}{20}$ and $\frac{16}{20}$

3. $\frac{15}{20}$ and $\frac{14}{20}$, $\frac{30}{40}$ and $\frac{28}{40}$

4. $\frac{16}{24}$ and $\frac{15}{24}$, $\frac{32}{48}$ and $\frac{30}{48}$

Page 67

1. > 2. < 3. = 4. < 5. <

Page 69

1. $\frac{1}{24}$

2. $\frac{20}{32}$, or $\frac{5}{8}$

3. $\frac{16}{48}$, or $\frac{4}{12}$, or $\frac{1}{3}$

4. $\frac{40}{48}$, or $\frac{10}{12}$, or $\frac{5}{6}$

5. $\frac{17}{24}$

Page 70

1. $3\frac{3}{4}$ 2. $11\frac{1}{12}$ 3. $9\frac{1}{2}$ 4. $5\frac{7}{8}$

Page 72

1. $1\frac{3}{5}$ 2. $\frac{5}{6}$ 3. $3\frac{3}{8}$ 4. $1\frac{5}{12}$

Page 73

1. 8 2. 24 3. 12

4. 40 5. Rita gets $16. Hunter gets $4.

Page 75

1. 44 2. 9 3. 20

Page 76

1. $\frac{1}{3}$ 2. $\frac{3}{10}$ 3. $\frac{3}{4}$ 4. $\frac{1}{12}$

Page 78

1. $\frac{3}{8}$ 2. 16 3. $8\frac{1}{2}$ 4. $2\frac{5}{12}$

Page 80

1. 14 2. 12 3. 16

Page 84

1. 0.75 2. 0.4 3. 3.5
4. 0.55 5. 0.8 6. 0.32

Page 85

1. 0.7 2. 0.375 3. 2.33
4. 0.75 5. 0.86

Page 87

1. 0.375 2. 0.167 3. 0.56

Page 88

1. 0.625 2. 0.1538461…
3. 0.875 4. 0.5714285…
5. $0.08\overline{3}$ 6. $0.5\overline{3}$

Answer Key

Page 90

Fraction	Decimal	Percent
$\frac{1}{4}$	0.25	25%
$\frac{8}{10}$ or $\frac{4}{5}$	0.80	80%
$\frac{1}{2}$	0.50	50%
$\frac{35}{100}$ or $\frac{7}{20}$	0.35	35%
$\frac{1}{10}$	0.10	10%
$\frac{5}{8}$	0.625	62.5%

Page 92

1. -2 **2.** -8 **3.** 1 **4.** -12

Page 94

1. -10 **2.** 12 **3.** -6 **4.** -10

Page 103

1. $\frac{6 \text{ dollars}}{1 \text{ hour}}$

dollars	6	12	18	24
hours	1	2	3	4

2. $\frac{8 \text{ pounds}}{1 \text{ gallon}}$

pounds	8	16	24	32
gallons	1	2	3	4

Page 104

1. 15 feet; 42 feet
2. 630 times; 1,400 times

Page 105

1. $6 **2.** $35 **3.** $1.20

Page 107

1. 12 to 20, $\frac{3}{5}$, 60%, or 12:20

2. 8 to 12, $\frac{8}{12}$, 67%, or 8:12 **3.** 40%

Page 109

1. $24 **2.** 20

Page 111

1. 20 cm **2.** 1,500 miles

Page 112

1. 6.28 inches **2.** 1.59 inches

Page 117

1.

Number of Hits	Number of Players
0	ЦHT
1	//
2	ЦHT
3	//
4	/

2.

Number of Players vs Number of Hits

Page 118

1.

Number of Points	Number of Games
10–19	/
20–29	ЦHT
30–39	///
40–49	///

2.

Number of Points Scored	
Stems (10s)	Leaves (1s)
1	7
2	9 6 8 7 1
3	5 5 5
4	4 6 5

Page 119

1. 0 **2.** 4 **3.** 4 **4.** 2 **5.** 2

Page 120

1. min. = 0; max. = 4; range = 4;
mode = 2 and 3; median: 2.5

2. 13.5

Page 121

Jason's mean (average), rounded to the nearest hundredth, is 80.91.

Page 122

Vacation Days per Year

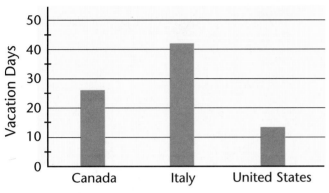

Page 118

Average Temperatures for Phoenix, Arizona

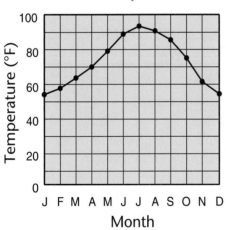

Page 125

3rd grade represents 62% − 45%, or 17%

4th grade represents 85% − 62%, or 23%

5th grade represents 100% − 85%, or 15%

Page 126

Hot Shots Game Points

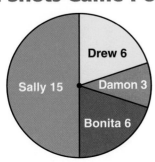

Page 127

Favorite Subjects

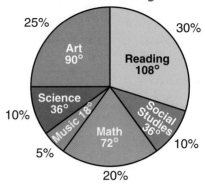

Page 133

1. $\frac{4}{13}$
2. $\frac{9}{13}$
3. $\frac{3}{13}$
4. $\frac{6}{13}$
5. $\frac{7}{13}$
6. $\frac{13}{13}$, or 1

Page 134

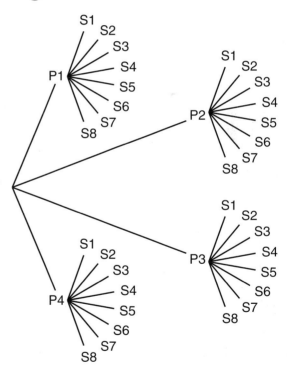

Page 139

1. Sample answer:
2. Sample answer:

3. **a.** $\angle 2 = 50°$ **b.** $\angle 1 = 130°$ **c.** $\angle 3 = 130°$

Answer Key

Page 140

Sample answers:

1.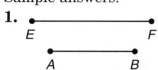
E   F

A   B

2.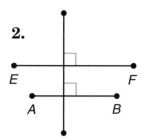
E   F

A   B

Page 141

Sample answers:

1.
H

2.
J   K

3.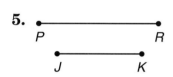
A   T   C

4.
T   U

5.
P   R

J   K

6.
F

E

Page 143

1. **a.** hexagon
b. quadrangle or quadrilateral
c. octagon

2. Sample answers:

a.

b.

3. The sides of the cover of the journal are not all the same length.

Page 144

Sample answers:

1.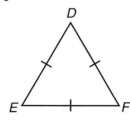
D

E   F

DFE
EDF
EFD
FDE
FED

2.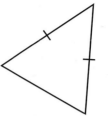

3.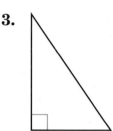

Page 145

1.
U   A

Q   D

2. No

3. *UADQ, ADQU, DQUA, QDAU, DAUQ, AUQD, UQDA*

Page 146

Sample answers:

1. All sides of a square have the same length. The sides of a rectangle may or may not all have the same length.

2. A rhombus is a parallelogram. A kite is not a parallelogram. All sides of a rhombus have the same length. The sides of a kite have two different lengths.

3. A trapezoid has exactly one pair of parallel sides. A parallelogram has two pairs of parallel sides.

Page 148

Sample answers:

1. **a.** They each have at least one circular face. They each have a curved surface.

   b. A cylinder has three surfaces; a cone has two. A cylinder has 2 circular bases, 2 edges, and no vertices. A cone has 1 circular base, 1 edge, and 1 vertex.

2. **a.** They each have at least one vertex. They each have a flat base.

   b. A cone has a curved surface; the surfaces of a pyramid are all flat surfaces (faces). A cone has 1 circular face; the faces of a pyramid are all shaped like polygons. A cone has only one vertex. A pyramid has at least four vertices.

Page 149

1. **a.** 5   **b.** 1

2. **a.** 5   **b.** 2

Page 150

1. **a.** 8   **b.** 18   **c.** 12

2. decagonal prism

Page 151

1. **a.** 4   **b.** 6   **c.** 4

2. pentagonal pyramid

3. Sample answers:

   a. The surfaces of each are all formed by polygons. Their base shape is used to name them.

   b. A prism has at least one pair of parallel faces; no two faces of a pyramid are parallel. The faces of a prism that are not bases are all parallelograms. The faces of a pyramid that are not the base are all triangles.

Page 152

1. regular tetrahedron, regular octahedron, regular icosahedron

2. **a.** 12   **b.** 6

3. Sample answers:

   a. Their faces are equilateral triangles that all have the same size.

   b. The tetrahedrons have 4 faces, 6 edges, and 4 vertices. The octahedrons have 8 faces, 12 edges, and 6 vertices.

Page 155

a, b, c, d, or all of these

Page 158

1.

2. C

Page 159

1.
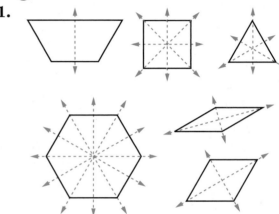

2. Infinite; any line drawn directly through its center is a line of symmetry.

Page 183

1. millimeter, gram, meter, centimeter

2. $\frac{1}{1,000}$ 3. 2,000 mg

Page 186

1. 27 feet 2 inches   2. 39 inches

Page 187

1. 24 mm   2. 75.4 mm

3. 44.0 in.

Page 189

1. 6 square units   2. 38 in.^{2}

3. 49 m^2

Page 191

1. 8 square units   2. 15 square units

3. 20 square units

Page 192

1. 768 ft^2 2. 80 in.^{2} 3. 2.2 cm^2

Page 193
1. 24 in.^{2} **2.** 27 cm^2 **3.** 10.8 yd^2

Page 194
1. 18 mm **2.** 9 mm **3.** 254.5 mm^2

Page 197
1. 42 yd^3 **2.** 1,000 cm^3 **3.** 288 ft^3

Page 199
1. 128 yd^3 **2.** 80 cm^3 **3.** 75 ft^3

Page 200
340 cm^2

Page 201
94.2 cm^2

Page 202
1. about 180 grams; 170.1 grams
2. 937 ounces

Page 206
1. 45° **2.** 210° **3.** 75°
Sample answers:
4.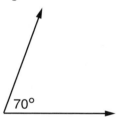
70°

5.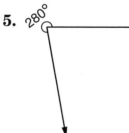
280°

6.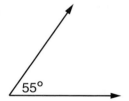
55°

Page 207
1. a. 2 **b.** 3 **c.** 6 **d.** 10
2. 540° **3.** 135°
4. number of triangles = number of sides − 2

Page 208
1–4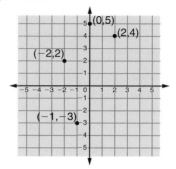

Page 217
1. $\frac{1}{2}x = 28$ or $\frac{x}{2} = 28$
2. $n = 15 * 4$
3. $c = \pi * d = 3.14 * 2$ cm $= 6.28$ cm
4. $c = \pi * d = 3.14 * 3$ in. $= 9.42$ in.

Page 218
1. Barbara's height $= A - 3$
2. miles run $= 2 * D$
3. 6 **4.** 16 **5.** 2 **6.** 18

Page 219
1. $y = 16$ **2.** $z = 6$ **3.** $m = 11$

Page 220
1. true **2.** false **3.** true
4. true **5.** false **6.** true

Page 221
1. false **2.** true **3.** false
4. < **5.** = **6.** >

Page 222
1. $x = 45$ **2.** $y = 600$
3. $w = 35$ **4.** $n = 56$
5. $25 - (15 + 10) = 0$
6. $100 = 10 * (9 + 1)$
7. $5 = 3 + (6 * 3) / (3 * 3)$
8. $26 = (7 + 6) * 2$

Page 223
1. 13 **2.** 25 **3.** 1

Page 225
1. $8 * (15 + 6) = (8 * 15) + (8 * 6)$
2. $(5 * 41) + (5 * 11) = 5 * (41 + 11)$
3. $16 * (10 - 8) = (16 * 10) - (16 * 8)$

Page 227
1. $12.95 − $9.50 = x; x = $3.45
2. 26 * $4.50 = n; n = $117

Page 229
1. One cube weighs the same as 2 marbles.
2. One cube weighs the same as 3 marbles.

Page 232

1.

in	out
v	2 * v + 1
0	1
1	3
2	5

2.

in	out
x	5x
5	25
9	45
20	100

3.

Rule

÷ 2

Page 234
1. $10.00   2. 360 miles

Page 243
1. The 15-ounce box is a better buy. It costs $0.24 per ounce; the 10-ounce box costs $0.25 per ounce.
2. a. 150 miles   b. 25 miles
   c. 125 miles   d. 600 miles

Page 245
1. 100 squares (Think: 1 + 3 + 5 + 7 + 9 + 11 + 13 + 15 + 17 + 19)
2. 2,500 squares (Notice the pattern:
   5 steps tall = 5^2 = 25 squares;
   10 steps tall = 10^2 = 100 squares;
   50 steps tall = 50^2 = 2,500 squares)

Page 248
1. Emily is not correct. The leading-digit estimate is 800 + 200 + 700 = 1,700. She should check her work.
2. Luis is not correct. The leading-digit estimate is 900 / 30 = 30. He should check his work.

Page 249
1. 25,800   2. 30,000   3. 25,800

Page 250
1. Sample estimate: 500 * 30 = 15,000; ten thousands
2. Sample estimate: 60,000/100 = 600; hundreds
3. Sample estimate: 3,000/1 = 3,000; thousands

Page 256
1. 82   2. 4   3. 60.5   4. 28

Page 258
1. 14 R4   2. 17 R0   3. 757 R25

Page 267
1. 0.3125   2. $\frac{77}{200}$, or $\frac{385}{1,000}$
3. 0.9%   4. 4.58
5. 21.875%   6. $\frac{29}{50}$, or $\frac{58}{100}$

Page 269
1. 0.8   2. 235
3. 1258.378   4. 1.00

Page 274
1. 0.00058   2. 76,000,000
3. 0.000004389, or 0.0000043   4. − 0.000011

Page 275
1. $6.0846 * 10^{15}$, or $6.085 * 10^{15}$
2. $5.3798 * 10^{11}$, or $5.38 * 10^{11}$
3. $1.5365 * 10^{16}$, or $1.537 * 10^{16}$
4. $2.7775 * 10^{17}$, or $2.778 * 10^{17}$
5. exactly $3.118752 * 10^8$ = 311,875,200 hands, or about $3.1187 * 10^8$ = 311,870,000 hands

Page 277
1. 25,447 ft^2 2. 83.9 cm

Page 280

about 620 ft^2

Page 281

1. $11.25; $86.25

Page 284

1. $63.50   2. $29.50

Page 286

1. 11, 18, 25, 32, 39, 46
2. 120, 107, 94, 81, 68, 55

Page 24A

1. 5,601   2. 10,316
3. 10,112   4. 45,152

Page 24B

1. 18   2. 262
3. 3,597   4. 2,085

Page 24D

1. 13,984   2. 43,108
3. 229,554   4. 332,514

Page 24F

1. $130   2. $243 R$1
3. $121 R$3   4. $103 R$4

Page 24H

1. 487   2. 2,263
3. 1,345   4. 1,076

Page 24J

1. 6 R17   2. 307 R15
3. 448 R10   4. 110 R31

Page 54A

1. 12.41   2. 81.23
3. 356.389   4. 95.54

Page 54B

1. 63.30   2. 0.91
3. 32.23   4. 3.59

Page 54D

1. 218.0   2. 1.7625
3. 0.0676   4. 80.8555

Page 54E

1. $1.25   2. $1.35
3. $0.60   4. $0.55

Page 54F

1. 1.5725   2. 2.61
3. 2.0925   4. 0.8425

Page 54H

1. 2290.0   2. $16.\overline{4}$
3. 79   4. 23.1

Page 54J

1. $0.8\overline{3}$ 2. $0.\overline{18}$
3. $0.\overline{5}$ 4. $2.\overline{3}$

Page 78D

1. $9 \div 24$ or 9/24   2. $\frac{111}{222}$
3. $1\frac{3}{5}$ 4. $3\frac{3}{8}$
5. $\frac{3}{4}$ of a pizza

Page 80B

1. $\frac{1}{15}$ 2. $\frac{1}{32}$
3. $\frac{1}{1000}$ 4. $\frac{1}{6}$ of a watermelon

Index

Index

Index

Index

Photo Credits

Cover (l)Steven Hunt/Stone/Getty Images, (c)Martin Mistretta/Stone/Getty Images, (r)Digital Stock/CORBIS, (bkgd)Pier/Stone/Getty Images; **Back Cover Spine** Martin Mistretta/Stone/Getty Images; **v** Jules Frazier/Photodisc/Getty Images; **vi** Burke/Triolo Productions/Brand X/Getty Images; **vii** PhotoDisc/Getty Images; **viii** Burke/Triolo Productions/Brand X Pictures/Getty Images; **ix** CORBIS; **xi** Medioimages/Photodisc/Getty Images; **1** Digital Vision/Getty Images; **2** (t)American Spirit Images/Fotosearch.com, (c)Stockbyte/Getty Images, (b)Kurt Amthor/CORBIS Edge/CORBIS; **3** The McGraw-Hill Companies; **8** Getty Images; **13** Erich Lessing/Art Resource, NY; **23** Science & Society Picture Library/SSPL via Getty Images; **25** Bob Martin/Sports Illustrated/Getty Images; **45** Adam Pretty/Getty Images Sport/Getty Images; **52** (t)Ryan McVay /Photodisc/Getty Images, (b)D. Hurst/Alamy; **53** Burgess Blevins/Taxi/Getty Images; **55** Kathryne Kleinman/FoodPix/Getty Images; **56** The McGraw-Hill Companies; **71** Sheila Terry/Photo Researchers, Inc.; **73** David Young Wolff/PhotoEdit; **75** The McGraw-Hill Companies; **078A** The McGraw-Hill Companies, Inc./Ken Karp photographer; **078B** Cre8tive Studios/Alamy; **078C** Peter Dazeley/Photographer's Choice RF/Getty Images; **81** The McGraw-Hill Companies; **95** (l)Bryan Allen/CORBIS, (c)NASA/Roger Ressmeyer/CORBIS, (r)Photodisc/Getty Images, (r bkgd) Siri Stafford/Lifesize/Getty Images; **96** (t)Detlev van Ravenswaay/Photo Researchers, Inc., (c)Adastra/Taxi/Getty Images, (b)David A. Hardy/Photo Researchers, Inc.; **97** (t)Roger Ressmeyer/CORBIS, (b)Gerard Lodriguss/Photo Researchers, Inc.; **98** (t)STScI/NASA/CORBIS, (c)Roger Ressmeyer/CORBIS, (b)Shigemi Numazawa/Atlas Photo Bank/Photo Researchers, Inc.; **99** (tl)Jason T. Ware/Photo Researchers, Inc., (tr)Chris Cook/Photo Researchers, Inc., (bl)Science Source/Photo Researchers, Inc., (br)Modern Technologies/Getty Images; **100** (tl)NASA/Roger Ressmeyer/CORBIS, (tr)The Stocktrek Corp/Brand X/CORBIS, (b)Nasa TV/Reuters/CORBIS; **101** Peter Dazeley/Photographer's Choice/Getty Images; **102** (t)D. Hurst/Alamy, (b)Ben Blankenburg/CORBIS; **103** Cadence Gamache; **104** Reuters/CORBIS; **107** Chev Wilkinson/Stone+/Getty Images; **108** Guy Motil/CORBIS; **110** (t bcl bcr bl br)The McGraw-Hill Companies, (c)Shannon Fagan/Photographer's Choice/Getty Images; **113** Edward A. Ornelas/San Antonio Express-News/Zuma Press, Inc.; **114** (t)Jeff Greenberg/Alamy, (b)Digital Vision/Getty Images; **115** Digital Vision/Getty Images; **116** Jim Arbogast/Digital Vision/Getty Images; **121** Tim Davis/Davis Lynn Wildlife/CORBIS; **122** CORBIS; **129 130** The McGraw-Hill Companies; **135** Graham Winterbottom/Getty Images; **136** (clockwise from top)C Squared Studios/Photodisc/Getty Images, (2)William Leaman/Alamy, (3)James Davis Photography/Alamy, (4) Charlie Schuck/Digital Vision/Getty Images, (5)The McGraw-Hill Companies, (6)Inti St Clair/Digital Vision/Getty Images, (7)SuperStock; **137** (t to b)Sharon Hoogstraten/Courtesy of Dave Wyman, (2)Jules Frazier/Photodisc/Getty Images, (3)Santokh Kochar/Photodisc/Getty Images, (4)Jim Wehtje/Photodisc/Getty Images; **140** F. Schussler/PhotoLink/Photodisc/Getty Images; **141** Benjamin Rondel Photography/Getty Images; **147** (tr)Erik Rank/Getty Images, (others)The McGraw-Hill Companies; **151** Digital Vision/Getty Images; **152** The McGraw-Hill Companies; **154** The McGraw-Hill Companies Inc./Ken Cavanagh Photographer; **157** Robert Stainforth/Alamy; **160** M.C. Escher's "Symmetry Drawing E38" © 2011 The M.C. Escher Company-Netherlands. All rights reserved. www.mcescher.com; **164** Jack Hollingsworth/Digital Vision/Getty Images; **165** Vincent Besnault/Stone/Getty Images; **175** (l to r)Alan Schein Photography/CORBIS, (2)Lewis Wickes Hine/Christie's Images/Corbis, (3)Sergio Dorantes/CORBIS, (4)Cory Langley/CORBIS; **176** (t)Hugh Sitton/Stone/Getty, (c)Danny Lehman/CORBIS, (b)Mark Karrass/CORBIS; **177** (t)Goodshot/Jupiterimages, (bc)Dave Bartruff/CORBIS, (bl)Photodisc/Getty Images, (br)Image Source/CORBIS; **178** (tl)Peter M. Wilson/Alamy, (tr)Dr. Robert Muntefering/The Image Bank/Getty Images, (bl) Ary Diesendruck/Stone/Getty, (br)Florian Monheim/Bildarchiv Monheim GmbH/Alamy; **179** (t)Keren Su/CORBIS, (c)1Apix/Alamy, (b)Marco Simoni/Robert Harding World Imagery/Getty Images; **180** (t)McDaniel Woolf/Photodisc Red/Getty Images, (c)Stuart Crump/Alamy, (b)Jan Jerabek/WoodyStock/Alamy; **181** Matthew May/Photonica/Getty Images; **183** Stockbyte/Getty Images; **185** (t c)The McGraw-Hill Companies, (b)Jupiterimages/Getty Images; **187 194 200** The McGraw-Hill Companies; **215** David R. Frazier Photolibrary/Alamy; **216** The Trustees of The British Museum/Art Resource, NY; **235** (t)2011 Herb Kawainui Kane/HawaiianEyes.com, (b)Dennis Kawaharada; **236** (t)The McGraw-Hill Companies, (bl)Big Bamboo Stock Photography, (br)Image 100/CORBIS; **237** (l)D. Nunuk/Photo Researchers, Inc., (r)First Light/CORBIS; **238** (t)2011 Herb Kawainui Kane/HawaiianEyes.com, (c)Jason Hosking/Corbis Edge/Corbis; **239** (t)Bill Varie/CORBIS, (c)2011 Herb Kawainui Kane/HawaiianEyes.com, (b)Frans Lanting/CORBIS; **240** (t)Jonathan Blair/CORBIS, (c)Mauritius/SuperStock, (b) David Hosking/flpa-images.co.uk; **241** Mel Curtis/Photodisc/Getty Images; **245** Ryan McVay/Photodisc/Getty Images; **247** TRBfoto/Digital Vision/Getty Images; **248** C Squared Studios/Photodisc/Getty Images; **250-285** The McGraw-Hill Companies; **287** (l)The Newark Museum/Art Resource, NY, (r)The Palma Collection/Photodisc/Getty Images, (b)Federal Reserve Bank of San Francisco; **288** (tl)American Numismatic Association, (tr)Charles O'Rear/CORBIS, (cl bl)Bettmann/CORBIS, (br)James Leynse/CORBIS; **289** (tl br)Charles O'Rear/CORBIS, (tr bl)James Leynse/CORBIS; **290** (tl)David Doubilet/National Geographic Stock, (tr)C Squared Studios/Photodisc/Getty Images, (bl)Barry Gregg/CORBIS, (br)Courtesy of the Burke Museum of Natural History and Culture, Catalog Number 2705.; **291** (tl)George Schweighofer/Currency Quest, (tr)Charles O'Rear/CORBIS, (bl)Danilo Calilung/CORBIS, (br)Federal Bureau of Investigation; **292** (t)Bruce Laurance/Digital Vision/Getty Images, (c)George Shelley/CORBIS, (b)Comstock Images/Alamy; **293** Brand X Pictures/Jupiterimages; **294** The McGraw-Hill Companies; **297** Dynamic Graphics/Punchstock; **299-320** The McGraw-Hill Companies; **329** Stockbyte/Getty Images; **330** The McGraw-Hill Companies; **337** Peter Dazeley/Stone/Getty Images; **338** (t)Scenics of America/PhotoLink/Photodisc/Getty Images, (c)Tetra Images/Alamy, (b)C. Borland/PhotoLink/Photodisc/Getty Images; **340** Peter Bostrom; **344** R. Morley/Photolink/Getty Images; **352** Nik Wheeler/CORBIS; **353** Robert Landau/CORBIS; **355** (t)Corbis Premium Collection/Alamy, (b)isifa Image Service s.r.o./Alamy; **356** (t)Gene Peach/Riser/Getty Images, (b)Image Source/Getty Images; **360** (t)Photodisc/Getty Images, (c)VisionsofAmerica/Joe Sohm/Getty Images, (b) Royalty-Free/CORBIS; **362** Purestock/Getty Images; **363** Tetra Images/CORBIS; **368** (t)Visions of America, LLC/Alamy, (b)Kim Kulish/CORBIS; **370** U.S. Bureau of the Census; **392** (t)Ryan McVay/Photodisc/Getty Images, (c) Reuters/CORBIS, (b)Sheila Terry/Photo Researchers, Inc.

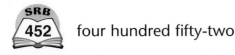